T0297634

Anti-Angiogenesis Drug Discovery and Development

Anti-Angiogenesis Drug Discovery and Development
Volume 2

Edited by

Atta-ur-Rahman, FRS
Kings College
University of Cambridge
Cambridge
UK

&

Muhammad Iqbal Choudhary
H.E.J. Research Institute of Chemistry
International Center for Chemical and Biological Sciences
University of Karachi
Karachi
Pakistan

AMSTERDAM • BOSTON • HEIDELBERG • LONDON • NEW YORK • OXFORD
PARIS • SAN DIEGO • SAN FRANCISCO • SINGAPORE • SYDNEY • TOKYO
ELSEVIER

Elsevier
Radarweg 29, PO Box 211, 1000 AE Amsterdam, Netherlands
The Boulevard, Langford Lane, Kidlington, Oxford OX5 1GB, UK
225 Wyman Street, Waltham, MA 02451, USA

Copyright © 2014 Bentham Science Publishers Ltd. Published by Elsevier Inc. All rights reserved.

No part of this publication may be reproduced or transmitted in any form or by any means, electronic or mechanical, including photocopying, recording, or any information storage and retrieval system, without permission in writing from the publisher. Details on how to seek permission, further information about the Publisher's permissions policies and our arrangements with organizations such as the Copyright Clearance Center and the Copyright Licensing Agency, can be found at our website: www.elsevier.com/permissions.

This book and the individual contributions contained in it are protected under copyright by the Publisher (other than as may be noted herein).

Notices
Knowledge and best practice in this field are constantly changing. As new research and experience broaden our understanding, changes in research methods, professional practices, or medical treatment may become necessary.

Practitioners and researchers must always rely on their own experience and knowledge in evaluating and using any information, methods, compounds, or experiments described herein. In using such information or methods they should be mindful of their own safety and the safety of others, including parties for whom they have a professional responsibility.

To the fullest extent of the law, neither the Publisher nor the authors, contributors, or editors, assume any liability for any injury and/or damage to persons or property as a matter of products liability, negligence or otherwise, or from any use or operation of any methods, products, instructions, or ideas contained in the material herein.

ISBN: 978-0-12-803963-2

British Library Cataloguing in Publication Data
A catalogue record for this book is available from the British Library

Library of Congress Cataloging-in-Publication Data
A catalog record for this book is available from the Library of Congress

For Information on all Elsevier publications visit our website at http://store.elsevier.com/

CONTENTS

PREFACE

Angiogenesis is one of the most important processes in the life cycles of higher animals, both in normal physiology, and in pathophysiology. The growth of new blood vessels modulates many processes including reproduction and development of cells, wound heating, etc. The molecular cascade of angiogenesis is tightly regulated by inhibitors and stimulators. Upregulation and disruption of angiogenic factors play key roles in the tumor growth and metastasis. Angiogenesis is also a key actor in cancer and other diseases. The discovery of angiogenic inhibitors is a promising approach for the treatment of various diseases, including cancers. The modern research in this area takes benefit of the understanding of angiogenisis at the molecular level. Key targets has been identified, and a large number of small molecular inhibitors have been discovered, which influence the angiogenic pathway in a very subtle manner. The developments in this field are fast and exciting, and deserve attention of both the drug discovery scientists, and the general public.

Volume 2 of this *e*book series entitled, "Anti-angiogenesis Drug Discovery and Development", is an outstanding collection of well written articles in this important field. The first volume of this series was greatly appreciated by the readers. The second volume is a continuation of the some high quality compilation of focused articles.

The first review by Parida and Mandal focuses on defining the complex mechanism of angiogenesis, and the recent advances in therapeutics which target the process at various points. These drugs and drugs candidates interact with various targets in the angiogenic cascade, and generate considerable disease response.

Akita in his review describes various *in vitro* assays / bioassays, used in the discovery of anti-angiogenic agents. These range from culturing techniques to capillary tube degradation and cell migration. The objective is to punitively assess the anti-angiogenic effects of drug candidates.

Hilmi and Pagès have reviewed various targets in angiogenesis in clear cell renal carcinoma (CCRCC). CCRCC is widely used for the study of the implications of angiogenesis in cancer. This has been taken as a model to understand the process of angiogenesis at molecular and cellular levels as well as to understand which of these targets can yield optimal therapeutic response. The success and failure of various antiagiogenic agents in clinical practices has also been reviewed.

Oxidative stress plays an important role in the on-set of many diseases. In cancer progression, oxidative stress is the primary cause of the up-regulation of several pro-angiogenic factors. Philips *et al* have contributed an excellent review on the role of antioxidants, both single molecules as well as plant extracts, and their combination, as supplement regimen to inhibit the invasiveness of several cancers.

Gude *et al* have reviewed studies on novel anti-cancer strategies, based on striking the balance between pro-angiogenic and anti-angiogenic molecules. They have also summarized various molecules of both synthetic and natural origins which inhibit the process of angiogenesis and can be used for the treatment of cancers.

Ovarian cancer is among the most lethal malignancies. Currently the available drugs are less than adequate for its treatment. The discovery of new anti-angiogenics can open the way for better treatment, and improve overall survival in ovarian cancer. Maile *at al* has attributed an excellent review on this topic.

Mc Murray and Klostergaard present a well referenced review on STAT3 (Signal Transducer and activator of Transcription 3) and its role in angiogenesis. STAT3 signaling activation has been identified as a valid target for drug discovery. Inhibition of STAT3 activation is achieved by inhibiting the interaction between the tumor cells and stromal compartment. The review summarizes decades of research work on small molecule-mediated inhibition of STAT3.

Katsi has written an excellent treatise on another aspect of angiogenesis, related to cardiovascular diseases such as an atherosclerosis. Historical prospects and details of the latest developments of angiogenesis therapies for cardiovascular diseases are reviewed.

This volume is the result of hard work of so many eminent contributors, for which we express our profound gratitude. We also like to acknowledge the commitment and untiring efforts of the outstanding team of the Bentham Science

Publishers, led by Mr. Mahmood Alam, Director Bentham Science Publishers. The efforts of Ms. Sara Yasir, Manager Publications, deserve special appreciation.

Atta-ur-Rahman, FRS
Kings College
University of Cambridge
Cambridge
UK

List of Contributors

Caroline Hilmi

University of Nice Sophia Antipolis, Institute for Research on Cancer and Ageing of Nice, UMR CNRS 7284, INSERM U1081, France

Christodoulos I. Stefanadis

1st Cardiology Department, Athens University Medical School, Greece

Costas T. Psarros

1st Cardiology Department, Athens University Medical School, Greece

Daphne Suzin

Massachusetts General Hospital, Boston, MA, USA

Dimitris Tousoulis

1st Cardiology Department, Athens University Medical School, Greece

Evelyn Y. T. Wong

Yong Loo Lin School of Medicine, Singapore

Georgia D. Vamvakou

Second Department of Cardiology, University of Athens, Attikon Hospital, Chaidari, Greece

Gilles Pagès

University of Nice Sophia Antipolis, Institute for Research on Cancer and Ageing of Nice, UMR CNRS 7284, INSERM U1081, France

Halyna Siomyk

School of Natural Sciences, Fairleigh Dickinson University, Teaneck, NJ 07666, USA

Harit Parakandi

School of Natural Sciences, Fairleigh Dickinson University, Teaneck, NJ 07666, USA

Hui Jia

School of Natural Sciences, Fairleigh Dickinson University, Teaneck, NJ 07666, USA

Ioannis E. Kallikazaros

Cardiology Department, Hippokration Hospital, Athens, Greece

Jim Klostergaard

Department of Molecular and Cellular Oncology, The University of Texas, MD Anderson Cancer Center, Houston, TX USA

John S. McMurray

Departments of Experimental Therapeutics and Molecular and Cellular Oncology, The University of Texas, MD Anderson Cancer Center, Houston, TX, USA

Madon M. Maile

Massachusetts General Hospital, Boston, MA, USA

Mahitosh Mandal

School of Medical Science and Technology, Indian Institute of Technology, Kharagpur, West Bengal, India

Marios G. Krokidis

1st Cardiology Department, Athens University Medical School, Greece

Masumi Akita

Division of Morphological Science, Biomedical Research Center, Saitama Medical University, 38 Moroyama, Iruma-gun, Saitama 350-0495, Japan

Mohammad Zahid Kamran

Gude Lab, Advanced Centre for Treatment Research and Education in Cancer (ACTREC), Tata Memorial Centre, Kharghar, Navi-Mumbai, India

Neena Philips

School of Natural Sciences, Fairleigh Dickinson University, Teaneck, NJ 07666, USA

Nicole E. Birrer

Massachusetts General Hospital, Boston, MA, USA

Peeyush N. Goel

Gude Lab, Advanced Centre for Treatment Research and Education in Cancer (ACTREC), Tata Memorial Centre, Kharghar, Navi-Mumbai, India

Prachi Patil

Gude Lab, Advanced Centre for Treatment Research and Education in Cancer (ACTREC), Tata Memorial Centre, Kharghar, Navi-Mumbai, India

Rajiv P. Gude

Gude Lab, Advanced Centre for Treatment Research and Education in Cancer (ACTREC), Tata Memorial Centre, Kharghar, Navi-Mumbai, India

Richard T. Penson

Massachusetts General Hospital, Boston, MA, USA

Sheetal Parida

School of Medical Science and Technology, Indian Institute of Technology, Kharagpur, West Bengal, India

Vasiliki K. Katsi

Cardiology Department, Hippokration Hospital, Athens, Greece

CHAPTER 1

Mechanism of Controlling Blood Vessel Growth and Development and Identification of Therapeutics Against Pathological Angiogenesis

Sheetal Parida and Mahitosh Mandal*

School of Medical Science and Technology, Indian Institute of Technology, Kharagpur, West Bengal, India

Abstract: Angiogenesis is a physiological process associated with development and repair of tissues. In embryonic stage, vasculogenesis occurs by *de novo* synthesis of a network of primitive blood vessels from precursors of endothelial cells called angioblasts which proliferate and coalesce to form the primary capillary plexus. The primary capillary plexus serves as a scaffold for further angiogenesis. It involves remodeling by sprouting and branching of preexisting vessels. In adults, angiogenesis occurs during ovarian cycles and in physiological processes like wound healing and tissue repair. Tumor induced angiogenesis is a pathological condition wherein angiogenesis is up regulated due to aberrant deployment of normal angiogenic machinery. In small tumors, the cells receive nutrition initially by passive diffusion. However, as the tumor grows in size, within the confinement of tumor the availability of nutrients is limited due to increasing competition between rapidly proliferating cells and the diffusion of nutrients is impeded by high interstitial pressure. In order to overcome this nutrient deprivation and for growth, invasion and subsequent metastasis, the tumor cells induce formation of new blood vessels from preexisting ones. This enables the survival of the tumor cells in a hostile microenvironment. Neo-angiogenesis is a complex process involving an extensive interplay between cells, soluble factors and extracellular matrix components. A critical equilibrium is regulated by anti and pro-angiogenic factors and the balance is shifted in favor of angiogenesis by hypoxia or inflammation. In tumor associated angiogenesis, the cancerous cells secrete or stimulate the secretion of various pro-angiogenic factors including Angiogenin, Vascular endothelial growth factor (VEGF), Fibroblast growth factor (FGF) and Transforming growth factor-β (TGF-β). The stimulation for neo-angiogenesis is called an angiogenic switch. The principal stimulation is thought to be oxygen deprivation possibly assisted by inflammation, oncogenic mutation, mechanical stress *etc.* VEGF is the most specific angiogenic factor for endothelial cells. VEGF binds to its receptors inducing signaling pathways that in turn bring about endothelial cell proliferation, differentiation,

***Corresponding author Mahitosh Mandal:** Associate professor, School of Medical Science and Technology, Indian Institute of Technology, Kharagpur, West Bengal-721302, India; Tel: 032222-83578; E-mail: mahitosh@smst.iitkgp.ernet.in

Atta-ur-Rahman and Muhammad Iqbal Choudhary (Eds)
Copyright © 2014 Bentham Science Publishers Ltd. Published by Elsevier Inc. All rights reserved.
10.1016/B978-0-12-803963-2.50001-6

migration, increased vascular permeability and release of endothelial cell precursors from the bone marrow. Sequentially, angiogenesis involves the degradation of basement by proteases, migration of endothelial cells (EC) into interstitial spaces and sprouting, ECs proliferation at the migrating tip and lumen formation, generation of new basement membrane with the recruitment of pericytes, formation of anastomoses and finally blood flow. Targeting angiogenesis for treatment of cancer has been an appealing concept among researchers for over three decades and recently many angiogenic inhibitors have moved from preclinical to clinical trials. Most of angiogenesis inhibitors have been found to be cytostatic rather than cytocidal. Hence, anti-angiogenic therapy is useful when administered in combination with conventional chemotherapeutic agents. Today there are more than 30 anti-angiogenic drugs in use showing considerable disease response. The development of anti-angiogenic drugs involves identification of new targets in the angiogenic pathway as well as identification and management of a new range of toxicities.

Keywords: Angiogenesis, angiogenesis inhibitors, endothelial cells, growth factors, hypoxia, intussusception, sprouting, tumor, vasculogenesis, VEGF.

INTRODUCTION

All metabolically active tissues require a continuous supply of oxygen to meet their energy requirements. Oxygen is carried to the tissues *via* blood capillaries extended throughout the body and these are essential for free exchange of nutrients and metabolites. No active tissue in the body is more than a few hundred micrometers from a blood capillary. Metabolic activity of a tissue is directly proportional to the extent of angiogenesis, hence, is also proportional to capillarity, oxygen being the key player in this regulation. Hemodynamic factors are critical for survival of vascular networks and for structural adaptations of vessel walls [1]. Angiogenesis is the physiological process of formation of new blood vessels form the pre-existing vasculature and is a continuous natural phenomenon vital in embryo development, during the ovarian cycle and wound healing. A precise balance of stimulatory and inhibitory factors maintains and regulates the process of angiogenesis [2] and any deviation from the balance results in either enhanced or retarded angiogenesis. It has now been established as common denominator underlying all cancers, cardiovascular disease, blindness, arthritis, complications of AIDS, diabetes, Alzheimer's disease, and more than 70 other major health conditions affecting children and adults in developed and developing nations [3].

Vasculogenesis, the formation of blood vascular system, is the earliest event in development of embryo. The precursors of both endothelial cells forming the blood vessels and hematopoietic cells forming the blood cells are a class of cells called the hemangioblasts [4]. The hemangioblasts further differentiate to form the angioblasts, which in turn aggregate to form the blood islands. In course of time the blood island start fusing together resulting in the formation of primary blood vascular plexus. The plexus is composed of a network of fine capillaries made up of the endothelial cells [4]. However, at the stage of primary plexus formation, the vessels have already acquired arterial or venous character [5]. Hence, it is evident that cell specificity is genetically programmed.The process of vasculogenesis is completed with formation of primary vascular plexus. Further transformations of the vascular net occur during angiogenesis, which is basically a physiological process involving formation of new vessels from the existing ones. Angiogenesis proceeds by extensive expansion of the primary vascular plexus. It undergoes numerous transformations to form a highly organized vascular net [4]. Initiation of angiogenesis is marked by the destruction of the walls of the existing blood vessels followed by an activation of the endothelial cell proliferation and migration. After proliferation, the endothelial cells assemble to form tube like structures around which the vessel walls are later formed [4]. The capillaries then fuse to form larger vessels, arteries and veins. The walls of capillaries and fine vessels consist of a single layer of cells (pericytes) but walls of arteries and veins are formed by several layers of smooth muscle cells [4]. Pericytes are known to have a mesenchymal origin, but the ontogeny is still not clear [6]. Pericytes are a heterogeneous group of cells capable of differentiating into a variety of mescenchymal cells like smooth muscle cells, fibroblasts and osteoblasts [6]. Hence, the vessels are known to be consisting of two main cell types, endothelial cells and mural cells. For understanding the mechanism underlying normal and tumor induced angiogenesis, it is important to understand the processes regulating the biological activity of these cells and their interaction among themselves. Formation and growth of new blood vessels is under stringent regulation activated only under the conditions of wound healing and in the uterine cycle.

Like all metabolically active cells and tissues, tumor cells require a continuous supply of oxygen and nutrients and removal of metabolic waste. Since cancerous cells have lost all check to growth and multiplication, they are metabolically far more active than their normal counterparts. Hence, they require an extensive vascular network within the tumor to fulfill the metabolic requirements. As solid tumors start developing, it is small in size and there is a free exchange of nutrients and oxygen and removal of wastes by simple passive diffusion [7, 8]. The tumor in this state is said to be in the pre-vascular condition. As the tumor starts enlarging and reaches up to a size of 2-3mm^3, exchange by diffusion is no more possible [7, 8]. A condition of hypoxia develops within the tumor which then stimulates the production of specific signaling molecules that further bring about the activation of a signaling cascade to induce the "angiogenic switch". Angiogenic switch can be defined as “a time-restricted event during tumor progression where the balance between pro- and anti-angiogenic factors tilts towards a pro-angiogenic outcome, resulting in the transition from dormant avascularized hyperplasia to outgrowing vascularized tumor and eventually to malignant tumor progression [9]”. Basically, there are two major forms of angiogenesis, sprouting angiogenesis and non-sprouting angiogenesis. Both the forms are observed during fetal development as well as in adults [3].

Interfering with the phenomenon of angiogenesis by targeting or blocking the process at intermediate steps has been an appealing strategy in cancer therapeutics for about four decades. However, even though the area has been so promising, most of the anti-angiogenic drugs and therapeutics are still in preclinical and clinical trial phase [10]. Only recently, two anti-angiogenic drugs, Avastin and Sutent have been approved by the FDA as solid tumor monotherapy [10]. Most of the anti-angiogenic compounds under preclinical and clinical trials are known to be cytostatic rather than cytotoxic. Hence, it would yield best results when administered in combination with conventional chemotherapeutic agents [10].

AN HISTORICAL OVERVIEW OF ANGIOGENESIS

First observations of angiogenesis date back to 1787.The word angiogenesis literally meaning “growth of new blood vessels” was first used by British surgeon Dr. John Hunter [11], to explain the blood vessels growing in antlers of reindeer. Vasculogenesis during reproduction in primates was first reported by Dr. Arthur Tremain Hertig [12], a pathologist from Boston, in the placenta of pregnant monkey in 1935. First observation of tumor angiogenesis was made by Dr. Judah Folkman [13] in 1971. He published his hypothesis in the "New England Journal of Medicine" which states that tumor growth is angiogenesis dependent. He put forward the concept that the tumor cells secrete certain diffusible molecules that induce the growth of new blood vessels towards the tumor. He also suggested that the neo-vascularization can be disrupted or inhibited by certain inhibitors of angiogenesis [13]. All solid and hematological tumors have an initial pre-vascular and subsequent vascular phase. The transition of pre-vascular to vascular phase depends on the release of pro-angiogenic factors. Acquiring angiogenic capabilities is the expression of progression from tumor growth to metastasis. The first angiogenesis inhibitor was discovered as a result of a joint effort by Dr. Henry Brem and Dr. Judah Folkman in 1975 [13]. In 1980s, the pharmaceutical industries started exploiting the field of angiogenesis in search of new therapeutic agents which could destroy or inhibit tumor formation by cutting off oxygen and nutrient supply. Basic fibroblast derived growth factor or bFGF was the first discovered angiogenic factor purified by Yuen Shing and Michael Klagsburn [14] in 1984 at the Harvard Medical School. The most important player of angiogenic mechanism, the vascular endothelial growth factor (VEGF), was discovered by Dr. Napoleone Ferrara [14] in 1989. However, an identical molecule called the vascular permeability factor (VPF) was reported by Dr. Harold Dvorak [14] earlier in 1983. It was in the same year of the discovery of VEGF that the first successful treatment of an angiogenesis-dependent benign tumor was carried out using Interferon 2α by a pediatric radiologist, Dr. Carl White [14]. The first anti-angiogenic candidate to undergo clinical trial on cancer patients was the drug TNP-470 which began in 1992. The great promises of anti-angiogenic therapy lead to the establishment of "The Angiogenesis Foundation" to improve the global efforts in this direction by facilitating development and trials of such therapeutic agents. An article published by Dr. Michael O'Reilly [14] in Nature in the year 1997 reports the complete regression of solid tumors following repeated cycles of anti-angiogenic therapy with Angiostatin and Endostatin [15-18]. The Food and Drug Administration (FDA) approved the use of angiogenesis stimulating laser treatment of severe, end stage coronary disease in 1998, and in 1999 for the treatment of age related muscular degeneration [14]. It was during this period that a massive wave of clinical trials for anti-angiogenic drugs started for cancer, muscular degeneration, diabetic retinopathy and psoriasis. Judah Folkman and Robert Kerbel [13] showed that cytotoxic chemotherapies may inhibit tumor angiogenesis at low doses. In the same year development of anti-angiogenic therapies for cancer was declared as a national priority by Dr. Richard Klausner [14], the director of U.S. National Cancer Institute. Avastin or Bevacizumab, entered into large scale clinical trials on cancer patients in 2003 [14]. It was first drug to attain such success in inhibiting blood vessel development in tumors and prolonging lives in cancer patients. The establishment of utility of Avastin as solid tumor monotherapy was further supported by the results of a pivotal phase III trial published in the New England journal of medicine in 2004 [14]. It revealed that the addition of this anti-VEGF monoclonal antibody to chemotherapy could significantly improve survival in metastatic colorectal cancer patients [19]. Based on these results, the FDA approved the use of Avastin for the treatment of advanced colorectal cancer and anti-angiogenic therapy was declared as the "fourth modality of cancer treatment" [14, 18, 20]. Later in 2004, Erlotinib or Tarceva, inhibitor of EGFR tyrosine receptor kinase was approved by the FDA for use in non-small cell lung carcinoma treatment (NSCLC) [14]. In 2005, Endostatin gained approval in China for the treatment of advanced lung cancer [14]. It is a potent inhibitor of angiogenesis by down regulation of multiple pro-angiogenic growth factors. Good results have been obtained in phase III trial with Sorafenib or Nexavar, a multi-tyrosine kinase inhibitor. It is now FDA approved second line therapy for advanced renal cancer [14]. Another multi-tyrosine kinase inhibitor, Sutent, obtained FDA approval as first line therapy for Gastrointestinal Stromal Tumor (GIST) and renal cancer in 2006. First time in 2006, a combination of Bevacizumab with Paclitaxel and Carboplatin was used and significantly improved the progression free survival, overall survival and response rates in NSCLC patients undergoing the treatment [14]. In 2007 results from a randomized phase III trial was published in the New England Journal of Medicine which showed that Sunitinib almost doubled the survival rate in metastatic renal cancer patients [14]. In November 2007, the FDA approved Sorafenib for the treatment of unresectable advanced hepatocellular carcinoma and was the first systemic agent which proved to be efficient in advanced liver carcinoma [14]. Based on the results from phase III trials, Avastin was approved for the treatment of breast cancer. It was the first anti-angiogenic agent to earn such approval. In the final trial, Avastin/Paclitaxel showed a double median progression free survival *versus* Paclitaxel alone in metastatic breast cancer [14].

TYPES OF ANGIOGENESIS

Angiogenesis is an intricate process that involves interactions between regulatory and effector molecules. Basically two types of angiogenesis are known; sprouting and non-sprouting type of angiogenesis. The classical form is the sprouting type of angiogenesis. It has been divided into a phase of sprouting and a phase of resolution by Pepper [21, 22].

Sprouting phase occurs in six steps:

- Increased vascular permeability and extravascular fibrin deposition.
- Vessel wall disassembly.
- Basement membrane degradation.
- Cell migration and extracellular matrix invasion.
- Endothelial cell proliferation.
- Capillary lumen formation.

The phase of resolution occurs in five steps:

- Inhibition of endothelial cell proliferation.
- Cessation of cell migration.
- Basement membrane reconstitution.
- Junctional complex maturation.
- Vessel wall assembly including recruitment and differentiation of smooth muscle cells and pericytes.

Apart from classical angiogenesis, some forms of non-sprouting angiogenesis have been described in tumors [23]. These include:

1. Intussusceptive vascular growth.
2. Co-option.
3. Formation of mosaic vessels.
4. Vasculogenic mimicry.

In intussusceptive type of vascular growth, a column of interstitial cells gets inserted into the lumen of a pre-existing blood vessel dividing the lumen to yield two blood vessels [21, 24, 25]. Fibroblasts and pericytes invade the column and extracellular matrix proteins get accumulated. Intussusception does not involve immediate proliferation of endothelial cells; instead there is a rearrangement and remodeling of existing vessels [24]. Since this process does not involve extensive cell proliferation, degradation of the basement membrane and invasion of surrounding tissue, it is metabolically economic over sprouting type of angiogenesis.

Co-option is a mechanism where tumors surround vessels in the organ or tissue of origin and incorporate these blood vessels. This process is important in cases where the tumors arise in or metastasize to vascular organs like the brain or lung [21, 26]. In yet another form of angiogenesis, the tumor cells along with the endothelial cells together form the luminal surface of the capillaries, thereby, giving rise to a mosaic vessel [21, 26]. About 15% of the vessels in human colon cancer xenograft models and in the biopsies of human colon carcinomas have been found to be mosaic channels lined by both endothelial cells and tumor cells [21, 26]. Vasculogenic mimicry was first observed in ocular melanoma. Here there is formation of vascular channels which are extracellular matrix rich tubular networks [27]. These tubular networks are devoid of endothelial cells but have circulating red blood cells. It has been observed in lung, prostate, ovarian and breast cancers [21, 28, 29]. However, the concept of vasculogenic mimicry is not widely accepted and only some investigators consider it as a type of non-sprouting angiogenesis [21, 22, 28]. Sprouting type of angiogenesis occurs under physiological and non-physiological conditions and the factors that determine the process that assist the sprouting mechanism has not yet been determined [21].

THE MECHANISM OF NORMAL ANGIOGENESIS

During initial phases of embryonic development, a network of blood vessels is formed by the process of vasculogenesis *i.e.,* the *de novo* synthesis of blood vessels from precursor endothelial cells called angioblasts, from which the adult vasculature is derived (Fig. **1**). The primitive network of blood vessels formed during embryogenesis is called the primary capillary plexus. It is formed by the conglomeration of the proliferating angioblasts. This endothelial cell lattice formed as a result of vasculogenesis then serves as the scaffold for angiogenesis [4, 30, 31] which is remodeled by sprouting and intussuseception. Most of the normal angiogenesis occurs in the developing embryo where the primary vascular tree is established with an adequate vasculature for the developing organs. In adults, angiogenesis occurs during the ovarian cycle and physiological repair processes like wound healing and tissue regeneration, but the turnover of endothelial cells occurring in adult vasculature is far lesser [30, 32].The newly formed microvessels then undergo maturation and remodeling in a step wise process (Fig. **1**). For the formation of new blood vessel sprouts, the pericytes or mural cells have to be first removed from the branching vessels followed by the degradation of endothelial cell basement membrane and extracellular matrix which are then remodeled by specific proteases like matrix metalloproteins [30, 33]. A newly formed matrix synthesized by stromal cells is then laid down. This newly formed matrix supports the proliferation and migration of endothelial cells with the help of soluble factors. After sufficient number of cell divisions, the endothelial cells get arrested in a monolayer forming a tube like structure by process called tubulogenesis. On the albumenal surface of the microvasculature, mural cells are recruited and the vessels uncovered by mural cells regress. Blood flow through the new blood vessels is then initiated. Angiogenesis under physiological conditions is a highly regulated process since it requires the induction of quiescent endothelial cells to get arranged into a monolayer forming a tube like structure. The tube then extends and branching occurs depending upon the extent of oxygenation required by the tissues for their metabolic activities. Regulation of angiogenesis depends on many positively and negatively influencing factors including soluble polypeptides, cell-cell and cell-matrix interactions and haemodynamic effects [4].

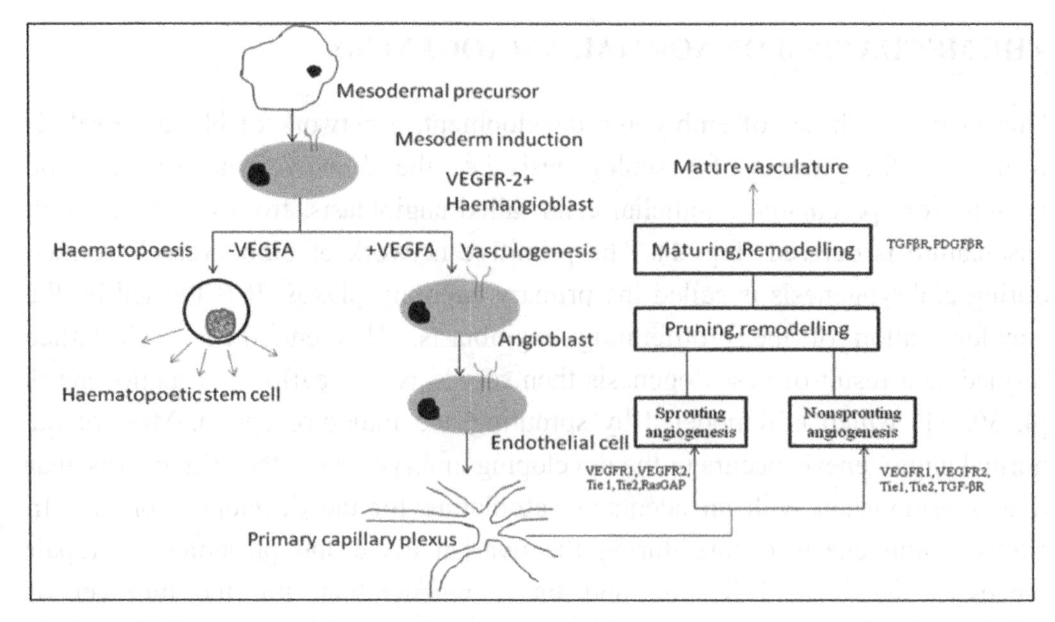

Figure 1: Development of blood vessel in embryogenesis and tissue repair. On mesoderm induction the precursor cells, haemangioblasts, form both angioblast and haematopoetic stem cells. Angioblasts aggregate to form blood islands and then the primary capillary plexus which ultimately forms the mature vasculature.

TUMOR INDUCED ANGIOGENESIS

A tumor is a population of host derived cells that has lost the capacity of self-regulating its growth and division leading to their aberrant proliferation. Like their normal counterparts, transformed cells also require a continuous supply of nutrients and oxygen for their metabolic functions and removal of wastes. Because of abnormally high rate of cell division, the cellular processes within the cancerous cells are accelerated causing an enhanced metabolic demand. To meet this enhanced metabolic demand and for the removal of cellular wastes, tumor cells require a network of blood vessels in their close proximity. Tumor angiogenesis is therefore the phenomenon of proliferation of a network of blood vessels that penetrates into the tumors, supplying nutrients and oxygen and removing waste products. Not only is the growth of solid tumor dependent on the process of angiogenesis, but also it regulates the establishment of precancerous microenvironment and metastasis in a hostile environment. In the pre-vascular stage the tumors grow up to 2-3 mm^3 in size containing about a million cells [34]. Oxygen and nutrients required for survival are taken up by simple passive diffusion. As the tumors start enlarging, diffusional transport of nutrients and wastes is no longer possible. A cascade of events leading to the proliferation of tumor vasculature then sets in. Several pro and anti-angiogenic factors are released and/or induced that are crucial for the regulation of endothelial cell proliferation, migration, induction or inhibition of apoptosis, cell to cell and cell to matrix adhesion. In response to hypoxia within the growing tumor, the tumor cells start releasing signaling molecules to the surrounding normal tissue. A signaling cascade sets in that activates certain genes in the host tissues which in turn induce the formation of proteins encouraging the growth of new blood vessels. An elaborate understanding of the mechanisms of tumor angiogenesis has led to the development of several effective anti-angiogenic and anti-metastatic agents which would affect the cancer angiogenic switch and are being tried and tested [34].

The Mechanism of Tumor Induced Angiogenesis

Like normal angiogenesis, tumor induced angiogenesis is a step wise process involving a complex interplay of cells, a variety of soluble factors and the Extracellular matrix (ECM) components. Sequentially, basic steps involved are [34]:

- Degradation of the basement membrane by proteases.
- Migration of the endothelial cells into the interstitial space and sprouting.
- Proliferation of the tip cells and lumen formation.
- Formation of anastomoses.
- Establishment of blood flow.

Angiogenic response in the microvasculature involves certain changes in the cellular adhesive interactions between adjacent endothelial cells, pericytes and the surrounding ECM.

The Angiogenic Switch

During neo-angiogenesis, the tumor cells exhibit an overexpression of one or more positive regulators of angiogenesis and promote the mobilization of angiogenic proteins from the ECM. Sometimes host cells like macrophages, which produce their own angiogenic proteins, are also recruited [35]. Neo-angiogenesis in tumors is driven by angiogenic growth factors secreted by the tumor cells that interact with their surface receptors expressed on ECs. VEGF and bFGF are the most commonly found angiogenic growth factors. They bind to the tyrosine kinase receptors on the membrane of ECs [36] (Fig. **2**). Binding brings about dimerization of the receptors and autophosphorylation of tyrosines on the receptor surface thus initiating several signaling proteins like PI3-kinase, Src, Grb2/m-SOS-1 (a nucleotide exchange factor for Ras) and signal transducers and activators of transcriptions (STATs) possessing src-homology-2 (SH-2) domains [37].

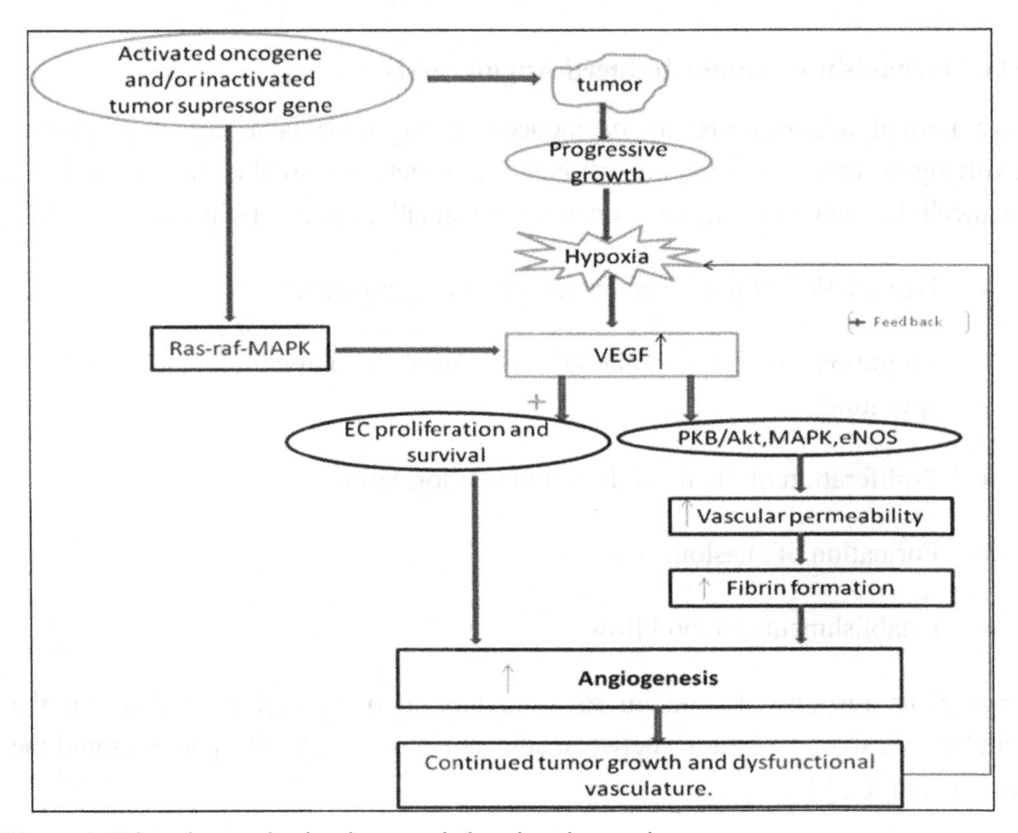

Figure 2: Triggering mechanism in tumor induced angiogenesis.

A number of signaling pathways crucial for the activation of cell cycle machinery are activated by the binding of SH-2 domains of these proteins to phosphotyrosine residues of the receptor tyrosine kinases [36, 37]. The most important and well-studied pathway passes through the GTP-binding protein Ras, activating the mitogen activated protein kinase MAPK cascade which in turn leads to the activation of transcription factors [37] (Fig. **2**). Activation of pro-angiogenic growth factors is necessary but not sufficient condition for the initiation of angiogenesis within tumors. There has to be a simultaneous downregulation of the negative regulators or inhibitors of angiogenesis. When there is a dominance of angiogenic factors within the tumor locality, neo-vascularization begins. It develops into capillaries or differentiates into mature venules and arterioles. If there is a dominance of the angiostatic factors, the newly forming vessels degenerate. The angiostatic factors cause regression of the vessels by either inducing apoptosis or by cell cycle arrest.

A complex interplay of many factors is involved in the induction and progress of both normal and tumor induced angiogenesis. These include the following factors

Soluble Factors

Vascular Endothelial Growth Factor

It is the most important and well characterized of all angiogenic growth factors also known as vascular permeability factor (VPF). There are six known isoforms of VEGF which are 121, 88, 165, 183, 189 and 206 amino acids long generated by the alternative splicing of a single gene. The most commonly expressed isoform is $VEGF_{165}$ [4, 31]. All isoforms of VEGF have similar biological effects but act in different microenvironments. $VEGF_{121}$ and $VEGF_{165}$ are known to act in extracellular environment while $VEGF_{189}$ and $VEGF_{206}$ are found to be cell or matrix associated owing to their affinity for heparin sulfates [33]. VEGF is a highly conserved, 34-45 kD dimeric glycoprotein. It loses activity in presence of a reducing agent because of its disulfide bond. Low levels of VEGF are constitutively expressed in many tissues but high levels of the protein are detected in the tissues which require angiogenesis like in placenta, corpus luteum and in healing wounds and almost all of human tumors [4, 31]. Many mescenchymal and stromal cells also produce VEGF.

VEGF receptors: VEGF is known to bind to three tyrosine kinase receptors *i.e.*, VEGFR1 or flt-1, VEGFR2 or kdr/flk-1 and VEGFR3 or flt-4 (Fig. **3**). Earlier these receptors were thought to be restricted to endothelial cells. However, functional VEGF receptors have recently been observed on normal cell types including vascular smooth muscle cells and macrophages. These receptors are members of the flt or 7-Ig family of genes characterized by seven extracellular immunoglobulin like domains, a membrane spanning domain and a conserved intracellular tyrosine kinase domain. The highest affinity receptor for VEGF is VEGFR1, K_d 10-30 pM. It is expressed in the endothelium of adult and embryonic mice, in case of healing wounds, vascular smooth muscle cells and monocytes. However, no direct migratory, proliferative or cytoskeletal effects are known to be mediated by VEGFR1. Instead, endothelial cell mitogenesis, chemotaxis and shape changes are mediated by the low affinity receptor VEGFR2 [17]. It is expressed on endothelial and hematopoietic precursors and in the proliferating endothelial cells of the embryo. But, in the quiescent endothelium of adult vasculature, VEGFR-2 RNA is dramatically reduced.VEGFR-3 is expressed on adult lymphatic endothelium. It is thought to be involved in lymphangiogenesis [4, 34] (Fig. **3**). It does not bind VEGF but complexes VEGF related proteins VEGF-C and VEGF-D. Another non flt family receptor of VEGF is neuropillin. It is found on surface of some tumors and endothelial cells. These receptors differ from VEGFR in both binding affinity and molecular mass. No detailed signal transduction is known after the binding of nuropillins to VEGF. But, expression of NRP1 and NRP2 has been found to be necessary for angiogenesis. Mice with damaged or defective Nrp1 die during embryonic development [18]. Neuropillins are the co-receptors of VEGF. They present VEGF to VEGFR2 receptors enhancing their binding efficiency [4, 32]. The neuropillins are precisely specific for arterial and venous blood vessels; Nrp1 is restricted to arteries while Nrp2 is found on veins and lymphatic vessels [4, 31]. Nrp2 expression is important for lymphangiogenesis (Fig. **3**). It interacts with VEGFR3 and binds VEGFC and VEGFD. It has been found that mouse deficient in VEGF and flk-1 gene are in most cases devoid of vascular structures and hence have defective events characterizing early vasculogenesis [38, 39]. But, on the contrary, flt-1 null mice form vascular structures having impaired vessel assembly [38, 39]. These facts indicate that flt-1 is involved in vascular remodeling in angiogenesis rather than *de novo* synthesis of blood vessels in vasculogenesis.

A number of factors regulate the response of cells to VEGF including the expression of neuropillin receptors. Various isoforms of VEGFA produced by the alternative splicing of a single gene is an important regulating factor. The accessibility of different VEGF isoforms for interaction with cells depends on the proteolytic release of isoforms attached to the extracellular matrix, hence depends on the activity of different proteases. Alternative splicing can also form VEGF isoforms having anti-angiogenic activity [31, 40]. These are called the b-isoforms which differ by their last six C-terminal amino acid residues [31, 40].

Functions of VEGF: VEGF, the mitogen specific to endothelial cells, after binding Flt-1 and Flk-1, induces the formation of a second messenger by the hydrolysis of inositol. This brings about autophosphorylation of the receptors in presence of heparin-like molecules. Phosphatidylinositol metabolic signal transduction pathways activate MAP kinases in EC and thereby VEGF exerts its mitogenic effect by promoting EC proliferation. It stimulates the production of urokinase-like plasminogen activator (uPA), tissue type plasminogen activator (tPA) and plasminogen activator inhibitor-1 (PAI-1) by the endothelial cells, proteolytic enzymes, tissue factors, and interstitial collagenase [36]. Plasminogen activators in turn activate plasminogen to plasmin, which then break down ECM components. In addition to remodeling the basement membrane, uPA also binds to uPAR, mediating intracellular signal transduction in the endothelial cells. The phosphorylation of focal adhesion proteins and MAP kinase activation follows, bringing about proliferation and migration of endothelial cells [34]. Hence, it leads to induction of a balanced system of proteolysis necessary for remodeling of extracellular matrix components for the purpose of angiogenesis.

In addition to aiding in endothelial cell proliferation, VEGF is also known to enhance cell migration during branching of endothelium from the existing vasculature and also acts as a survival factor by inhibiting apoptosis in endothelial cells [27, 41]. It enhances endothelial cell permeability by increasing the activity of vesicular-vacuolar organelles in endothelial cells lining the small vessels. These facilitate transport of metabolites between luminal and ablumenal plasma membranes [31]. VEGF may also enhance endothelial cell permeability by bringing about the activation of PKB/Akt, endothelial nitric oxide synthase or eNOS and mitogen activated protein kinase (MAPK) dependent signal transduction cascade [36]. It brings about the rearrangement of cadherin/catenin complexes thereby loosening the adherent junction between the endothelial cells in a monolayer [42]. As a result of enhanced vascular permeability, there occurs an extravasation of plasma proteins and an extracellular matrix is formed to support the migration of endothelial and stromal cells. VEGF also induces the expression of $\alpha_1\beta_1$, $\alpha_2\beta_2$ and $\alpha_v\beta_3$-integrins which promote the migration and proliferation of cells and reorganization of the matrix. It is a potent pro-survival factor for the endothelial cells of the newly formed vessels. Endothelial survival factors VEGF, angiopoetin and $\alpha_v\beta_3$ integrin suppress p53, p21, p16 and p27 and the pro-apoptotic protein Bax, at the same time variably activating PI3k/Akt, p42/44 MAPK, bcl2, A1 and the survivin pathways [36, 37]. The recruitment of the activator protein-2 complex, AP-2/Sp1, on the -88/-66 position of the VEGF promoter by p42/p44 MAP kinases and direct phosphorylation of the HIF-1α induces and promotes VEGF expression by activating its transcription [43, 44]. Ras dependent signaling pathways have also been speculated to synergistically act with PI3K/Akt pathway, in VEGF signaling. Downregulation of H-rasV12G brings about a profound tumor regression characterized by an initial massive apoptosis of tumor as well as host derived endothelial cells. The resistance to induction of apoptosis on enforced expression of VEGF indicates that an intact Ras dependent signaling pathway is indispensible for mediating the apoptosis inhibitory effect of VEGF. Additionally, it has also been observed that VEGF induces an enhanced expression of the serine threonine protein kinase Akt *via* the KDR/Flt-1, which is a downstream target of PI3-kinase. It interferes with the various apoptosis signaling pathways thereby blocking apoptosis and at the same time promoting endothelial cell migration and enhancing HIF expression (Fig. **4**). This mechanism ensures oxygenation and vascularization of hypoxic environment in healthy as well as pathological conditions. Hence, VEGFR2 and the PI3K/Akt signal transduction pathway have been identified as crucial elements in VEGF induced endothelial cell survival. On the other hand, downstream effectors mediate the anti-apoptotic effect of VEGF including Akt-dependent activation of the endothelial nitric oxide synthase(eNOS) (Fig. **4**). Moreover, the VEGF-induced activation of the MAPK/extracellular signal regulated kinase (ERK) pathway and inhibition of the stress activated protein kinase/c-Jun amino-terminal kinase pathway is also implicated in the anti-apoptotic effect mediated by VEGF

[34]. In the tumor cells around necrotic areas the level of expression of VEGF mRNA is the maximum, and it is partially mediated by the binding of HIF1α to a corresponding binding site on the VEGF promoter and partially by the stimulation of stress induced PI3K/Akt pathway [34]. As tumor growth progresses, the level of hypoxia within the tumor keeps on increasing thereby inducing the up-regulation of pro-angiogenic factors like VEGF, bFGF, IL-8, TNF-α, TGF-β *etc.* These factors in turn enhance vessel hyper permeability, release of plasma proteins, protease induction, formation of fibrin, proliferation and migration of endothelial cells *etc.* An enhanced angiogenesis and fibrinolysis results in continuous growth of dysfunctional vasculature which then acts as a positive feedback loop creating a continuous hypoxia within the tumors [34, 37]. Enhanced HIF expression is known to stimulate VEGF expression (Fig. **4**). Under hypoxic conditions, Hypoxia inducing factor (HIF) binds to the *cis* element in the VEGF promoter, enhancing VEGF gene transcription and mRNA stability [45]. VEGF not only stimulates the entire angiogenic process but also functions as a physiological sensor that stimulates angiogenesis under low oxygen conditions [45].

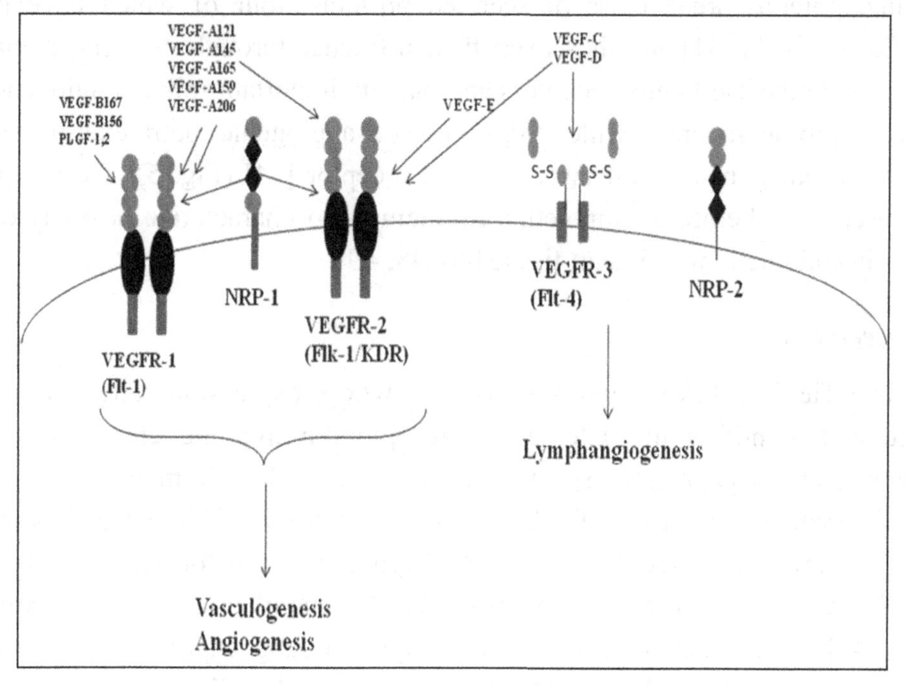

Figure 3: Interaction of members of VEGF family with their receptors.

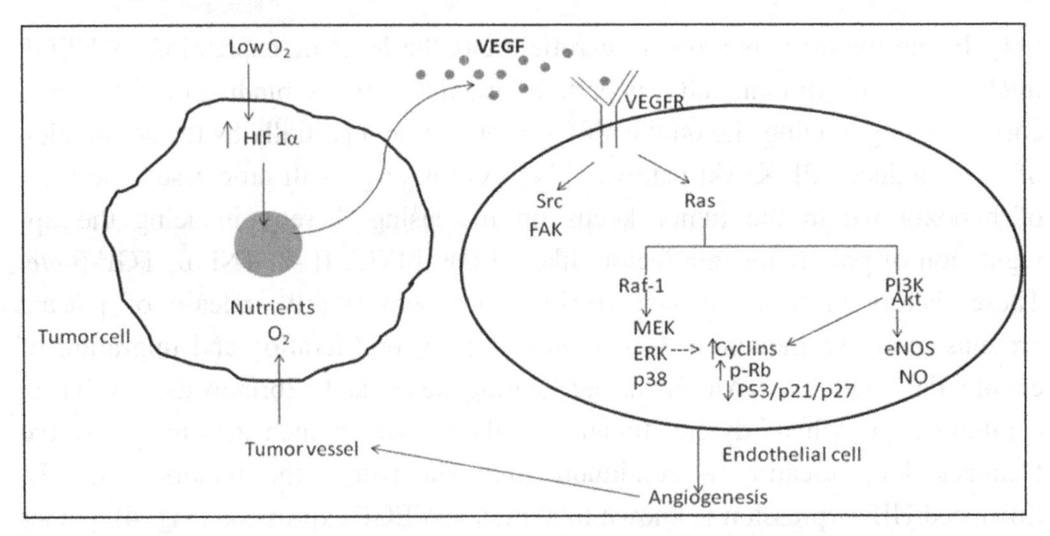

Figure 4: Tumor induced angiogenesis controlled by VEGF. Under conditions of hypoxia the tumor cells secrete VEGF which activates KDR/Flk-1 in tumor associated endothelial cells thereby inducing Ras dependent angiogenic signaling pathway Raf1/MEK/ERK/p38MAPK and PI3K/Akt/eNOS together controlling progression of cell cycle and angiogenesis.

Angiopoietins and Tie Receptors

The angiopoietins are a class of secreted proteins, four of which have been identified so far [4, 31] and they exert their influence through the TIE receptors. These bind to the Tie family receptors and play an important role in angiogenesis. Ang-1 is pro-angiogenic while Ang-2 is anti-angiogenic both exerting their agonist and antagonist action through Tie-2 receptor [34] (Fig. **5**). Tie receptors were discovered before angiopoietins in attempts to characterize novel tyrosine kinases in endothelium and heart tissue [46, 48, 49].

Tie Receptors

Tie 1 and Tie 2 or Tek are two tie receptors whose expression pattern has been studied and identified till date. These receptors mimic the effect of VEGF receptors and are apparently specific for vascular endothelium [48]. However, cells of hematopoetic lineage, like K562 which is a tumor cell line, have also been found to express these receptors [31, 48]. High expression for Tie-1 mRNA has been observed in embryonic angioblasts, vascular endothelium and endocardium [4, 31, 46]. In adults, the expression is high on lung capillaries but weakly expressed in the endocardium [31]. Tie 2 is expressed earlier during embryonic development but is detected on the same sites as Tie 1 receptors. It is lesser evident in adult endocardium and vasculature endothelium. Such an expression pattern of the Tie receptors gives a clue about its involvement in developmental angiogenesis. Extensive edema, hemorrhage and prenatal death in 14.5 days have been observed in Tie deficient mice [46]. Mice deficient in Tie 2 receptors show premature embryonic death and have a reduced number of endothelial cells in blood vessels compared to wild variety, underdeveloped hearts, vasodilation and abnormal network formation with a lack of sprouting and branching blood vessels. These evidences suggest that even though both Tie 1 and Tie 2 are important for vascular integrity, Tie 2 is specifically important for sprouting and branching of blood vessels.

Angiopoietins

Angiopoetins are secretory protein ligands, 70 kDa, specific for Tie 2, but ligand specific to Tie 1 has not yet been recognized. Angiopoetin1 (Ang1) mRNA is the best studied member of this family, found in embryonic heart myocardium surrounding the endocardium and in later stages in the mesenchyme surrounding the blood vessels [4, 31]. *In situ* localization shows that Ang1 is produced not only by the mescenchymal cells closely associated with endothelium but also by human neuroepithelioma and mouse myoblast cell lines [31]. Ang1 is not known to induce endothelial cell proliferation or tube formation under experimental conditions but induces sprout formation from confluent endothelial cells in a three dimensional experimental setup. Hence, Ang1 deficient mice show defect in vasculature similar to Tie 2 deficient mice. The primary vasculature is normally developed but the mice die in 12-13 days of embryonic development because of inefficient vascular remodeling. The heart development is most defective with severely impaired endocardial and trabecular development [31, 47-51]. In addition, branching and organization of the vascular network into large and small vessels is hampered and the blood vessels are dilated. Moreover, periendothellial cells known to have significant role in normal vessel branching, are absent from the underdeveloped tissue folds. Overexpression of Ang1 in transgenic mice induces numerous, extensively branched blood vessels, larger in diameter than the wild type mice. These vessels are more resistant to leakage induced by inflammatory agents or co-expressed VEGF [31]. Ang1 improves the strength and stability of endothelium from the newly developed sprouts which can be related to vessel branching and remodeling induced by the Ang1 signaling cascade. Angiopoetin 2 or Ang2 is a structural homologue of Ang1 which binds to Tie 2 and also antagonizes Ang1 [31, 46]. Hence, the vessel branching and remodeling stimulated by Ang1 signaling appears to be mechanistically related to Ang1's ability to increase the girth and stability of endothelium in newly formed angiogenic sprouts. Ang2 inhibits Ang1 mediated Tie2 autophosporylation in endothelial cells which expresses endogenous Tie2 but at the same time it also stimulates Tie2 autophosphorylation in NIH 3T3 ectopically expressing Tie2, suggesting that Ang2 is an Ang1 antagonist conditionally only in the context of vasculature [31]. In embryonic condition, Ang2 is expressed in the dorsal aorta and in the punctuate regions of the vasculature and in adults it is expressed in placenta, ovary and uterus that are the primary sites of angiogenesis. At the same time it has been observed that an overexpression of Ang2 in vascular structures is lethal to embryos with similar but much more severe defects as those in Ang1 and Tie2 deficient mice. It is evident from the above mentioned experimental facts that Ang2 antagonizes Ang1 in vasculature development *in vivo* and acts as a check on the Ang1/Tie2 mediated angiogenesis thereby preventing excessive branching and sprouting of blood vessels by promoting destabilization of blood vessels [31, 49-51]. Destabilization of vessels due to Ang2 causes the sprouts to be plastic and susceptible to the action of remodeling factors. There exists a crucial balance maintained by the angiogenic processes stimulated by VEGF and Ang1 and inhibited by Ang2 which regulates the extent of normal blood vessel remodeling [50].

Fibroblast Growth Factor (FGF)

This is a large family of structurally related growth factors having both acidic (pI = 5) and basic (pI=9.6) members which play a major role in angiogenesis [4, 31] (Fig. **5**). Acidic fibroblast growth factors were the first growth factor found having a role in angiogenesis. Both aFGF and bFGF are known to stimulate angiogenesis in endothelial cells *in vitro*. In addition to stimulating endothelial cell proliferation and migration, FGF also induces the cell to produce plasminogen activator and collagenase [52]. In three dimensional collagen matrix, bFGF induces tubulogenesis in endothelial cells. However, FGF is not selectively mitogenic for endothelial cells as is the case with VEGF. It induces proliferation and migration in most of the embryonic cells of mesodermal and neuroectodermal origins like pericytes, fibroblasts, myoblasts, chondroblasts and osteoblasts [31]. Like VEGF, FGF induces sprouting in pre-existing blood vessels towards an implanted bolus *in vivo* in the cornea and chick chorioallantoic membrane [52, 53]. However, development of vasculature in mice deficient in both aFGF and bFGF is not affected indicating lack of role of FGF *in vivo*. A signal sequence is absent in aFGF and bFGF, hence it is not a secretory protein. Most of it is cytoplasmic or remains bound to extracellular matrix because of its strong affinity for heparin. This suggests that FGF may be produced as a result of cell disruption due to injury having a regenerative role in tissue repair and is deposited locally in the extracellular matrix. It is quite evident that bFGF does not play a role in angiogenic responses but is necessary for blood vessel remodeling in tissue repair. FGF along with their specific receptors are overexpressed in most of the cancers and are important for development of tumor vasculature as well as endothelial cell survival. It has been precisely observed by Pardo *et al.,* [54] that bFGF induces the expression of bcl-XL and bcl-2, the anti-apoptotic proteins, *via* MEK/ERK signal transduction. An overexpression of bFGF significantly enhances VEGF mRNA expression while the bFGF induced tumor development is hindered by KDR/Flk-1 mAb. This suggests a synergistic role with VEGF in tumor progression [34, 55]. bFGF also induces an increased expression of VEGF mRNA in vascular smooth muscle cells and increased VEGF receptors in microvascular ECs. Both aFGF and bFGF are mitogenic for endothelial cells and promote their migration and the production of plasminogen activator (PA) and collagenase responsible for the degradation of the basement membrane [36]. These are the key players in production of the extracellular matrix and the release of matrix metalloproteinases (MMPs) for selective degradation and reorganization of the extracellular matrix. FGFs bind to their high affinity receptors thus activating the intrinsic tyrosine kinase followed by a cascade of events which in turn induce early gene transcription and cell proliferation. In order to induce a proliferative signal through the tyrosine kinase receptors, these factors need to interact with heparin sulfate [34, 36]. The receptors on binding to the ligand dimerize and transphosphorylate at the tyrosine residue.

Platelet Derived Growth Factor (PDGF)

Initially isolated and purified from platelets, the Platelet derived growth factor (PGDF) is a protein produced by multiple sources including fibroblasts, keratinocytes, myoblasts, astrocytes, epithelial cells and macrophages [31, 52]. It is a 45kDa protein which is either homodimeric or heterodimeric composed of two PGDF chain A and B. Most of the cells express both the isoforms of the protein but in a few only one isoform is expressed. Even the receptors are dimeric in nature composed of complexes between α and β subtypes. Of the two known receptors, α subtype can bind both the A and B chains of PGDF molecule while the β receptor can only bind to the PGDF B chain, hence only the α receptor can bind all the three PGDF isoforms [52, 53].

Transforming Growth Factor-β

It is a family of highly conserved, 25kDa proteins. These multifunctional polypeptides regulate cell growth and differentiation, matrix composition, cellular adhesion, angiogenesis as well as immune functions [34]. These are homodimeric disulfide linked cytokines, a typical member of which is the TGF-β. Prior to secretion from the cell, a furin peptidase cleaves the precursor to generate a C-terminal peptide chain, 112 amino acid long, that noncovalently associates with the N-terminal pro region called the latency associated peptide or LAP and then dimerizes to form the mature TGF-β [31, 34]. Secretory TGF-β is biologically inactive, incapable of binding to TGF-β receptors. The latent complex is activated by proteases such as plasmin and cathepsin D, low pH, chaotropic agents such as urea, and heat [56, 57]. Under living conditions, latent TGF-β is known to be activated by an exposure to low pH or on cleavage by proteases. Constitutive expression of TGF-β has been observed on a wide variety of normal and transformed cells. Its receptors are expressed on all mammalian and avian cells. It is found in the extracellular matrix of many tissues and even the microvasculature. Both pericytes and endothelial cells produce TGF-β. Hence, the effect of TGF-β is diverse influencing an array of cell types including those of the vasculature. Its various functions include the stimulation and inhibition of cell proliferation, controlling cell adhesion by regulating the extracellular matrix production, production of protease inhibitors and integrins and the induction of cellular differentiation.

Several *in vitro* studies indicate the importance of TGF-β in vascular cells. One study claims that it acts as a growth stimulating factor at low dosage but inhibits growth at higher doses. TGF-β influences endothelial cell migration and tube formation in a collagen gel, processes which characterize angiogenesis *in vitro*. As in case of endothelial cell proliferation, TGF-β may either stimulate or inhibit endothelial tube formation *in vitro*. At doses as low as 0.5ng/ml it stimulates tube formation, but at higher dose about 1-5ng/ml tube formation is inhibited [31].

The angiogenic activity of TGF-β is isoform specific. TGF-β2 has no effect on tube formation *in vitro* at low concentrations unlike TGF-β1 but does induce tube formation at higher doses [58-60]. The proteolytic activity of TGF-β may be responsible for its effect on vascular tube formation. It can inhibit the synthesis of proteases like transin and can also stimulate synthesis of protease inhibitors like tissue inhibitors of metalloproteinase (TIMP). It is capable of producing an overall anti-proteolytic activity by modulation of the urokinase-like plasminogen activator uPA [59, 61] and plasminogen activator inhibitor PAI levels. Cumulatively, these effects of TGF-β prevent the remodeling of the matrix and hence inhibit angiogenesis. At the same time some other studies suggest that TGF-β is a promoter of angiogenesis by another mechanism. On co-culturing endothelial cells with pericytes or vascular smooth muscle cells, plasmin cleaves the latent TGF-β to yield the active TGF-β influencing both the endothelial cells and pericytes. Active TGF-β mediates the inhibition in endothelial cell growth observed upon endothelial cell: mural cell contact [60, 62, 63]. This growth factor is also known to bind to the pericytes and induce vascular smooth muscle actin (VSMA) expression and myogenic determination.

Hence, from these studies it can be inferred that TGF-β might have an important role in establishment of the structural integrity of the newly sprouting capillaries during angiogenesis. It aids the formation and strengthening of vessel walls by promoting the inhibition of endothelial cell proliferation and mural cell differentiation while tube assembly is stimulated by its matrix modulating effects.

Depending upon the experimental conditions, varying results have been obtained *in vivo* suggesting a role of TGF-β in angiogenesis. It induces robust angiogenesis when subcutaneously administered into mice or implanted into rabbit cornea. However, in most of the cases new blood vessel formation is always accompanied by inflammation [31, 58]. Since TGF-β is known to induce chemotactic response in a variety of cells including monocytes and fibroblast, its effect on angiogenesis may be indirect *i.e.*, it might mediate angiogenesis by recruiting these cells which then produce the factors and enzymes which in turn mediate the process of angiogenesis directly.

TGF-β's failure to induce angiogenesis *in vivo* may be because the latent form of the factor has not been activated by enzymatic cleavage. Possible role of TGF-β in angiogenesis has also been suggested by some genetic studies. In mice embryos deficient in TGF-β1, mesodermal cells normally differentiate into endothelial cells but premature death is observed in the embryonic stage due to a defective yolk sac vasculature as well as hematopoetic system [63, 64]. The walls of the vessels are weak and frail because of disruption of cell to cell contact between the endothelial cells. Similar response has been observed in mice deficient in TGF-β receptor I. All these observations indicate a role of TGF-β in establishment of vessel wall integrity. Altogether it can be summarized that TGF-β has a role in development of vasculature *in vitro* as well as *in vivo*.

Other Soluble Factors

Some other soluble factors have been observed to be involved in the process. However, the effects are not much wide spread. Some of these factors exhibiting angiogenic properties are the Tumor necrosis factor-alpha (TNF-α), epidermal growth factor (EGF), transforming growth factor-alpha (TGF-α), and the colony stimulating factors (CSFs) [31]. TNF-αβγ primarily involved in inflammation and immunity, has many properties similar to TGF-β like stimulation of angiogenesis *in vivo* and tubulogenesis *in vitro* and the inhibition of endothelial cell growth. EGF and TGF-α, both bind the EGF receptor and both are mitogenic for endothelial cells *in vitro* and induce angiogenesis *in vivo*. Granulocyte-colony stimulating factor (G-CSF) and granulocyte macrophage-colony stimulating factor (GM-CSF), proteins involved in growth and differentiation of hematopoietic precursors, to some extent, induce migration and proliferation of endothelial cells. Apart from these growth factors, there are other proteins which to some extent influence angiogenesis like angiogenin, a polypeptide obtained from human adenocarcinoma cell line. Strangely, it induces angiogenesis in chorioallantoic membrane (CAM) and rabbit cornea but is neither mitogenic nor chemotactic for endothelial cells *in vitro* [31]. It can also bind to the extracellular components thereby supporting the adhesion and spreading of endothelial cells *in vitro*. Angiogenin also binds to extracellular matrix components and can support the adhesion and spreading of endothelial cells *in vitro*. But the synthesis of angiogenin is negligible in fetus where angiogenesis mainly takes place; it is mostly expressed in adult stage. Another polyribonucleopeptide, angiotropin, induces random capillary endothelial cell migration and tube formation but is non-mitogenic for the endothelial cells. It is also known to stimulate angiogenesis in the CAM, cornea, and ear lobe accompanied by epidermal and stromal cell proliferation. In addition to the above mentioned proteins, some non-peptide low molecular weight molecules like prostaglandins, nicotinamide and monobutyrin, appear to have some contribution to the process of angiogenesis, but their roles are controversial and their mechanisms of action are still unknown. These soluble factors have been derived from more than one source and have a direct or indirect effect on the endothelial and mural cells. The involvement of such large number of proteins and non-protein factors in angiogenesis gives a clear idea of its complex nature and also of the numerous modes positive and negative regulations of the process.

Membrane-Bound Factors

Many membrane bound molecules requiring cell-cell or cell-matrix contact have pronounced effect on angiogenesis. Integrins, cadherins, and ephrins are few such endothelial membrane proteins that mediate many functions involved in blood vessel assembly, particularly, integrin a_vb_3, VE cadherin, and ephrin-2B [65]. These are known to have important roles in normal angiogenesis.

Integrin

Integrins are a group of heterodimeric complexes composed of αα and ββ subunits [31]. These are the receptors for extracellular matrix proteins and membrane-bound polypeptides on other cells. There exists a diverse array of as many as 20 different integrins formed by a combination of 16α and 8β subunits [4, 31]. The matrix substrates for integrins are polysaccharide glycosaminoglycans and fibrous proteins like fibronectin, vitronectin, collagen, laminin, and elastin. Some integrins bind short peptide sequences, such as the RGD (Arg-Gly-Asp) sequence found in fibronectin and vitronectin, but others recognize three dimensional conformations. The process of angiogenesis involves the invasion of the ECM and migration of endothelial cells through it. Integrins are important for the process because cell to matrix interactions are mediated by integrins and play an important role for vascular remodeling. Its involvement in neovascularization is suggested by the fact that it mediates attachment, spreading, and migration of endothelial cells *in vitro* and is localized to the tip cells of the capillary sprouts during wound healing. $\alpha_v\beta_3$ is abundant on angiogenic blood vessels in granulation tissue but not on normal skin vessels [64]. It has also been observed that the neutralizing antibodies inhibit bFGF-induced vessel sprouting in the CAM while it does not affect the pre-existing vessels which again suggest a role of these integrins in angiogenesis. $\alpha_v\beta_3$ is also highly expressed on angioblasts before and during vasculogenesis. Lack of $\alpha_v\beta_3$ expression leads to an abnormal patterning of blood vessels in the developing embryos [31]. Therefore, $\alpha_v\beta_3$ appears to mediate endothelial cell functions *in vitro* and important role in angiogenesis *in vivo*. In addition to binding to the ECM components, $\alpha_v\beta_3$ also binds to matrix metalloproteinase-2 localizing the active form of the enzyme to the tips of the developing vessels. Hence, $\alpha_v\beta_3$ may be involved in the regulation of localized degradation of the ECM and migration of the endothelial cells by adhering to the modulated matrix. $\alpha_v\beta_3$ ligation also leads to MAP kinase activation thereby suppressing apoptosis in endothelial cells [31, 60, 64, 66]. $\alpha_v\beta_3$ integrin may therefore mediate endothelial cell survival in many ways, by activating intracellular pathways promoting proliferation and activation indirectly mediating the process of angiogenesis (Fig. **5**). Additionally, other integrins have also been implicated in regulation of angiogenesis to a lesser extent [31].

VE-Cadherin

Cadherins are a family of Ca^{+2}-binding transmembrane protein molecules responsible for homotypic cell to cell interactions and also mediate a wide range of other functions in many cell types [31]. The intracellular domain of cadherins binds to β-catenin and plakoglobulin both of which are anchored to cortical actin by β-catenin. These proteins also help in maintaining the intracellular levels of beta catenin and plakoglobulin [67]. These molecules when released from the cadherins, translocate into the cell nucleus and regulate transcription of specific gene thereby mediating intracellular signaling. Basically, two types of cadherins are expressed on the endothelial cells, VE-cadherin, localized to adherens junctions exclusively in endothelial cells and N-cadherin, not found at the cell-cell contacts [60]. The importance of VE cadherins have been implicated from a number of studies. Enhancement of endothelial cell permeability by VEGF is accompanied by the phosphorylation of the tyrosine residue and dissociation of VE-cadherins suggesting a role of these proteins and passage of molecules across the endothelium. Contact inhibition of endothelial cell growth is regulated by VE-cadherins [68]. Inhibition of cell proliferation helps in maintaining a stable monolayer of cells in the vessel walls. It has been observed that VE-cadherin deficient mice show seriously abnormal vasculature. Differentiation of angioblasts to endothelial cells and primary capillary network formation is normal but the later stages of vascular development are severely impaired [69]. The overall function of cadherins is not only to establish a junctional stability between the endothelial cells in vessel wall but also to enhance the endothelial cell survival by helping in transmission of anti-apoptotic signal of VEGF to the nucleus. Even though VE-cadherin is not involved in vasculogenesis [60], it plays a crucial role in maturation and remodeling of vessels during development [31].

Eph-B4/Ephrin-B2

These are a class of receptor ligand pairs having important role in blood vessel development. The Eph receptors are the largest known family of receptor tyrosine kinases, which include at least 14 membrane bound proteins and eight transmembrane ligands identified till date [31]. The mechanism of ephrin signaling is interesting because on one hand ephrin expressed on a cell surface binds and activates its cognate receptors on another cell surface and on the other hand, through a reciprocal signaling system, the ephrin is activated upon receptor engagement. The activation and engagement of these signaling molecules requires a direct cell to cell contact which suggests that their major function is associated with formation of spatial boundaries establishing the basic body plan in developing embryo [69-71].

Interleukin-8 and Matrix Metalloproteinase-2

An upregulation of MMP activity results in an increased proteolytic degradation of the basement membrane and extracellular matrix and is directly associated with tumor growth, metastasis and tumor induced angiogenesis [34, 72]. An enhanced expression of IL-8 mRNA has been observed in neoplastic tissues and is correlated with the extent of neovascularization, tumor progression and survival (Fig. **5**). The main mechanism by which the interleukins induce angiogenesis is by an IL-8 induced production of MMP-2. It is evident from the observation that in tumor cells transfected with IL-8, there is an increased expression of MMP-2 while the levels of VEGF and bFGF remain constant [73-75].

Biomechanical Forces

In addition to soluble and insoluble factors, some mechanical forces also play important role in pruning and remodeling process of normal angiogenesis [31]. Fluid shear stress has been shown to be an important regulator of vascular structure and function through its effect on the endothelial cell. The flow of blood which is a viscous fluid is associated with shear stress, the tractive force per unit area [76]. It acts in the direction of blood flow on the surface of the inner wall of the blood vessel at the endothelial cell [76]. In the simplifying case of fully laminar flow, the wall shear stress is proportional to the flow velocity and medium viscosity and inversely proportional to the third power of the internal radius [76]. Endothelial cells of the atrial vascular system contain bundles of straight actin filaments or stress fibers. *In vitro*, in case of a perfectly laminar flow, there is a dramatic increase in expression of these stress fibers in the endothelial cells where as when the flow is turbulent, division of endothelial cells begins and the transcription of genes for PDGF and TGF-β is stimulated which promotes angiogenesis [31, 76]. *In vivo*, from studies in rabbit ear vessels, enhanced blood flow is known to associate with an increase in microvascular area.

Hence, blood vessel morphogenesis is modulated by blood flow induced shear stress [76]. In case of laminar flow, the vessels are stabilized and protected by the increased expression of stress fiber in endothelial cells and in case of turbulent flow there is further growth of blood vessels [31, 76]. Hence, remodeling of the primary capillary plexus is efficiently maintained as vasculogenesis may result in overproduction of blood vessels. The capillaries that remain unperfused by blood regress, apparently by apoptosis of endothelial cells those in which blood flow is established form a stable part of the vasculature.

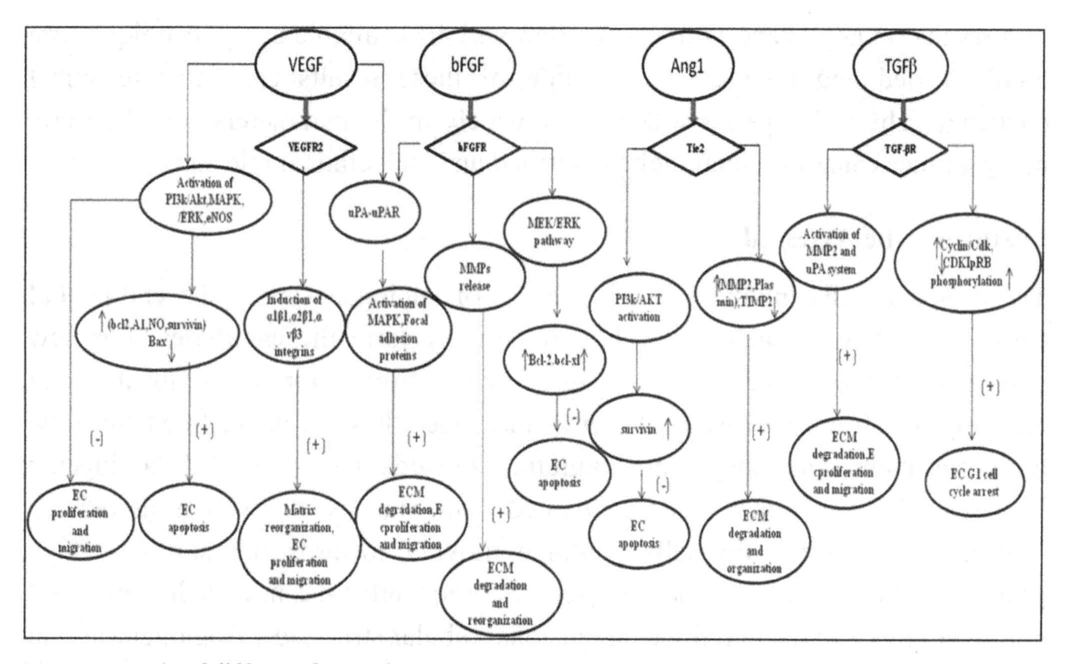

Figure 5: Role of different factors in the process of angiogenesis.

Oncogene and Tumor Suppressor Genes

Under normal conditions the oncogenes remain dormant while the tumor suppressor genes are constitutively activated. In tumors, the tumor suppressor genes lose their functions while the oncogenes are activated thereby promoting tumor growth and angiogenesis by different mechanisms. VEGF is introduced by Kor, H-ras mutant gene, v-src and v-raf in transformed fibroblast and ECs while TNF-a, TGF-β etc are up-regulated by mutant ras [34]. Possibly, these effects are mediated through a ras-raf-MAP kinase signal transduction pathway resulting in the activation of promoter regions of genes of the angiogenic growth factors [75,

77, 78]. More importantly, the expression of ras is a critical suppressor gene, inhibiting the angiogenesis by inducing thrombospondin-1, downregulation of VEGF and NOS and also the down-regulation of hypoxia-induced angiogenesis, either inducing apoptosis or enhancing anti-angiogenic factors [79, 80]. The biological function of p53 is to act at a G1/S check point, halting DNA under conditions of stress like DNA damage and activates the apoptotic pathway [81].

ASSAYS FOR ANGIOGENESIS

The results on angiogenesis obtained from various available assay techniques are highly varied and precise interpretation of these results can be a technical challenge. The techniques available vary widely in the parameters they measure, many a times cannot be accurately quantified and lack clinical relevance.

Features to be Assessed

The principal cells involved in the process of angiogenesis are the endothelial cells, which line the blood vessels and virtually form the capillaries. For new vessels to form, endothelial cells first escape from their stable location by breaking free from the basement membrane. Once these cells are freed from the basement membrane, they start migrating towards the source of angiogenic stimulus like the tumor cells, activated lymphocytes or wound associated macrophages. As the tip cells continue moving towards the stimulus, they proliferate thereby forming new vessels. Subsequently, the new endothelial cell outgrowth reorganizes into three dimensional tubular structures. Angiogenesis can be inhibited by targeting and blocking any of these events *i.e.*, disruption of the basement membrane, cell migration, cell proliferation and tube formation which can be evaluated *in vitro*. However, the critical assays for angiogenesis require a more holistic approach, which have to be further supported by *in vivo* data for a more realistic appraisal of the angiogenic response [82, 83].

In Vitro Assays

Cell Proliferation

Of the various well established methods available for evaluating cell proliferation, the most reliable and frequently used is the thymidine incorporation assay. It measures cell proliferation by directly measuring the DNA. In this commonly used assay, a radioactive nucleoside, 3H-thymidine is incorporated into the newly synthesized strands of chromosomal DNA during mitotic division. The radioactivity of the DNA recovered from the cells is measured using a scintillation β counter. However, more recently techniques have been developed which no longer require the use of radioisotopes. For example, 5-bromo-2'deoxyuridine (BrdU), a nucleoside analog can be used that is detectable during active DNA synthesis using immunohistochemistry. Another useful nucleoside analog is 5-ethynyl-2'-deoxyuridine.Its detection is not antibody based and therefore does not require DNA denaturation. However, additional confirmatory tests need to be done since proliferation or cell viability is not necessarily a measure of angiogenic activity. There are several problems in validation of *in vitro* angiogenesis assays. While testing the potential of a potent agent to inhibit angiogenesis, cell proliferation assays do not hold good because some substances may be merely toxic for the cells, hence, inhibit cell proliferation rather than being anti angiogenic. Cytotoxicity can result from simply diluting the culture media too much using distilled water or by adding excess of sodium chloride. For research usage, endothelial cells are obtained from two major sources, the human umbilical vein endothelial cells and the bovine aorta endothelial cells. But all endothelial cells are not uniform in their properties. There exists a considerable difference in the nature of large blood vessel derived cells and cells of the microvascular origin. In addition to that cells derived from different sites even within same organ have some unique features, naturally species differences are much more pronounced. Cells *in vitro* both gain and lose attributes found *in vivo*, and it is generally not feasible to use truly primary (not passaged) endothelial cells in angiogenesis assays [1, 82]. Another fact to consider is the environmental conditions under which the endothelial cells are cultured. Cells *in vivo* are subjected to shear stress and other hydrodynamic forces leading to induction of multiple signaling pathways. Cells *in vitro* are cultured in room air which constitutes 21% oxygen. Such a condition is hyperoxic when compared to *in vivo* oxygen tension, more so in the microcirculation. With the advancement of technology, two and three dimensional scaffolds have been in use recently, composed of matrigel, collagen or fibrin, which to some extent mimic the three dimensional structure *in vivo* but it is practically impossible to completely stimulate the complex interactions between endothelial cells and their complex physical environment. Equally challenging is to stimulate the *in vivo* interactions between the endothelial cells and the other cell types [82].

Cell Migration Assays

Migration of the endothelial cells is key step in the process of angiogenesis and there are a number of assays available which focus on the migratory response of endothelial cells to inducers or inhibitors of angiogenesis. One of the most used techniques is the blind-well chemotaxis chamber assay or modified Boyden chamber assay. In this assay, the endothelial cells are placed on the upper layer of a cell permeable filter and permitted to migrate in response to test factor placed in the medium below the filter. For accurate measurements, cells are enumerated after the retained cells are separated from those that have migrated across the membrane. This system is based on the principle of concentration gradients and describes the conditions *in vivo* well. But one practical problem is that microvascular endothelial cells are less resistant as compared to the standard large vessel endothelial cells or neutrophils [82]. Some labs use an advanced two dimensional cell migration assay that measures cell motility and can be readily quantified [82, 84]. A monolayer of 1-μm beads is deposited on the bottom of 96-well plates. Endothelial cells (100 cells/well) are then placed in the well, along with test medium. Cell movement is scored after 24 h (phagokinetic track assay), and the assay lends itself to computer-assisted quantification [82].

Tube Formation Assay

The tube formation assay is one of the most reliable and specific methods to evaluate the potential of endothelial cells to form three dimensional network or tubulogenesis. Endothelial cells, irrespective of their origin, have an inherent capability of forming tubules spontaneously when subjected to conditions *in vitro* where they can lay down the appropriate extracellular matrix components given proper time lag. Tube formation can be further improved using collagen or fibrin clots to coat the culture dishes. Using such systems supplemented with clots better mimics the situation *in vivo* and the formation of tight junctions can be verified with the help of an electron microscope. One of the best discoveries so far in this regard is the matrigel. It is a matrix rich product prepared from Engleberth-Holm-

Swarm (EHS) tumor cells, the principal component of which is laminin [1, 82]. The matrigel can induce tube formation by the endothelial cells within a time as short as 24 hours. Since it is a faithful means of testing angiogenesis, tube formation assays have attained a prominent place in the array of angiogenic measures. However, it must be kept in mind that sometimes cultured cells, which do not have endothelial origin, may also show a positive response to matrigel like the fibroblasts according to some unpublished sources. Another important criterion to be careful about is to control the protein concentration since it may affect the tube forming capacity *in vitro*.

Organ Culture Assays

The Aortic Ring Assay

Organ culture method, which has come into picture recently, is far more accurate and reliable than earlier techniques because under *in vivo* conditions, angiogenesis involves not only endothelial cells but also other cells from the surroundings. One of such techniques quite widely used is the rat aortic ring assay. The isolated rat aorta is cut into segments and then placed in culture medium, generally in a matrix-containing environment such as matrigel. Over the next 7-14 days, the explants are monitored for the outgrowth of endothelial (and other) cells as this is affected by the addition of test substances. Quantification is achieved by measurement of the length and abundance of vessel-like extensions from the explants [82]. Quantification by pixel count is done using endothelium selective reagents like fluroscein labled BSL-I. With recent modifications, the assay can be done using defined culture media in shorter time periods as opposed to earlier culture conditions which required complex media and the culture had to be monitored for at least a week's time. This *in vitro* assay system is the closest replica of *in vivo* conditions. In addition to involving the surrounding non endothelial cells, the endothelial cells have not been preselected by passaging and thus are in a non-proliferative stage at the time of experimentation, hence are closer to real life situation. However, the system is not ideal since angiogenesis is primarily a microvascular event which is not the case with the aorta.

The Chick Aortic Arch Assay

This assay is basically a major modification of the rat aortic ring assay. It was originally developed specifically for testing thalidomide as it had limited effect on rodents but strong effect in chick embryo. This assay is more convenient because it avoids the use of laboratory animals, is rapid which can be completed in a short period of 1-3 days, and can be carried out in serum free media. Aortic arches are dissected from day 12-14 chick embryos and cut into rings similar to those of the rat aorta. On placing the rings on matrigel, substantial outgrowth of cells occurs within 48h, with the formation of vessel-like structures which can be easily made out. Both growth-stimulating factors, such as FGF-2, or inhibitors, such as endostatin, can be added to the medium, where their effect can be easily measured. The endothelial cell outgrowth can be quantified using fluorescein-labeled lectins like BSL-I and BSL-B4 or by staining the cultures with labeled antibodies to CD31. Standard imaging techniques are useful both for the enumeration of endothelial cells and for delineating the total outgrowth area. One more advantage of this system is that embryonic arch endothelial cells are very close to microvascular endothelial cells in their basic properties, but since they are obtained from growing embryos, they undergo rapid cell division before explantation and exposure to angiogenic mediators.

In Vivo Assays

The best way to study angiogenesis and role of external factors on angiogenesis is best done *in vivo*. There is always a greater emphasis on results obtained from experimentation on intact animals because a complex interaction among complex cell types, the extracellular matrix, the hydrodynamic factors and metabolic factors are often difficult to mimic *in vitro*. The oldest methods use diffusion chambers made up of milipore filters. And other techniques designed for visual monitoring of the progress of neo-vascularization of the implanted tumor. However, the most accurate details about angiogenesis *in vivo* are still provided by histological studies. The standard information available from histological studies is further augmented by new improved techniques like blood flow monitoring by Doppler or radiologic approaches. *In vivo* systems developed over years which are easier to perform and permit better quantification are the chick chorioallantoic membrane assay(CAM), Matrigel plug assay *etc.* [82].

The CAM Assay

The CAM (Chick Chorioallantoic Membrane) assay is a age old technique which has been used by embryologists to study embryonic organ development.

Originally, the CAM of day 7-9 chick embryos was exposed by making a window in the egg shell, and tissue or organ grafts were then placed directly on the CAM. The window was sealed, eggs were re-incubated, and the grafts were recovered after an appropriate length of incubation time. The grafts were then scored for growth and vascularization. Assessment of angiogeneisis was done by ranking the vascularization on a 0 to 4 scale. However, these days because of the use of better imaging techniques it is possible to analyze parameters like the measurement of a bifurcation point in a certain designated area around the test material, hence the quantification of the assay has greatly improved. In a modification of this *in ovo* method, the entire egg contents were transferred to a plastic culture dish (whole embryo culture) after 72h of incubation [84, 85]. Although technically this may be considered an *in vitro* assay, it is a whole-animal assay. After 3-6 additional days of incubation, during which time the CAM develops, grafts can be made more readily than within the egg shell and can be monitored throughout the time of subsequent development. Angiogenic stimulation by tumors or immunocompetent allografts or xenografts (graft *vs.* host reaction) yields extensive angiogenesis over the next several days [84]. Test substances can be administered by placing them on membranes or on the underside of cover slips. Inhibitors can be assessed by their effect either on the normal development of the CAM vasculature itself or on induced angiogenesis such as the FGF-2-evoked angiogenesis. CAM assay is highly valuable technique owing to its ease of procedure, ready availability of the experimental material and for the ex-plant method, the feasibility of carrying out multiple tests on individual CAMs as well as of monitoring the reaction throughout the course of the assay. However, there still exist some aspects which limit the utility of this assay. The test is conducted on chicken cells which in itself is a limiting factor. In addition to this, there is always an underlying concern that that the CAM itself undergoes rapid changes morphologically and in terms of the gradual change in the rate of endothelial cell proliferation during the course of embryonic development.

The Corneal Angiogenesis Assay

This assay although has been in use for long is the best *in vivo* assay because the cornea itself is an avascular structure. Hence, any vessel that starts appearing on being stimulated by an angiogenesis inducing substance or tissue are newly formed vessels. It was originally conducted on rabbit eye but is now being modified for use in mice since it is the standard animal used in any research. In this setup, a pocket is made in the cornea, and test tumors or tissues, when introduced into this pocket, elicit the in-growth of new vessels from the peripheral limbal vasculature. Slow-release materials such as ELVAX (ethylene vinyl copolymer) or Hydron have been used to introduce test substances into the corneal pocket [82]. Sponge is the most commonly used material to hold the test cell suspension or the angiogenesis inducing substances because the slow releasing substances are likely to be toxic. To evaluate the effect of angiogenesis inhibitors, it can be monitored on the locally induced angiogenic reaction in the cornea like by FGF, VEGF or tumor cells The test inhibitors can be administered orally or systemically, the latter either by bolus injection or, more effectively, by use of a sustained release method such as implantation of osmotic pumps loaded with the test inhibitor [82, 84, 85].The vascular response can be monitored by direct observation throughout the course of the experiment using a slit lamp for the rabbit or a simple stereomicroscope in mice. Definitive visualization of the mouse corneal vasculature was earlier done by injecting India ink. These days fluorochrome labeled high molecular weight dextran is preferably used. Quantification is done by measuring the area of vessel penetration, progress of vessels towards angiogenic stimulus over a period of time, or in case of fluorescence, by histogram analysis or pixel counts above a specific threshold. Corneal angiogenesis assay scores a point of advantage over other processes because of better ability to monitor the progress of angiogenesis, absence of existing vasculature in cornea and mice can be used as the experimental model. At the same time the surgical procedures are demanding. A few animals can be grafted at a single setting. Apart from that, space available for introduction is limited, inflammatory reactions occur very frequently and are often difficult to avoid and the site, although ideal for visualization, is atypical precisely because the cornea is avascular.

The Matrigel Plug Assay

Matrigel plug assay is a simple to administer assay procedure which does not require much technical skill. In this setup, matrigel containing test cells or substances are subcutaneously administered where it solidifies forming a plug.

The plug is obtained after 7-21 days and histologically examined in order to determine the extent to which the blood vessels have penetrated. Quantification of the vessels in histologic sections is tedious but accurate [82]. Fluorescence measurement of plasma volume can be achieved using fluorescein isothiocyanate (FITC)-labeled dextran 150 [84, 85]. Quantification can also be achieved by measuring the amount of hemoglobin contained in the plug [84]. However, the hemoglobin assay may be misleading because blood content is much affected by the size of vessels and by the extent of stagnant pools of blood. In a modified matrigel assay [82], matrigel alone is first introduced into the test animal, sponge or tissue fragment is then inserted into the plug. New vessels can then be measured by injection of FITC-dextran as described for the corneal assay. The greatest disadvantage of the sponge/Matrigel assay is that it is more time-consuming than the standard Matrigel plug assay.

MECHANISM OF ANGIOGENESIS INHIBITION

In order to control angiogenesis, specific targets within the signaling cascade have to be identified and some of the major targets identified so far are [34]:

- Growth factors promoting endothelial cell proliferation.
- Proteases required for endothelial cells for penetration into the basement membrane and formation of new blood vessels.
- Interference with specific intracellular signal transduction pathway.
- Induction of apoptosis in the endothelial cells or inhibition of endothelial cell survival.
- Inhibition of endothelial bone marrow precursor cells.
- Inhibition of $\alpha_v\beta_3$-integrin-vitronect interaction that is the most important player in adhesion of endothelial cells to extracellular matrix during neo-vascularization.

INHIBITORS OF GROWTH FACTORS AND RECEPTORS

This is one major class of inhibitors of angiogenesis which target growth factors like VEGF and bFGF. The bioavailability of these factors depends on their ability to bind heparin which helps them to remain trapped within the extracellular matrix. The first generation drugs of this kind, the heparin like molecules, *e.g.*, pentosan polysulfate, have multiple negative charges, thereby promoting the binding of growth factors. The receptor targeting agents may also interfere with tumor growth and metastasis by blocking specific receptors like those of FGFs and VEGF hence inhibiting the transduction of angiogenic stimulus into intracellular response. The receptors are in almost all cases transmembrane tyrosine kinases, wherein the ligands bind to the extracellular domain inducing an autophosphorylation of an intracellular kinase domain which in turn activates the downstream signals. In order to disrupt the signaling cascade the inhibitor might compete for receptor binding and block the autophosphorylation of receptor tyrosine kinase (Fig. **6**).

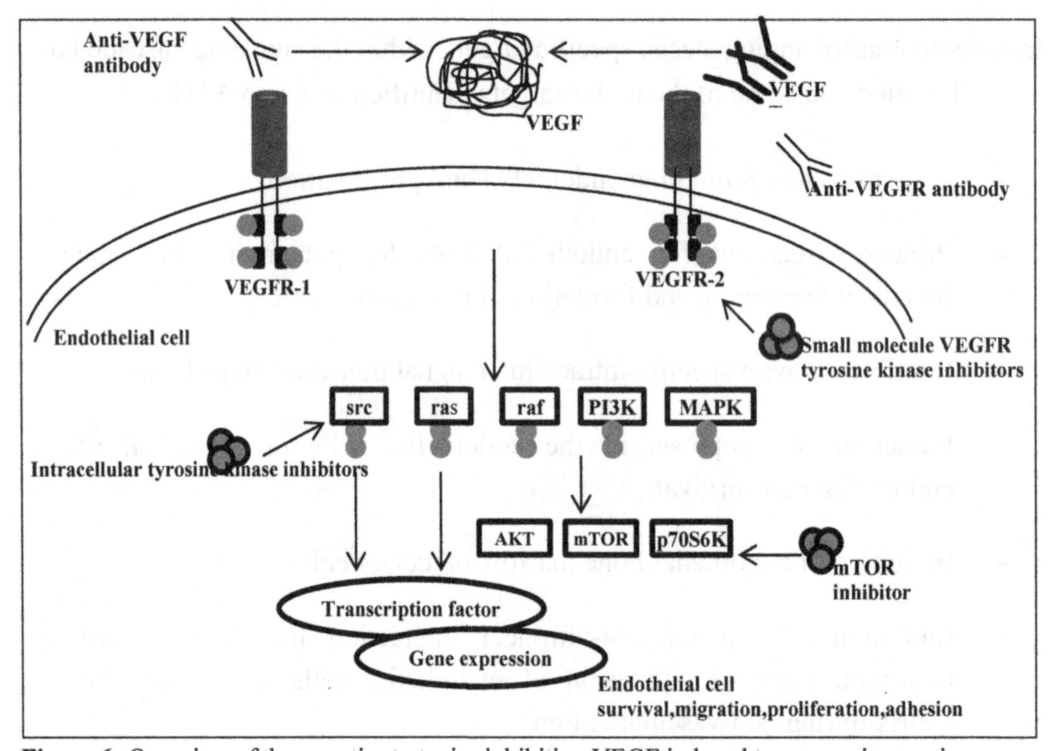

Figure 6: Overview of therapeutic strategies inhibiting VEGF induced tumor angiogenesis.

Various known VEGF inhibitors include the anti-VEGF monoclonal antibody that directly neutralizes the VEGF protein thereby inhibiting its biological functions, some soluble VEGF receptors which bind to the VEGF molecule thus indirectly inhibiting VEGF function, VEGF signaling inhibitors which inhibits the autophosphorylation of VEGF thereby interfering with a series of signal transduction pathways and the VEGF antisense which interferes with VEGF mRNA translation and protein formation by binding to the VEGF mRNA. Recently, another strategy has come into picture *i.e.*, VEGFR2 DNAzyme which has been found to be highly efficient in cleaving is substrate in a time and concentration dependent manner thereby inhibiting the proliferation of endothelial cells with a marked reduction in VEGFR2 mRNA inhibition of tumor growth *in vivo*.

Endogenous Inhibitors of Angiogenesis

There are over 40 known and well characterized molecules which inhibit angiogenesis endogenously and can be grouped into four broad categories, the interferons or IFNs, proteolyic fragments, interleukins and tissue inhibitors of matrix metalloproteinase or TIMPs [31, 86-88]

Interferons

Interferons are a group of secreted glycoproteins, INF-α, -β, and -γ which directly or indirectly inhibit tumor growth and angiogenesis. Studies in different cancers have revealed that IFN-α/β down-regulate the expression of pro-angiogenic factor MMP-9 mRNA and down regulate IL-8 expression in bladder cancer [31, 86, 89]. It has also been observed that the administration of an optimal dose of IFNα/β induced apoptosis in endothelial cells and reduced the bFGF mRNA and protein expression. It also reduced the microvessel density in tumors. However, results from some studies ruled out the possibility of IFN-γ/αβ treatment in reducing bFGF and VEGF levels in serum from carcinoid tumor and leukemia patients [90, 91]. Hence, anti-angiogenic effect of IFNs treatment is possibly dependent on the regulation and activity of different angiogenic factors in different tumors. It shows both dose as well as time dependence. In addition, IFN-αβ is thought to induce anti-angiogenic effects through the secretion of inducible protein 10 (IP-10) and monokine, both induced by IFN-γ [92]. IFNs are also known to have antitumor properties, which are apparently mediated through a direct cytotoxic effect on tumor cells, an elevation of immunogenicity of tumor by up-regulation of major histocompatibility complex(MHC)classes I and II and tumor associated antigens, and activation of macrophages, T lymphocytes and natural killer cells [92].

Interleukins

It has been observed that the interleukins having an ELR motif *i.e.*, Glu-Leu-Arg motif at the NH2 terminus like that in IL-8, cause an increase in angiogenesis while those members lacking the sequence as in IL-4 inhibit it [87, 93]. *In vivo* studies on rat cornea show that bFGF induced neo-vascularization is inhibited by IL-4. IL-12, a cytokine of activated macrophages, causes a marked increase in expression of angiogenic growth factor VEGF, IL-8 and FGF resulting in induction of angiogenesis [35]. In addition, IL-6 also counteracts p53 mediated apoptosis. It has been reported that IL-12 suppresses the expression of VEGF mRNA, bFGF, and MMP-9 mRNA as well as stimulates mRNA expression of IFN-α,β,γ. It also stimulates the expression of anti-angiogenic chemokine IFNγ-inducible protein (IP-10) in cultured endothelial cells [35, 94]. It significantly promotes apoptosis of the endothelial cells and inhibits the proliferation human tumors and induces severe necrosis in the murine, hence reduces the tumor vessel density. *In vivo*, it has been observed that inhibition of neo-vascularization in tumors secreting IL-10 occurs possibly by down-regulation of VEGF, IL-12, TNF-α, IL-6, and MMP-9 synthesis in the tumor-associated macrophages or TAMs. IL-10 is also known to inhibit tumor metastasis through a natural killer (NK) cell-dependent mechanism [87].

Tissue Inhibitors of Metalloproteinases

Remodeling of the extracellular matrix leads to the formation of a scaffold upon which endothelial cells can adhere, migrate, and form tubes. Deposition of the matrix components forms the basal lamina which sheaths the endothelial cells and the mural cells [34]. The migration of endothelial cells through gelatin is impeded by the overexpression of TIM-1 *in vitro*. TIMP-2 inhibits the proliferation of endothelial cells induced by bFGF but TIMP-1 does not [95, 96]. TIMP-2 also inhibits soluble FGFR1 released by MMP-2. It is evident from studies that TIMP-3 overexpression induces apoptosis in a number of cancer cell lines and rat vascular smooth muscle cells by stabilizing the TNF-alpha surface receptors possibly by the inhibition of a receptor shedding MMP. Apparently, the anti-angiogenic and antitumor effects of TIMP-3 is mediated partially by a decreased expression of the VE-cadherins in the presence of TIMP-3 *in vitro* and its overexpression in tumors. To summarize, TIMP-1, TIMP-2, TIMP-3 and TIMP-4 inhibit neo-vascularization by the inhibition of MMP-1, MMP-2, and MMP-9 induced breakdown of surrounding matrix [97, 98].

Proteolytic Fragments

These fragments are the derivatives of the matrix components like collagen or fibronectin, or of enzymes such as plasminogen and MMP-2 responsible for the matrix remodeling. The most well studied and well characterized inhibitors in this class are angiostatin and endostatin [16, 99].

Angiostatin

It is an internal, 38kDa fragment of plasminogen which is known to down regulate the expression of VEGF within the tumor thereby influencing the process of angiogenesis negatively [34]. It is also known to exert an anti-angiogenic effect by disruption of the HGF/c-met signaling by inhibiting the phosphorylation of c-met, Akt and ERK1/2 which is induced by hepatocyte growth factor [100]. Observations reveal that the ATP metabolism on endothelial cell surface is hindered on binding of angiostatin to the α/β-subunits of plasma membrane-localized ATP synthase thereby inhibiting angiogenesis by the down regulation of endothelial cell proliferation and migration. In addition, transfer of angiostatin gene by adenovirus causes selective inhibition of endothelial cell proliferation and disruption of the G2/M transition which is induced by the M-phase promoting factors and because of the down regulation of M-phase phosphoproteins, endothelial cells show a significant mitotic arrest [100, 101]. It also decreases tumor blood vessel density and increase in endothelial cell apoptosis. Angiostatin is also related to actin stress fiber reorganization, detachment and death by a transient increase in ceramide and a treatment with angiostatin or ceramide also in activation of RhoA which is an important effector of cytoskeletal structure. It selectively regulates E-selectin expression hence inhibits endothelial cell proliferation.

Endostatin

This collagen derivative known to inhibit endothelial cell proliferation, angiogenesis and tumor growth is produced by the hemangioendothelioma cells. It is a 20 kDa fragment of the type XVIII collagen [34]. Firstly, VEGF induces tyrosine phosphorylation of KDR/Flk-1 in the endothelial cells which is inhibited by endostatin thereby exerting its anti-angiogenic effect; secondly endostatin suppresses the VEGF-driven activation of ERK, p38MAPK, and p125FAK, downstream of the KDR/Flk-1 signaling and associated with the mitogenic activities of VEGF in endothelial cells. Thirdly, it hinders the binding of VEGF to ECs and to its cell surface receptor, KDR/Flk-1 and lastly it directly binds to KDR/Flk-1 but not to VEGF [102]. Endostatin also exerts an anti-migratory effect by reducing VEGF-induced phosphorylation of endothelial NOS (eNOS). Rhen *et al.,* in heir elegant experiments demonstrated that soluble endostatin is capable of binding αv- and α5-integrins hence inhibit integrin function like the migration of endothelial cells [103]. It may also exert its anti-proliferative and anti-angiogenic effects by competing with bFGF to bind to the cell surface heparan sulphate proteoglycans thus impairing mitogenic growth factor signaling. On being stimulated by neuropeptide Y and bombesin and in the basal state, endostatin significantly reduces endothelial cell proliferation. It can also block the invasiveness of endothelial cells and the tumor cells by interfering with the extracellular activation of proMMP-2 by inhibiting the membrane-type1 MMP (MT1-MMP) and masking the catalytic activity ofMMP-2.

Somatostatin and Its Analogs

The anti-proliferative action of somatostatin is regulated *via* five G-protein coupled receptors called somatostatin receptors or SSTRs(SSTR1-SSTR5) [104, 105]. SS and analogs are known to inhibit the receptor positive endocrine neoplasms [34]. Somatostatin are capable of maintaining high levels of CDKIs, p27 (Kip1) and p21 and inactivation of cyclinE-CDK2 complexes thereby hypophosphorylating pRb, thus the activation of SSTR leads to G1 arrest of cell cycle. Growth inhibition of normal and cancer pancreatic acinar cells mediated by somatostatin is also triggered by an inhibition of the PI3-kinase signaling pathway [104]. SS can even induce apoptosis directly in the tumor cells *via* the SSTR3 dependent G protein signaling which stimulates the tumor suppressor gene p53 and Bax [104, 105]. In addition to the above described mechanisms, somatostatin and analogs can also exert their biological responses by communicating with specific receptors coupled to a number of signaling pathways which involve major players like adenylate cyclase, guanylate cyclase, ionic conductance channels, phospholipase C-b, phospholipase A2, and tyrosine phosphatase and by protein dephosphorylation, hence directly influence cell growth [104, 105]. The best characterized pathway involves the inhibition of adenylate cyclase, leading to a reduction in intracellular cAMP levels. SSTR1 and SSTR2 exert their influence by stimulating tyrosine phosphatases while SSTR5 couples to the inositol phospholipid/calcium pathway. Therefore to summarize, the mechanisms involved in inhibition of angiogenesis by somatostatin and its analogs are [34, 104-118]:

- By inhibiting the synthesis and secretion of hormones like GH, insulin and/or gastrointestinal hormones.
- By directly or indirectly inhibiting IGF-1 and/or other growth factors which stimulate tumor growth. SS analogs selectively stimulate the formation of IGF-binding protein 1, hence interfere with IGF-1 action at the receptor level.
- By direct anti-mitotic effects of growth factors, which act on tyrosine kinase receptors such as EGF and FGF, *via* SSTRs on the tumor cells.
- Immune modulation.

Thrombospondin-1

Thrombospondin-1(TSP-1), a naturally occurring angiogenesis inhibitor, restricts tumor growth and limits the density of vasculature in normal tissues. TSP-1 induces an apoptotic signaling pathway *via* CD36 receptor binding thereby inhibiting angiogenesis [34]. As TSP-1 binds to CD36 receptor, the Src-related kinase p59-fyn is recruited leading to the activation of p38 MAPK. Activation of the p38 MAPK is p59-fyn-dependent and to requires a caspase-3-like proteolytic activity. Activation of p38 MAPK further activates caspase-3 and leads to apoptosis. Strangely, induction of apoptosis by TSP-1 is restricted to the endothelial cells already activated to participate in angiogenic process but not in quiescent vessels. It acts through CD36 modulating activity of focal adhesion kinase (FAK) and thereby inhibiting endothelial cell migration and proliferation. It also inhibits chemotaxis *in vitro* and neovascularization *in vivo*, induced by several angiogenic stimuli *i.e.*, the protein acting *via* tyrosine kinase receptors (VEGF, bFGF, aFGF, PDGF), *via* G proteins (IL-8), *via* serine/threonine kinase receptors (TGF-b), and also lipids (PGE-1) [34].

AN OVERVIEW OF ANTI-ANGIOGENIC THERAPY

When a tumor starts developing, since they are small in size, they can obtain oxygen and nutrients by simple passive diffusion. But as they enlarge, they become devoid of the essential nutrients and oxygen and need to develop new collateral blood vessel for oxygenation and nutrient supply for invasion, growth and subsequent metastasis. Neo-vascularisation, the angiogenesis driven process, has been an appealing target for anticancer drugs for over three decades and recently some hopes have come true. More than 30 inhibitors of angiogenesis are in clinical trials and it is apparent that these are cytostatic and not cytotoxic leading to tumor dormancy. It is therefore clear that these drugs would yield best results when administered in conjugation with conventional chemotherapy.

Most of the anti-angiogenic substances being investigated presently can be classified as direct inhibitors and indirect inhibitors. The direct inhibitors are agents that directly target the endothelial cell recruitment, endothelial cell proliferation as well as tube formation whereas the indirect inhibitors affect the production of pro-angiogenic growth factors by the tumor cells or interfere with their receptors or intracellular signaling pathways. Anti-angiogenic treatment has been particularly appealing for researchers because these strategies have certain theoretical advantages over conventional cytotoxic chemotherapy directed against malignant tumor cells. To begin with, the treatment is not restricted to any particular histologic tumor entity as all solid tumors depend on angiogenesis for the maintenance of functional microvasculature. The tumor microvasculature is well accessible to systemic treatment. Moreover, unlike conventional chemotherapy, no endothelial barrier has to be crossed by these therapeutic substances. Adult angiogenesis is induced only under certain physiologic conditions of stress, like during the reproductive ovarian cycle or wound healing.

An antagonism of angiogenesis is, therefore, a highly selective therapy promising less serious side effects. One of the very important advantages of anti-angiogenic therapy is that since the endothelial cell, as a target is genetically quite stable, it is expected to be less prone to development of drug resistance. Some of the leading anti-angiogenic agents known are inhibitors of VEGF-mediated endothelial cell functions during angiogenesis, thereby inhibiting proliferation, migration and/or survival [11]. The six isoforms of VEGF and their receptors *i.e.*,VEGFR-1/Flt1, VEGFR-2/Flk1/KDR, and VEGFR-3/Flt4 are crucial for angiogenesis. Prominent inhibitors of VEGF signaling are monoclonal antibodies against VEGF protein or receptors [119].

Bevacizumab or Avastin is a humanized monoclonal VEGF antibody against soluble VEGF and has been investigated in numerous preclinical and studies. High tolerability has been proven in clinical phase I studies and encouraging results from clinical phase II and III trials resulted in clinical approval of the drug in 2004 [88, 121, 122] (Table **1**). IMC-1C11 is another antibody against the extracellular domain of the VEGF receptor Flk-1. It has been found to exhibit efficient anti-tumor activity in several preclinical animal studies and has passed in clinical phase I testing. Further advances in this field include the development of a soluble decoy receptor incorporating both VEGFR-1 and VEGFR-2 domains (VEGF-Trap), binding VEGF with significantly higher affinity than previously reported VEGF antagonists [123].VEGF-Trap is currently investigated in a phase III study in advanced ovarian cancer patients with recurrent symptomatic malignant ascites [123].

Second leading class of anti-angiogenic drugs which have undergone extensive investigation are the small receptor tyrosine kinase inhibitors (Table **1**). Some of these have undergone numerous pre-clinical and clinical trials. SU5416 and SU6668 with additional inhibitory effects on bFGF and PDGF receptor tyrosine kinase, PTK787/ZK22854, a VEGFR-1 and VEGFR-2 tyrosine kinase inhibitor, SU11248/Sunitinib (Sutent) [88, 124], a broad spectrum orally available tyrosine kinase inhibitor of VEGF, PDGF, c-kit, and Flt-3 kinase activity as well as BAY-43-9006/Sorafenib (Nexavar), an orally available small-molecule inhibitor of VEGFR-2 and -3, PDGF receptor β and Raf-1 kinase have shown significant improvement in progression-free survival in metastatic renal cell cancer in clinical phase III studies [88, 124-129]. In addition to the above-mentioned anti-angiogenic agents, numerous other drugs with anti-angiogenic activity have been developed and characterized *in vitro* and *in vivo*. They include matrix metalloproteinase inhibitors, natural peptide inhibitors (*e.g.*, angiostatin, endostatin), and angiogenesis inhibitors with unknown mechanism, *e.g.*, thalidomide and inhibitors of integrin signaling (RGD peptides to $\alpha_v\beta_3$ integrin).

In addition to the above, there are some drugs that have been identified and chemically designed to block important intracellular pro-angiogenic signaling pathways more downstream. A promising strategy in this regard is the blocking of the mammalian target of rapamycin (mTOR), a threonine kinase involved in intracellular pro-survival and pro-angiogenic signaling [123] (Fig. **6**).

Even though the concept of anti-angiogenic therapy is quite a promising and results from the numerous pre-clinical and clinical trials have been sufficiently encouraging, till date there are only few compounds that have been successful in getting approved as anti-angiogenic monotherapy for treatment of different cancers. Nexavar or sorafenib has been approved by the FDA in December 2005 for treatment of patients with advanced metastatic renal cell carcinoma [88, 124, 129]. Originally developed as a Raf inhibitor, it is a small molecule kinase inhibitor which is administered orally. However, on further laboratory characterization it was found that it is a multikinase inhibitor to many targets which also include VEGFR-2, VEGFR-3 and PDGF receptor β. It is an inhibitor of angiogenesis since it binds VEGFR. In a randomized, placebo controlled phase III study on more than 900 patients suffering from advanced renal cell carcinoma who had a previous session of systemic therapy but had failed the treatment, an overall survival was observed. The overall response rate, quality of life and safety have also been assessed and data published (http://www.onyxpharm.com/wt/page/nexavar) and a continuous improvement in overall survival was observed. Some side effects have been noticed, but they are mild and temporary. These include rash, diarrhea, increases in blood pressure, and redness, pain swelling, or blisters on the palms of the hands or soles of the feet. The second approved drug of the category is Sutent or Sunitinib. It is a broad-spectrum, orally administered tyrosine kinase inhibitor approved for the treatment of patients with gastrointestinal stromal tumors (GIST) at an advanced stage of the disease or for patients unable to tolerate Gleevec, the current treatment for GIST. During clinical trials on patients with this rare type of cancer, it was found that Sutent delayed the time taken for tumors or new lesions to grow in GIST from a median time period of 6 weeks to 27 weeks. It has also been approved by the FDA for treatment of advanced renal cell carcinoma based on the observation that it could reduce the size of the tumor. The first anti-angiogenic therapy to gain approval for use in combination was Avastin in 2006 based on a phase III clinical trial on patients suffering from metastatic colorectal carcinoma in 2004. The study evaluated first-line therapy of 815 patients with metastatic colorectal carcinoma. The humanized, recombinant, anti-VEGF antibody Avastin (Bevacizumab; 5 mg/kg q 2 week) was added to the IFL standard first-line chemotherapy. The overall response rate was 44.9% in the Bevacizumab group in comparison to 34.7% in controls. Median progression-free survival (10.6 *vs.* 6.2 months) and median survival of the patients (20.3 *vs.* 15.6 months) were significantly improved by the addition of Bevacizumab to standard chemotherapy. Some minor side effects of VEGF-antibody therapy were observed which included hypertension. However, more serious side effect includes six gastrointestinal perforations observed only in Avastin treated patients. Avastin was approved for treatment metastatic colon cancer. Some adverse side effects observed were neutropenia, fatigue, hypertension, infection and hemorrhage. Patients who underwent chemotherapy in combination with Avastin showed 25% better overall survival. Although few anti-angiogenic drugs could prove to be efficient enough for being considered as candidates for monotherapy, ongoing preclinical and clinical trials are providing growing evidence that this therapeutic approach would yield best result when conjugated with conventional therapy, precisely with standard chemotherapy (Table **1**).

Visualisation of a synergistic effect of the anti-angiogenic drugs with chemotherapeutics would be a bit complicated initially. This is because, it generally appears that anti-angiogenic drugs would retard intratumoral transport and delivery of chemotherapeutics by radically decreasing the number of blood vessels leading to impaired intratumoral blood flow and drug transport. Furthermore, it would create hypoxia within tumors thereby suppressing the proliferation of tumors. Since proliferating tumor cells are most sensitive to chemotherapy, the anti-angiogenic compounds could retard the effect of chemotherapeutic agents on tumor.

The first possible mechanism is normalization of tumor microvessels by anti-angiogenic therapy. The concept was first put forward in 2001.The immature and inefficient angiogenic tumor blood vessels could be pruned by eliminating excess endothelial cells resulting in a more "normal" and more conductive tumor microcirculation for delivery of oxygen, nutrients, and therapeutics. Improved delivery of cytotoxic drugs and oxygen during this "normalization" window explains partially the synergistic effects of chemo-and anti-angiogenic therapy. In addition, a reduction in interstitial fluid pressure in solid tumors by anti-angiogenic therapy may further improve drug transport and thus contribute to the success of a combination therapy [119].

The second possible mechanism suggests that anti-angiogenic drugs prevent rapid tumor cell repopulation after maximum tolerated dose of chemotherapy during the break periods of successive courses. Thirdly, anti-angiogenic drugs are capable of augmenting antivascular effects of chemotherapy [119].

In addition to cytotoxic effects on tumor cells, several chemotherapeutic agents are known to have anti-angiogenic effects, including Cyclophosphamide, 5-Fu, Taxans, and Camptothecins [11]. VEGF is a potent survival factor for angiogenic endothelial cells. Hence, anti-VEGF therapy can amplify cytotoxic effects of chemotherapy on proliferating endothelial cells. Additionally, the myelosupressive effects of these compounds seem to inhibit the mobilization of the precursors of endothelial cells thereby influencing synthesis of new vessels subsequent to homing at the site of angiogenesis. To improve the anti-angiogenic effect of chemotherapy, the concept of dose-dense chemotherapy called metronomic or anti-angiogenic chemotherapy was developed. Metronomic chemotherapy which can be defined as the chronic administration of chemotherapy at a relatively low, minimally toxic doses on a frequent schedule of administration at close intervals, with no prolonged drug free breaks, works basically through anti-angiogenic mechanisms by directly killing the tumor endothelial cells and suppressing the circulation endothelial progenitor cells [119]. Undesirable side effects can be eliminated or reduced to negligible by using minimally toxic doses, hence metronomic chemotherapy seems to be very useful for long term combinatorial use of anti-angiogenic drugs [130].

Table 1: Anti-angiogenic drugs, mode of action and their stage of development [125-136]

Drug	Trade Name/Code Name	Category	Target	Stage of Development	References
Bevacizumab	Avastin	Monoclonal antibody	VEGF-A	Approved for metastatic colon cancer in combination with flouropyrimidine based chemotherapy,Lung cancer Phase III	[88, 120, 121, 122]
Ramucirumab	IMC-1121B	Monoclonal antibody	VEGFR-2	Phase I clinical trial	[88, 124]
2C3		Monoclonal antibody	VEGF-A	Preclinical phase	[88]
Vatalanib	PTK-787	Receptor tyrosine kinase inhibitor	VEGFR-1,-2	Phase III clinical trial	[88, 124]
Pazopanib hydrochloride	GW786034/Votrient/Armala	Receptor tyrosine kinase inhibitor	VEGFR-1,-2,-3	Approved for advanced soft tissue carcinoma, Lung cancer Phase II,III	[88, 122]
AEE 788		Receptor tyrosine kinase inhibitor	VEGFR-2,EGFR	Preclinical phase	[88]
Vandetanib	ZD6474	Receptor tyrosine kinase inhibitor	VEGFR-1,-2,-3,EGFR	Approved for unresectable, locally advanced or metastatic medullary thyroid cancer.	[88]
Sunitinib	SU11248/Sutent	Receptor tyrosine kinase inhibitor	VEGFR-1,-2,PDGFR	Approved for unresectable,locally advanced or metastatic pancreatic neuroendocrine tumors.	[88, 124]
Sorafenib	Nexovar/BAY 43-9006	Receptor tyrosine kinase inhibitor	VEGFR-1,-2,PDGFR	Approved for unresectable hepatocellular carcinoma	[88, 124, 129]
Motesanib	AMG-706	Receptor tyrosine kinase inhibitor	VEGFR-1,-2,-3	Phase II/Phase III clinical trials.	[88]
VEGF-Trap		Soluble receptor chimeric protein	VEGF-A,PIGF	Advanced ovarian cancer Phase III	[88, 123]
ABT-510		Inhibitor of endothelial cell proliferation	Endothelial CD36	Phase I clinical trial in combination with Bevacizumab	[88]

Table 1: contd...

Thalidomide	Thalomid	Inhibitor of endothelial cell proliferation	Reduction of TNF-α	Multiple myeloma Approved	[88, 129, 134]
Etaracizumab	Etaratazumab/Abegrin/MEDI522	Inhibitor of integrin's pro-angiogenic activity	Integrin αV	Phase II clinical trial for metastatic melenoma	[133]
Marimastat	Marimastat	Matrix metalloproteinase inhibitor	MMP-1,-2,-3,-7,-9	Non small cell Lung cancer Phase III	[122]
Prinomastat	AG3340	Matrix metalloproteinase inhibitor	MMP-2,-9	Brain and CNS Phase II	[135]
Rebimastat	BMS-275291	Matrix metalloproteinase inhibitor	MMP-1,-2,-8,-9,-13,-14	Prostate cancer	[136]
Neovastat	Neovastat/AE941	Matrix metalloproteinase inhibitor	MMP-2,-9,-12,VEGF	Stage III clinical trial in combination with chemotherapy and radiotherapy for NSCLC, Refractory multiple myeloma and metastatic kidney cancer.	[88]
Combrestatin		Vascular targeting	Endothellin tubulin	Head and Neck cancer Phase II	[93]
N-Acetylcolchicinol dihydrogenphosphate	ZD6126	Vascular targeting	Endothelin tubulin	Reached stage IV clinical trial for metastatic renal cell carcinoma but has been suspended.	[88]
DMXAA	AS1404	Vascular targeting	Endothelin tubulin	Clinical trials completed for solid tumors	[88]

Anti-Angiogenic Therapy in Surgical Oncology

Because angiogenesis is indispensible for wound healing, anti-angiogenic therapy could impair or inhibit wound healing or tissue repair which has even been demonstrated by numerous clinical studies. The timing of surgical and adjuvant intervention should therefore be an important consideration while anti-angiogenic drugs are being administered with surgery. For example, in the pivotal phase III clinical trial, comparing bolus FU/Leucovorin/Irinotecan with or without Bevacizumab as first-line therapy for advanced colorectal cancer, patients who underwent surgery while receiving Bevacizumab had a trend towards higher complication rates than those receiving chemotherapy alone [11, 40]. Even though the comparative results were not statistically significant, this trend was confirmed by an analysis of pooled data. However, treatment with the anti-angiogenic drug Avastin has been found to be feasible and safe when started more than 28 days after the primary surgery because from a study including 230 patients undergoing surgery prior to Avastin treatment, less than 1.3% patients developed complications compared to 0.5% in controls. In addition to its important role in wound healing, VEGF plays a critical role in liver regeneration following liver resection [11, 40]. Apart from regulating angiogenesis in liver regeneration, VEGF is also a mediator of a paracrine pathway by which other cytokines can be upregulated in liver sinusoidal endothelial cells. Potential inhibition of normal hepatic regeneration by anti-VEGF containing neo-adjuvant treatment regimens has to be kept in mind when patients are being considered for curative resection of liver metastasis. At present, sufficient preclinical and clinical data are missing to make evidence-based recommendations in terms of surgical interventions during neo-adjuvant or adjuvant anti-angiogenic therapy. Due to different half lives, one has to keep in mind that recommendations cannot be generalized for different anti-angiogenic molecules. With the use of tyrosine kinase inhibitors to VEGF receptors, such as Nexavar or Sutent, waiting periods might be significantly shorter due to their short half-life time. New clinical studies are necessary to determine if and how surgery is feasible and safe after administration of novel targeted anti-angiogenic agents in a neoadjuvant and adjuvant setting.

CONCLUSION

The establishment of a new vascular network by the process of neo-angiogenesis is of utmost importance for development of solid tumors as well as for disease progression and metastasis. However, the ongoing research on the present field has impressively indicates that the same could be exploited for more efficient drug delivery and administration. Angiogenesis is a vast arena and a lot is yet to be unearthed which could be highly useful as adujuvant therapy in conjugation with conventional chemotherapy. The ongoing research in this area holds in itself a promising future for therapeutics as well as diagnostics.

ACKNOWLEDGEMENTS

The authors are extremely thankful to all the members of The Cancer Biology Lab, School of Medical Science and Technology, IIT Kharapur, India for their constant support and valuable suggestions.

CONFLICT OF INTEREST

The authors confirm that this chapter contents have no conflict of interest.

REFERENCES

[1] Adair TH, Montani JP. Angiogenesis. Chapter 1, Overview of Angiogenesis. San Rafael (CA): Morgan & Claypool Life Sciences; 2010. Available from http://www.ncbi.nlm.nih.gov/books/NBK53238/.

[2] Angioworld.com [Homepage on the internet] Investigational Drug Brunch, Cancer Therapy Evaluation Program, Division of Cancer Treatment, Diagnosis, and Centers, National Cancer Institute, Rockville, MD [Cited 29th September, 2012]. Available from http://www.angioworld.com/angiogenesis.htm.

[3] Angio.org [Homepage on the internet] The angiogenesis foundation: understanding angiogenesis. Cambridge, MA 02142 USA. [Updated 7 September 2011; Cited 29th September 2012]. Available from http://www.angio.org/understanding/disease.php.

[4] Karamysheva AF. Mechanism of angiogenesis. Biochemistry(Mosc) 2008; 73(7): 751-62.

[5] Coultas L, Chawengsaksophak K, Rossant J. Endothelial cells and VEGF in vascular development. Nature 2005; 438(7070): 937-45.

[6] Gerhardt H, Betsholtz C. Endothelial-pericyte interactions in angiogenesis. Cell Tissue Res. 2003; 314(1): 15-23.

[7] Folkman J. What is the evidence that tumors are angiogenesis dependent? J Natl Cancer Inst 1990; 82(1): 4-6.

[8] Folkman J. Clinical applications of research on angiogenesis. N Engl J Med 1995; 333(26): 1757-63.

[9] Baeriswyl V, Christofori G. The angiogenic switch in carcinogenesis. Semin Cancer Biol 2009; 19(5): 329-37.

[10] Grepin R, Pages G. Molecular Mechanisms of Resistance to Tumor Anti-Angiogenic Strategies. J Oncol 2010; (2010):835680.

[11] Chen S, Hanna G. Use of antiangiogenic therapy in clinical oncology. UTMJ 2008; 859(3): 164-66.

[12] Hertig AT. Angiogenesis in the Early Human Chorion and in the Primary Placenta of the Macaque Monkey. Issue 459. Carnegie Institution of Washington publication Contributions to embryology, Contributions to embryology 1935; pp: 1-81.

[13] Folkman J, Klagsbrun M. Angiogenic Factors. Science 1987; 235(4787):442-7.

[14] Angio.org [Homepage on the internet]. The angiogenesis foundation: Historical highlights. Cambridge, MA 02142 USA. [Updated 23 June 2009; Cited 29th September 2012]. Available from http://www.angio.org/understanding/highlight.php.

[15] Ozawa S, Shinohara H, Kanayama HO, Bruns CJ, Bucana CD, Ellis LM, *et al.* Suppression of angiogenesis and therapy of human colon cancer liver metastasis by systemic administration of interferon-alpha. Neoplasia 2001; 3(2); 154-64.

[16] Hohenester E, Sasaki T, Olsen BR, Timpl R. Crystal structure of the angiogenesis inhibitor endostatin at 1.5 angstrom resolution. EMBO J 1998; 17(6): 1656-64.

[17] Kim YM, Jang JW, Lee OH, Yeon J, Choi EY, Kim KW, *et al.* Endostatin inhibits endothelial and tumor cellular invasion by blocking the activation and catalytic activity of matrix metalloproteinase. Cancer Res 2000; 60(19); 5410-13.

[18] Dixelius J, Larsson H, Sasaki T, Holmqvist K, Lu L, Engstrom A, *et al.* Endostatin-induced tyrosine kinase signaling through the shb adaptor protein regulates endothelial cell apoptosis. Blood 2000; 95(11); 3403-11 .

[19] Jayson GC, Zweit J, Jackson A, Mulatero C, Julyan P, RansonM, *et al.* Molecular imaging and biological evaluation of HuMV833 anti-VEGF antibody: implications for trial design of antiangiogenic antibodies. J Natl Cancer Inst 2002; 94(19): 1484-93.

[20] Dhanabal M, Ramchandran R, Waterman MJ, Lu H, Knebelmann B, Segal M, Sukhatme VP. Endostatin induces endothelial cell apoptosis. J Biol Chem 1999; 274(17): 11721-26.

[21] Roskoski R Jr, Vascular endothelial growth factor (VEGF) signaling in tumor progression. Crit Rev Oncol Hematol 2007; 62(3): 179-213.

[22] Pepper MS. Manipulating angiogenesis. From basic science to the bedside. Arterioscler Thromb Vasc Biol 1997; 17(4): 605-19.

[23] Auguste P, Lemiere S, Larrieu-Lahargue F, Bikfalvi A. Molecular mechanisms of tumor vascularization. Crit Rev Oncol Hematol 2005; 54(1): 53-61.

[24] Djonov V, Schmid M, Tschanz SA, Burri PH. Intussusceptive Angiogenesis: Its Role in Embryonic Vascular Network Formation. Circ Res 2000; 86(3): 286-92.

[25] Patan S, Munn LL, Jain RK. Intussusceptive microvascular growth in a human colon adenocarcinoma xenograft: a novel mechanism of tumor angiogenesis. Microvasc Res 1996; 51(2): 260-72.

[26] Holash J, Maisonpierre PC, Compton D, Boland P, Alexender CR, Zagzag D, Yancopoulos GD, Wiegand SJ.Vessel cooption, regression and growth in tumors mediated by angiopoietins and VEGF. Science 1999; 284(5422): 1994-1998.

[27] Dvorak HF, Brown LF, Detmar M, Dvorak AM.Vascular permeability factor/vascular endothelial growth factor, microvascular hyperpermeability, and angiogenesis, Am J Pathol 1995; 146(5): 1029-39.

[28] Gossmann A, Helbich TH, Kuriyama N, Ostrowitzki S, Roberts TP, Shames DM, *et al.* Dynamic contrast-enhanced magnetic resonance imaging as a surrogate marker of tumor response to anti-angiogenic therapy in a xenograft model of glioblastoma multiforme. JMagn Reson Imaging 2002; 15(3); 233-240.

[29] Hendrix MJ, Seftor EA, Hess AR, Seftor RE. Vasculogenic mimicry and tumor-cell plasticity: lessons from melanoma. Nat Rev Cancer 2003; 3(6): 411-21.

[30] Risau W. Mechanisms of Angiogenesis. Nature 1997; 386(6626): 671-74.

[31] Michael Papetti, Ira M. Herman. Mechanism of normal and tumor derived angiogenesis. Am J Physiol Cell Physiol 2002; 282(5): C947-70.

[32] Folkman J. Angiogenesis in cancer, vascular, rheumatoid and other disease. Nat Med 1995; 1(1): 27-31.

[33] Klagsbrun M, D'Amore PA. Regulators of angiogenesis. Annu Rev Physiol 1991; (53): 217-39.

[34] Gupta MK, Qin RY. Mechanism and its regulation of tumor-induced angiogenesis. World J Gastroenterol 2003; 9(6): 1144-55.

[35] Torisu H, Ono M, Kiryu H, Furue M, Ohmoto Y, Nakayama J,*et al*. Macrophage infiltration correlates with tumor stage and angiogenesis in human malignant melanoma. Int J Cancer 2000; 85(2); 182-8.

[36] Bussolino F, Mantovani A, Persico G. Molecular mechanisms of blood vessel formation. Trends Biochem Sci 1997; 22(7): 251-6.

[37] Sebti SM, Hamilton AD. Design of growth factor antagonists with antiangiogenic and antitumor properties. Oncogene 2000; 19(56): 6566-73.

[38] Ferrara N, Houck K, Jakeman L, Leung DW. Molecular and biological properties of the vascular endothelial growth factor family of proteins. Endocr Rev 1992; 13(1):18-32.

[39] Conn G, Soderman DD, Schaeffer MT, Wile M, Hatcher VB, Thomas KA. Purification of a glycoprotein vascular endothelial cell mitogen from a rat glioma-derived cell line. Proc Natl Acad Sci USA 1990; 87(4): 1323-7.

[40] Denekamp. Endothelial cell proliferation as a novel approach to targeting tumor therapy. Br J Cancer 1982; 45(1): 136-9.

[41] Robinson CJ, Stringer SE. The splice variants of vascular endothelial growth factor (VEGF) and their receptors. J Cell Sci. 2001; 114(5): 853-65.

[42] Klagsbrun M, Moses MA. Molecular angiogenesis. Chem Biol 1999; 6(8): R217-24.

[43] Tang H, Kerins DM, Hao Q, Inagami T,Vaughan DE. The urokinase-type plasminogen activator receptor mediates tyrosine phosphorylation of focal adhesion proteins and activation of mitogen-activated protein kinase in cultured endothelial cells. J Biol Chem 1998; 273(29): 18268-72.

[44] Lobov IB, Brooks PC, Lang RA. Angiopoietin-2 displays VEGF dependent modulation of capillary structure and endothelial cell survival *in vivo*. Proc Natl Acad Sci USA 2002; 99(17): 11205-10.

[45] Beck L, D'Amore PA. Vascular development: cellular and molecular regulation. FASEB J 1997; 11(5): 365-73.

[46] Yancopoulos GD, Davis S, Gale NW, Rudge JS, Wiegand SJ, Holash J. Vascular-specific growth factors and blood vessel formation. Nature 2000; 407(6801): 242-48.

[47] Korhonen J, Partanen J, Armstrong E, Vaahtokari A, Elenius K, Jalkanen M, *et al*. Enhanced expression of the tie receptor tyrosine kinase in endothelial cells during neovascularization. Blood 1992 ; 80(10); 2548-55.

[48] Korhonen J, Polvi A, Partanen J, Alitalo K. The mouse tie receptor tyrosine kinase gene: expression during embryonic angiogenesis. Oncogene 1994; 9(2): 395-403 .

[49] Kim I, Kim HG, So JN, Kim JH, Kwak HJ, Koh GY. Angiopoietin-1 Regulates Endothelial Cell Survival Through the Phosphatidylinositol 3-Kinase/Akt Signal Transduction Pathway. Circ Res 2000; 86(1): 24-9.

[50] Sato TN, Tozawa Y, Deutsch U, Buchholz KW, Fujiwara Y, Maguire MG, *et al*. Distinct roles of the receptor tyrosine kinases Tie-1 and Tie-2 in blood vessel formation 1995; Nature 376(6535); 70-4.

[51] Kim I, Kim HG, Moon SO, Chae SW, So JN, Koh KN *et al*. Angiopoietin-1 induces endothelial cell sprouting through the activation of focal adhesion kinase and plasmin secretion. Cir Res 2000; 86(9); 952-9.

[52] Folkman J. Tumor angiogenesis. Adv Cancer Res 1985; (43): 175-203 .

[53] Carmeliet P, Jain RK. Angiogenesis in cancer and other diseases. Nature 2000; 407(6801): 249-257.

[54] Pardo OE, Arcaro A, Salerno G, Raguz S, Downward J, Seckl MJ. Fibroblast growth factor-2 induces translational regulation of Bcl-XL and Bcl-2 via a MEK-dependent pathway: correlation with resistance to etoposide-induced apoptosis. J Biol Chem 2002; 277(14): 12040-6.

[55] Kim CK, Choi YK, Lee H, Ha KS, Won MH, Kwon YG, *et al.* The Farnesyltransferase Inhibitor LB42708 Suppresses Vascular Endothelial Growth Factor-Induced Angiogenesis by Inhibiting Ras-dependent Mitogen-Activated Protein Kinase and Phosphatidylinositol 3-Kinase/Akt Signal Pathways. Mol Pharmacol 2010; 78(1); 142-50 .

[56] Lawrence DA, Pircher R, Jullien P. Conversion of a high molecular weight latent beta-TGF from chicken embryo fibroblasts into a low molecular weight active beta-TGF under acidic conditions. Biochem Biophys Res Commun. 1985; 133(3): 1026-34.

[57] Derynck R, Jarrett JA, Chen EY, Goeddel DV. The murine transforming growth factor-beta precursor. J Biol Chem 1986; 261(10): 4377-9. .

[58] Bar RS, Boes M, Booth BA, Dake BL, Henley S, Hart MN. The effects of platelet-derived growth factor in cultured microvessel endothelial cells. Endocrinology. 1989 ; 124(4):1841-8.

[59] Massague J. The transforming growth factor-beta family. Annu Rev Cell Biol 1990; (6): 597-641.

[60] Merwin JR, W Newman, LD Beall, A Tucker, J Madri. Vascular cells respond differentially to transforming growth factors β1 and β2 *in vitro*. Am J Pathol 1991; 138(1): 37-51.

[61] Piek E, Heldin CH, Dijke PT. Specificity, diversity, and regulation in TGF-β superfamily signaling. FASEB Journal 1999; 13(15): 2105-24.

[62] Roberts AB, Lamb LC, Newton DL, Sporn MB, De Larco JE, Todaro GJ. Transforming growth factors: isolation of polypeptides from virally and chemically transformed cells by acid/ethanol extraction. Proc Natl Acad Sci USA 1980; 77(6): 3494-8.

[63] Myoken Y, Kan M, Sato GH, McKeehan WL, Sato JD. Bifunctional effects of transforming growth factor-beta (TGF-beta) on endothelial cell growth correlate with phenotypes of TGF-beta binding sites. Exp Cell Res 1990; 191(2): 299-304.

[64] Pepper MS. Transforming growth factor-beta: vasculogenesis, angiogenesis, and vessel wall integrity. Cytokine Growth Factor Rev 1997; 8(1): 21-43.

[65] Gerety SS, Wang HU, Chen ZF, Anderson J. Symmetrical mutant phenotypes of the receptor eph B4 and its specific transmembrane ligand ephrin-B2 in cardiovascular development. Mol Cell 1999; 4(3): 403-414.

[66] Wang HU, Anderson DJ. Eph family transmembrane ligands can mediate repulsive guidance of trunk neural crest migration and motor axon outgrowth. Neuron 1997; 18(3): 383-96.

[67] Behrens J, von der Kries JP, Kühl M, Bruhn L, Wedlich D, Grosschedl R, Birchmeier W. Functional interaction of β-catenin with the transcription factor LEF-1. Nature 1996; 382(6592): 638-42.

[68] Orlidge A, D'Amore PA. Inhibition of capillary endothelial cell growth by pericytes and smooth muscle cells. J Cell Biol 1987; 105(3): 1455-62.

[69] Swerlick RA, Brown EJ, Xu Y, Lee KH, Manos S, Lawley TJ. Expression and modulation of the vitronectin receptor on human dermal microvascular endothelial cells. J Invest Dermatol 1992; 99(6): 715-22.

[70] Senger DR, Ledbetter SR, Claffey KP, Papadopoulos-Sergiou A, Peruzzi CA, Detmar M. Stimulation of endothelial cell migration by vascular permeability factor/vascular endothelial

growth factor through cooperative mechanisms involving the $\alpha_v\beta_3$ integrin, osteopontin, and thrombin. Am J Pathol 1996; 149(1): 293-305.

[71] Gumbiner BM. Cell adhesion: the molecular basis of tissue architecture and morphogenesis. Cell 1996; 84(3): 345-57.

[72] Laping NJ, Grygielko E, Mathur A, Butter S, Bomberger J, Tweed C, *et al.* Inhibition of transforming growth factor (TGF)-beta1-induced extracellular matrix with a novel inhibitor of the TGF-beta type I receptor kinase activity: SB-431542. Mol Pharmacol 2002; 62(1); 58-64.

[73] Yuan A, Yang PC, Yu CJ, Chen WJ, Lin FY, Kuo SH, *et al.* Interleukin-8 messenger ribonucleic acid expression correlates with tumor progression, tumor angiogenesis, patient survival, and timing of relapse in non-small-cell lung cancer. Am J Respir Crit Care Med 2000; 162(5); 1957-63.

[74] Bar-EliM. Role of interleukin-8 in tumor growth and metastasis of human melanoma. Pathobiology 1999; 67(1): 12-8.

[75] Kitadai Y, Takahashi Y, Haruma K, Naka K, Sumii K, Yokozaki H, *et al.* Transfection of interleukin-8 increases angiogenesis and tumorigenesis of human gastric carcinoma cells in nude mice. Br J Cancer 1999; 81(4); 647-53.

[76] Malek AM, Gibbons GH, Dzau IVJ, Izumo S. Fluid Shear Stress Differentially Modulates Expression of Genes Encoding Basic Fibroblast Growth Factor and Platelet-derived Growth Factor B chain in Vascular Endothelium. J Clin Invest 1993; 92(4): 2013-21. .

[77] Rak J, Filmus J, Finkenzeller G, Grugel S, Marme D, Kerbel RS. Oncogenes as inducers of tumor angiogenesis. Cancer Metastasis Rev 1995; 14(4): 263-77.

[78] Relf M, Le Jeune S, Scott PA, Fox S, Smith K, Leek R, *et al.* Expression of the angiogenic factors vascular endothelial cell growth factor, acidic and basic fibroblast growth factor, tumor growth factor beta-1, platelet-derived endothelial cell growth factor, placenta growth factor and pleiotrophin in human primary breast cancer and itrelation to angiogenesis. Cancer Res 1997 57(5); 963-9.

[79] Volpert OV, Lawler J, Bouck NP. A human fibrosarcoma inhibits systemic angiogenesis and the growth of experimental metastases via thrombospondin-1. Proc Natl Acad Sci USA 1998; 95(11): 6343-8.

[80] Panetti TS, Chen H, Misenheimer TM, Getzler SB, Mosher DF. Endothelial cell mitogenesis induced by LPA: inhibition by thrombospondin-1 and thrombospondin-2. J Lab Clin Med 1997; 129(2): 208-16.

[81] Mazure NM, Chen EY, Yeh P, Laderoute KR, Giaccia AJ. Oncogenic transformation and hypoxia synergistically act to modulate vascular endothelial growth factor expression. Cancer Res 1996; 56(15): 3436-40.

[82] Auerbach R, Lewis R, Shinners B, Kubai L, Akhtar N. Angiogenesis assays: a critical overview. Clin Chem 2003; 49(1): 32-40.

[83] Kim SW, Park SS, Ahn SJ, Chung KW, Moon WK, Im JG, *et al.* Identification of angiogenesis in primary breast carcinoma according to the image analysis. Breast Cancer Res Treat 2002; 74(2); 121-129.

[84] Auerbach R, Akhtar N, Lewis RL, Shinners BL. Angiogenesis assays: problems and pitfalls. Cancer Metastasis Rev 2000; 19(1-2): 167-72.

[85] Murray JC. Angiogenesis protocols. Methods in Molecular Medicine. Volume 46. In: Totowa, NJ. Humana Press 2001; 267 .

[86] Oehler MK, Bicknell R. The promise of anti-angiogenic cancer therapy. Br J Cancer 2000; 82(4): 749-52.

[87] Duda DG, Sunamura M, Lozonschi L, Kodama T, Egawa S, Matsumoto G, *et al.* Direct in vitro evidence and in vivo analysis of the antiangiogenesis effects of interleukin-12. Cancer Res 2000; 60(4); 1111-6.

[88] Stephen J. Clarke, Rohini Sharma. Angiogenesis inhibitors in cancer - mechanisms of action. Experimental and clinical pharmacology 2006; 29(1): 9-12.

[89] Slaton JW, Perrotte P, Inoue K, Dinney CP, Fidler IJ. Interferon alpha-mediated down-regulation of angiogenesis-related genes and therapy of bladder cancer are dependent on optimization of biological dose and schedule. Clin Cancer Res 1999; 5(10): 2726-34.

[90] Bauvois B, Dumont J, Mathiot C, Kolb JP. Production of matrix metalloproteinase-9 in early stage B-CLL: suppression by Interferons. Leukemia 2002; 16(5): 791-8.

[91] Marler JJ, Rubin JB, Trede NS, Connors S, Grier H, Upton J, *et al.* Mulliken JB, Folkman J. Successful antiangiogenic therapy of giant cell angioblastoma with interferon $\alpha_2\beta$: report of 2 cases. Pediatrics 2002; 109(2); E37.

[92] Nilsson A, Janson ET, Eriksson B, Larsson A. Levels of angiogenic peptides in sera from patients with carcinoid tumors during alpha-interferon treatment. Anticancer Res 2001; 21(6A): 4087-90.

[93] Volpert OV, Fong T, Koch AE, Peterson JD, Waltenbaugh C, Tepper RI, *et al.* Inhibition of angiogenesis by interleukin 4. J Exp Med 1998; 188(6), 1039-46.

[94] Clinicaltrials.gov [Home page on the internet]. National Cancer Institute Fact Sheet. National Cancer Institute, U.S. National Institutes of Health; c1971-2013 [Updated 10th June 2010; Cited 25th April 2013]. Available from http://clinicaltrials.gov/show/NCT00060242.

[95] Murphy AN, Unsworth EJ, Stetler-Stevenson WG. Tissue inhibitor of metalloproteinases-2 inhibits bFGF-induced human microvascular endothelial cell proliferation. J Cell Physiol 1993; 157(2): 351-8.

[96] Valente P, Fassina G, Melchiori A, Masiello L, Cilli M, Vacca A, *et al.* TIMP-2 overexpression reduces invasion and angiogenesis and protects B16F10 melanoma cells from apoptosis. Int J Cancer 1998; 75(2); 246-53.

[97] Smith MR, Kung H, Durum SK, Colburn NH, Sun Y. TIMP-3 induces cell death by stabilizing TNF-alpha receptors on the surface of human colon carcinoma cells. Cytokine 1997; 9(10): 770-80.

[98] Baker AH, Zaltsman AB, George SJ, Newby AC. Divergent effects of tissue inhibitor of metalloproteinase-1, -2, or -3 overexpression on rat vascular smooth muscle cell invasion, proliferation, and death *in vitro*. TIMP-3 promotes apoptosis. J Clin Invest 1998; 101(6): 1478-87.

[99] Marion-Audibert AM, Nejjari M, Pourreyron C, Anderson W, Gouysse G, Jacquier MF *et al.* Effects of endocrine peptides on proliferation, migration and differentiation of human endothelial cells. Gastroenterol Clin Biol 2000; 24(6-7); 644-48.

[100] Hajitou A, Grignet C, Devy L, Berndt S, Blacher S, Deroanne CF, *et al.* The antitumoral effect of endostatin and angiostatin is associated with a downregulation of vascular endothelial growth factor expression in tumor cells. FASEB J 2002 16(13): 1802-4.

[101] Cao Y, Ji RW, Davidson D, Schaller J, Marti D, Sohndel S, *et al.* Kringle domains of human angiostatin. Characterization of the anti-proliferative activity on endothelial cells. J Biol Chem 1996; 271(46); 29461-7.

[102] Gilmore AP, Romer LH. Inhibition of focal kinase protein (FAK) signaling in focal adhesion decreases cell motility and Proliferation. Mol Biol Cell 1996; 7(8): 1209-24.

[103] Rehn M, Veikkola T, Kukk-Valdre E, Nakamura H, Ilmonen M, Lombardo C, *et al.* Interaction of endostatin with integrins implicated in angiogenesis. Proc Natl Acad Sci USA 2001; 98(3); 1024-9.

[104] Pages P, Benali N, Saint-Laurent N, Esteve JP, Schally AV, Tkaczuk J, *et al.* sst2 somatostatin receptor mediates cell cycle arrest and induction of p27(Kip1). Evidence for the role of SHP-1. J Biol Chem 1999; 274(21); 15186-93.

[105] Sharma K, Patel YC, Srikant CB. C-terminal region of human somatostatin receptor 5 is required for induction of Rb and G1 cell cycle arrest. Mol Endocrinol 1999; 13(1): 82-90 .

[106] Charland S, Boucher MJ, Houde M, Rivard N. Somatostatin inhibits Akt phosphorylation and cell cycle entry, but not p42/p44 mitogen-activated protein (MAP) kinase activation in normal and tumoral pancreatic acinar cells. Endocrinology 2001; 142(1): 121-8.

[107] Sharma K, Patel YC, Srikant CB. Subtype-selective induction of wild-type p53 and apoptosis, but not cell cycle arrest, by human somatostatin receptor 3. Mol Endocrinol 1996; 10(12): 1688-96.

[108] Albini A, Florio T, Giunciuglio D, Masiello L, Carlone S, Corsaro A, *et al.* Somatostatin controls Kaposi's sarcoma tumor growth through inhibition of angiogenesis. FASEB J 1999; 13(6); 647-55.

[109] Woltering EA, Watson JC, Alperin-Lea RC, Sharma C, Keenan E, Kurozawa D, *et al.* Somatostatin analogs: angiogenesis inhibitors with novel mechanisms of action. Invest New Drugs 1997; 15(1); 77-86.

[110] Gulec SA, Gaffga CM, Anthony CT, Su LJ, O'Leary JP, Woltering EA *et al.* Antiangiogenic therapy with somatostatin receptor-mediated in situ radiation. Am Surg 2001; 67(11); 1068-71.

[111] Patel YC. Molecular pharmacology of somatostatin receptor subtypes. J Endocrinol Invest 1997; 20(6): 348-67.

[112] Lewin MJ. The somatostatin receptor in the GI tract. Annu Rev Physiol 1992; (54): 455-68.

[113] Buscail L, Esteve JP, Saint-Laurent N, Bertrand V, Reisine TO'Carroll AM, Bell GI, *et al.* Inhibition of cell proliferation by the somatostatin analogue RC-160 is mediated by somatostatin receptor subtypes SSTR2 and SSTR5 through different mechanisms. Proc Natl Acad Sci USA 1995; 92(5); 1580-4.

[114] Mentlein R, Eichler O, Forstreuter F, Held-Feindt J. Somatostatin inhibits the production of vascular endothelial growth factor in human glioma cells. Int J Cancer 2001; 92(4): 545-50.

[115] Lawnicka H, Stepien H, Wyczolkowska J, Kolago B, Kunert-Radek J, Komorowski J. Effect of somatostatin and octreotide on proliferation and vascular endothelial growth factor secretion from murine endothelial cell line (HECa10) culture. Biochem Biophys Res Commun 2000; 268(2): 567-571.

[116] Wu PC, Liu CC, Chen CH, Kou HK, Shen SC, Lu CY, *et al.* Inhibition of experimental angiogenesis of cornea by somatostatin. Graefes Arch Clin Exp Ophthalmol 2003; 241(1); 63-9.

[117] Lamberts SW, Krenning EP, Reubi JC. The role of somatostatin and its analogs in the diagnosis and treatment of tumors. Endocr Rev 1991; 12(4): 450-82.

[118] Jimenez B, Volpert OV, Crawford SE, FebbraioM, Silverstein RL, Bouck N. Signals leading to apoptosis-dependent inhibition of neovascularization by thrombospondin-1. Nat Med 2000; 6(1): 41-8.

[119] Renaud Grepin, Gilles Pages. Molecular mechanisms of resistance to tumor anti-angiogenic strategies. J Oncol 2010; (2010): 835680.

[120] Clinicaltrials.gov [Home page on the internet]. National Cancer Institute Fact Sheet. National Cancer Institute, U.S. National Institutes of Health; c1971-2013 [Updated January 2012; Cited 24th April 2013]. Available from http://www.cancer.gov/clinicaltrials/search/results?protocolsearchid=4586362.

[121] Clinicaltrials.gov [Home page on the internet]. National Cancer Institute Fact Sheet. National Cancer Institute, U.S. National Institutes of Health; c1971-2013 [Updated January 2012; Cited 24th April 2013]. Available from http://www.cancer.gov/clinicaltrials/search/results?protocolsearchid=8706973.

[122] Clinicaltrials.gov [Home page on the internet]. National Cancer Institute Fact Sheet. National Cancer Institute, U.S. National Institutes of Health; c1971-2013 [Updated December 2012; Cited 24th April 2013]. Available from http://clinicaltrials.gov/ct2/show/NCT00002911.

[123] Eichhorn ME, Kleespies A, Angele MK, Jauch KW, Bruns CJ. Angiogenesis in cancer: molecular mechanisms, clinical impact. Langenbecks Arch Surg 2007; 392(3): 371-9.

[124] Clinicaltrials.gov [Home page on the internet]. National Cancer Institute Fact Sheet. National Cancer Institute, U.S. National Institutes of Health; c1971-2013 [Updated 10th July 2011; Cited 24th April 2013]. Available from http://www.cancer.gov/cancertopics/factsheet/Therapy/angiogenesis-inhibitors.

[125] White CW, Sondheimer HM, Crouch EC, Wilson H, Fan LL. Treatment of pulmonary hemangiomatosis with recombinant interferon alpha-2a. N Engl J Med.1989; 320(18): 1197-200.

[126] Izawa JI, Sweeney P, Perrotte P, Kedar D, Dong Z, Slaton JW, *et al.* Inhibition of tumorigenicity and metastasis of human bladder cancer growing in athymic mice by interferon-beta gene therapy results partially from various antiangiogenic effects including endothelial cell apoptosis. Clin Cancer Res 2002; 8(4); 1258-70.

[127] Sasamura H, Takahashi A, Miyao N, Yanase M, Masumori N, Kitamura H, *et al.* Inhibitory effect on expression of angiogenic factors by antiangiogenic agents in renal cell carcinoma. Br J Cancer 2002; 86(5); 768-73.

[128] Strieter RM, Polverini PJ, Kunkel SL, Arenberg DA, BurdickMD, Kasper J, *et al.* The functional role of the ELR motif in CXC chemokine-mediated angiogenesis. J Biol Chem 1995; 270(45); 27348-57.

[129] Clinicaltrials.gov [Home page on the internet]. National Cancer Institute Fact Sheet. National Cancer Institute, U.S. National Institutes of Health; c1971-2013 [Updated 10th July 2011; Cited 24th April 2013]. Available from http://www.cancer.gov/clinicaltrials/search/results?protocolsearchid=4586421.

[130] Boehm T, Folkman J, Browder T, O'Reilly MS. Antiangiogenic therapy of experimental cancer does not induce acquired drug resistance. Nature 1997; 390(6658): 404-7.

[131] Clinicaltrials.gov [Home page on the internet]. National Cancer Institute Fact Sheet. National Cancer Institute, U.S. National Institutes of Health; c1971-2013 [Updated 3rd March 2012; Cited 24th April 2013]. Available from http://www.cancer.gov/clinicaltrials/search/results?protocolsearchid=4586416.

[132] Clinicaltrials.gov [Home page on the internet]. National Cancer Institute Fact Sheet. National Cancer Institute, U.S. National Institutes of Health; c1971-2013 [Cited 24th April 2013]. Available from http://www.cancer.gov/clinicaltrials/search/results?protocolsearchid=4586442.

[133] Clinicaltrials.gov [Home page on the internet]. National Cancer Institute Fact Sheet. National Cancer Institute, U.S. National Institutes of Health; c1971-2013 [Cited 24th April 2013]. Available from http://www.cancer.gov/drugdictionary?cdrid=38491.

[134] Clinicaltrials.gov [Home page on the internet]. National Cancer Institute Fact Sheet. National Cancer Institute, U.S. National Institutes of Health; c1971-2013 [Cited 24th April 2013]. Available from http://www.cancer.gov/cancertopics/druginfo/fda-thalidomide.

[135] Clinicaltrials.gov [Home page on the internet]. National Cancer Institute Fact Sheet. National Cancer Institute, U.S. National Institutes of Health; c1971-2013 [Cited 24th April 2013]. Available from http://clinicaltrials.gov/ct2/show/NCT00004200.

[136] Clinicaltrials.gov [Home page on the internet]. National Cancer Institute Fact Sheet. National Cancer Institute, U.S. National Institutes of Health; c1971-2013 [Cited 24th April 2013]. Available from http://clinicaltrialsfeeds.org/clinical-trials/results/term=rebimastat.

CHAPTER 2

Development of *In Vitro* Method for Assaying Anti-Angiogenic Effect of Drugs

Masumi Akita*

Division of Morphological Science, Biomedical Research Center, Saitama Medical University, 38 Moroyama, Iruma-gun, Saitama 350-0495, Japan

Abstract: Culture techniques using matrix structures have been improved for *in vitro* studies of angiogenesis. Collagen gel culture was used for studying the biological process of angiogenesis. During angiogenesis, electron microscopic and immunohistochemical studies were performed. DNA micro-array gene expression was also conducted. Capillary tubes in the collagen gel were positive for Tnf-α, Nrp-1 and CD133. To test anti-angiogenic drugs, the collagen gel culture was applied. Thalidomide induced the inhibition of cell migration and suppression of Tnf-α. Thalidomide-induced inhibition of angiogenesis involves apoptosis. Cell migration was inhibited by lovastatin. Lovastatin caused the capillary tube degradation. The collagen gel culture provides a useful method for assaying anti-angiogenic effect of drugs.

Keywords: Angiogenesis, Aorta, CD133, collagen gel, culture, DNA micro-array, Fgf, Hematopoietic progenitor cell, Hematopoietic stem Cell, Integrin, laser, lovastatin, nrp, SEM, skeletal muscle, TEM, thalidomide, Tnf-α, Vascular injury, Vegf.

INTRODUCTION

Angiogensis is indispensable in the state of a normal developmental process and pathological condition like tumor growth. Solid tumors recruit new blood vessels from the outer side. Recently brain tumors produce endothelial cells inside the tumor [1]. In any case, the inhibition of angiogenesis prevents the extensive growth of blood vessels that tumors require to survive. Suitable assays are essential for studying the process of angiogenesis and testing new agents with angiogenic or anti-angiogenic potential [2].

***Corresponding author Masumi Akita:** Division of Morphological Science, Biomedical Research Center, Saitama Medical University, 38 Moroyama, Iruma-gun, Saitama 350-0495, Japan; Tel: +81-49-276-1422; Fax: +81-49-276-1422; E-mail: makita@saitama-med.ac.jp

Atta-ur-Rahman and Muhammad Iqbal Choudhary (Eds)
Copyright © 2014 Bentham Science Publishers Ltd. Published by Elsevier Inc. All rights reserved.
10.1016/B978-0-12-803963-2.50002-8

Numbers of angiogenesis assays are increasing. The chorioallantoic membrane (CAM) assay [3] and the corneal micro pocket assay [4, 5] are commonly used *in vivo* assays. The CAM assay is used for studying the effectiveness of anti-angiogenic drugs. The corneal micropocket assay is applied within an avascular environment to study the effectiveness of angiogenic drugs. These assays have proven useful and have dramatically advanced the knowledge of angiogenesis. However, they are also limited in several respects (for a detailed review see Jain *et al.,* [6]: (a) skillful surgical techniques are needed; (b) only a limited number of drugs allows to be assayed (in the case of the micro pocket assay, for example); and (c) simultaneous evaluation of both angiogenic exogenous growth factors. The effectiveness of anti-angiogenic drugs is not attainable without the addition of the advantages and limitations of the assays in use [7]. Staton *et al.,* [7] reviewed *in vitro* assays (endothelial cell proliferation, migration and differentiation, vessel outgrowth from explant) and *in vivo* assays (implantation of sponge and polymers, corneal angiogenesis assay, chamber assay, zebrafish assay, chick CAM assay and tumour angiogenesis models).

Although the endothelial cell plays an important role during angiogenesis, it is not an only cell type related to angiogenesis. Pericytes and smooth muscle cells surrounding the endothelial cells are involved. The circulating blood and an extracellular matrix (ECM) produced by mesenchymal cells, endothelial cells and nonendothelial cells are also involved. Staton *et al.,* [7] noted that now *in vitro* assay is not available to simulate this complex process.

Among the methods *in vitro* assay, vessel outgrowth from explant using collagen gel can simulate considerably the *in vivo* situation. The aortic ring assay is an example of this approach. Small pieces of rodent aorta are explanted into collagen gels. Cell migration from the aorta and three-dimensional (3D) growth were achieved in this culture [8].

In this chapter, I would like to review the collagen gel culture method, and describe the application of collagen gel culture for assaying angiogenesis and the effect of anti-angiogenic drugs (thalidomide and lovastatin).

COLLAGEN GEL CULTURE OF AORTIC EXPLANT

The culture technique was modified from the previous works [9-11]. Fig. **1** shows a schematic diagram of 3D collagen gel culture.

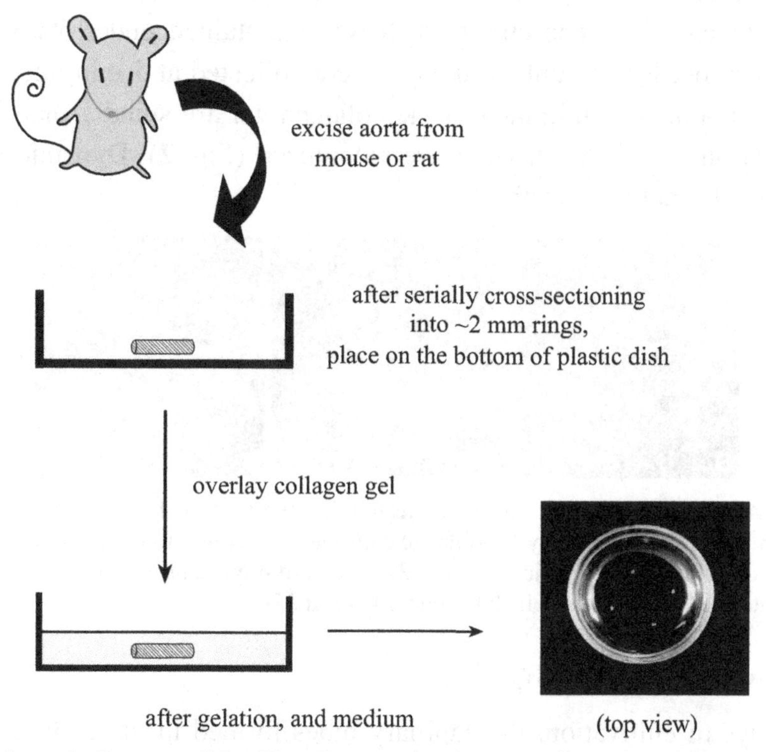

Figure 1: Schematic diagram of the 3D collagen gel culture. Medium was added just to cover the gel. Four pieces of aortic explant were embedded in the gel (top view).

Thoracic aortas from mice or rats were used as a material. Blood and adipose tissue were removed under a stereoscopic microscope. The thoracic aortae were then serially cross-sectioned into ~2mm rings. Four pieces of aortic ring were placed at the bottom of a plastic dish (35 mm). Type I collagen gel matrix (0.3% Cellmatrix type-IA, Nitta Gelatin, Japan) was overlaid. After gelation, culture medium (Ham's F-12, Invitrogen Corp., USA) supplemented with 20% fetal bovine serum (FBS), 1% non-essential amino acids, 100 mg/ml streptomycin, and 100 units/ml penicillin (Invitrogen Corp., USA) were added. The medium was changed three times per week until day 14 (95% air/5% CO_2). Phase contrast microscope was utilized to observe the capillary tube formation.

CAPILLARY TUBE FORMATION

Time-Lapse Imaging

Imaging was performed by a phase-contrast microscope (Nikon TE2000, Japan) with a stage preheated to 37 °C and an imaging system (Aquacosmos, Hamamatsu Photonics, Japan) [11]. The culture dish was maintained under 5% CO_2 during image acquisition. Phase-contrast images were collected at 5 min intervals. After 2 days incubation, cells migrated in the collagen gel are spindle-shaped. After 7 days incubation, capillary sprouts were recognized (Fig. **2**). Dynamic process of capillary tube formation was observed.

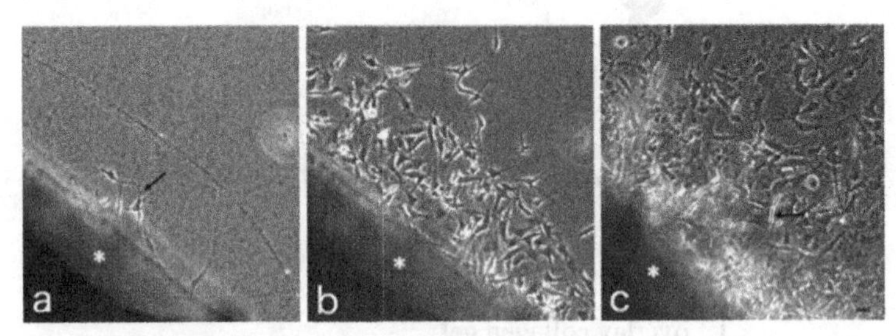

Figure 2: a: After 2 days of culture, arrow indicated fibroblastic cells from a mouse aortic explant (*) b: After 5 days of culture, many fibroblastic cells were outgrown from an aortic explant (*). c: After 7 days of culture, arrow indicated a tubular structure protruding (*). a, b, c: Scale bar = 20 µm. (See Akita *et al.,* [11] Stem Cells Int, Epub 2013 Jun 23).

Histology and Histochemistry

After 14 days of cultivation, the capillary tubes formed in the collagen gel were observed. The cultured aortic rings were fixed in 4% paraformaldehyde/0.1 M phosphate buffer (pH 7.2) and stained with Giemsa and toluidine blue (Fig. **3**).

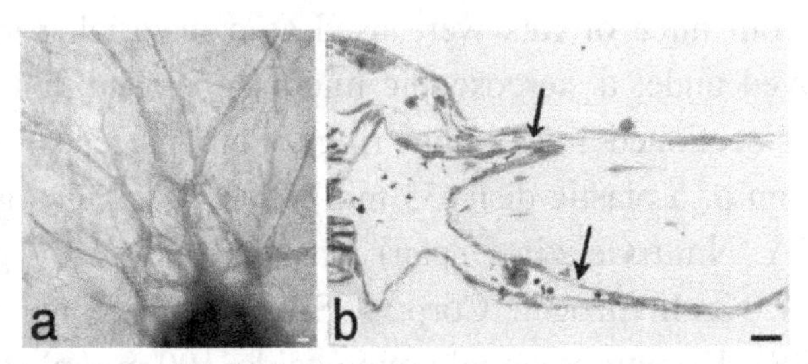

Figure 3: a: Giemsa staining. Scale bar = 10 µm b: Toluidine blue staining. A longitudinal section showing of the aortic explant and two capillary sprouts (arrows). The sprouting endothelial cells extended into the collagen gel. Cells migratingfrom the endothelium of a mouse aortic explant form the luminal inside of the capillary tube, whereas adventitial cells of aortic explant migrated around the capillary sprouts. Scale bar = 10 µm.

To determine the presence of F-actin, rhodamine-phalloidin (Invitrogen Corp., USA) was used. FITC-conjugated tomato lectin (Lycopersicon esculentum; EY Labo, USA) was also used [11]. Tomato lectin recognized the fucose residues on the endothelial cell membrane [12]. Acetylated low-density lipoprotein (Ac-LDL) labeled with 1,1'-dioctadecyl-3,3,3',3'-tetramethylindocarbocyanine perchlorate (DiI-Ac-LDL; Biomedical Technol. Inc., USA) was added to the growth medium and incubated 6 hours. Fig. **4** showed the results of rhodamine-phalloidin, DiI-Ac-LDL and tomato lectin staining.

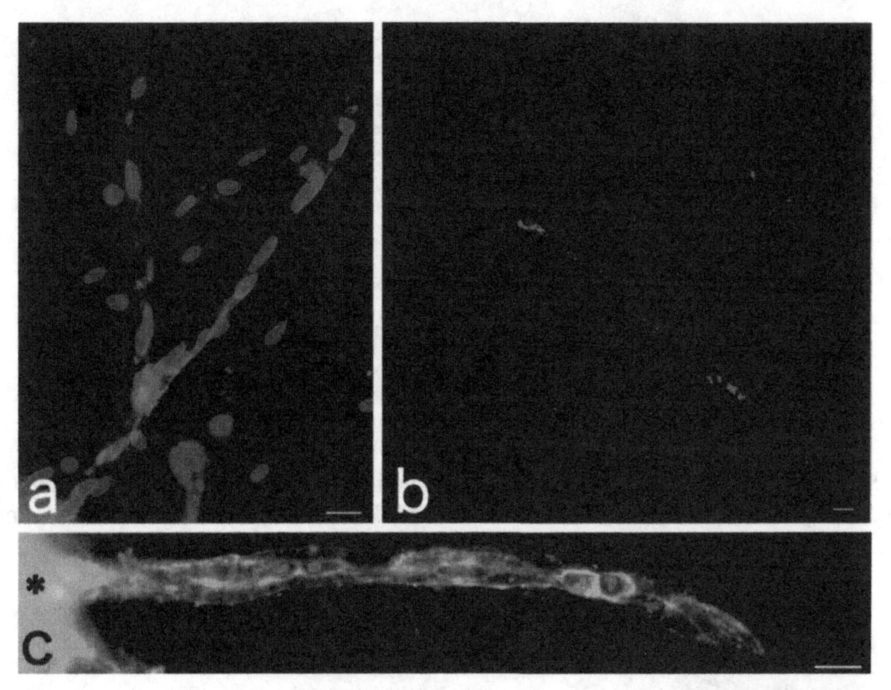

Figure 4: a: After 11 days of culture: the capillary tubes were stained with Rhodamine-phalloidin for F-actin. Scale bar = 20 µm. b: After 11 days of culture: DiI-Ac-LDL, Scale bar = 10 µm. c: After 11 days of culture: the capillary tubes were strongly positive for FITC-conjugated tomato lectin. The asterisk (*) showed a mouse aortic explant. Scale bar = 20 µm. a, c (See Akita *et al.*, [11] Stem Cells Int, Epub 2013 Jun 23).

Transmission Electron Microscopy (TEM)

Samples were fixed in 2.5% glutaraldehyde/0.1 M phosphate buffer (pH 7.2) for 1 hour and subsequently fixed in 1% OsO_4/0.1 M phosphate buffer (pH 7.2) for 1 hour, dehydrated in a graded ethanol series, and embedded in epoxy resin. Ultrathin sections were prepared and treated with uranyl acetate and lead citrate.

A transmission electron microscope (JEM-1400, Japan) was used for observation. Capillary tubes with lumen were demonstrated by a cross-section. Pericyte-like cells were observed on the outside of the lumen (Fig. **5a**) [13]. The cells formed lumen did not show typical gap and tight junctions. Fig. **5b** showed a longitudinal section of the capillary tube [11].

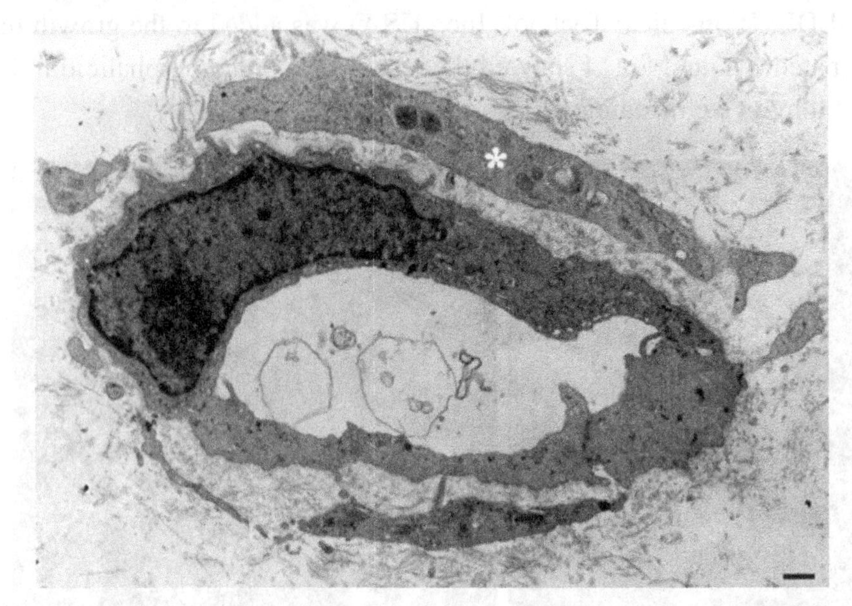

Figure 5a: Transmission electron microphotograph indicated a capillary tube with lumen (a cross section). Pericyte-like cells (*) were observed around the capillary tube. Scale bar = 1 μm. (See Akita *et al.*, [13] J Jpn Coll Angiol (1997) 37 ; 331–6).

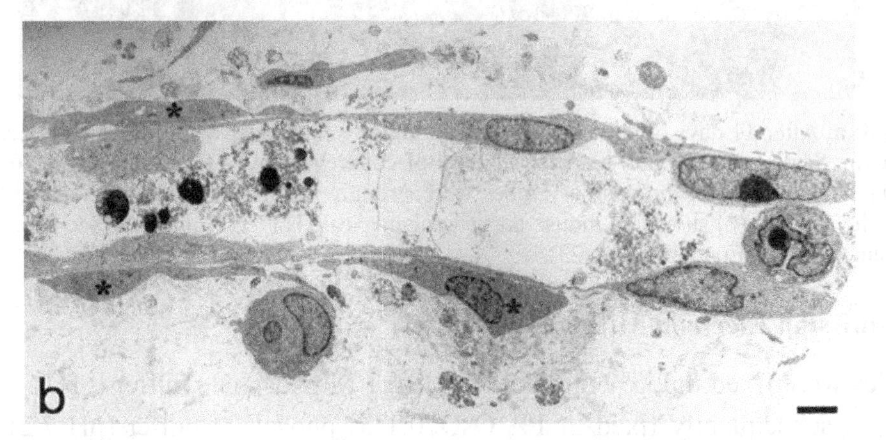

Figure 5b: Transmission electron microphotograph of a capillary tube (a longitudinal section). Pericyte-like cells (*) were observed around the capillary tube. Scale bar = 2 μm. (See Akita *et al.*, [11] Stem Cells Int, Epub 2013 Jun 23).

Scanning Electron Microscopy (SEM)

Samples were fixed in glutaraldehyde and OsO_4 in the same manner described above. The samples were dehydrated in a graded ethanol series, and critical-point air-dried after treatment with isoamyl acetate. The samples were sputter coated with OsO_4 and a scanning electron microscope (Hitachi S-4800, Japan) was used for observation. Before fixation, some samples were treated with 2% collagenase/Hank's balanced salt solution for 10 to 20 min at 37 °C [9, 10]. Figs. **6** and **7** show the capillary tubes.

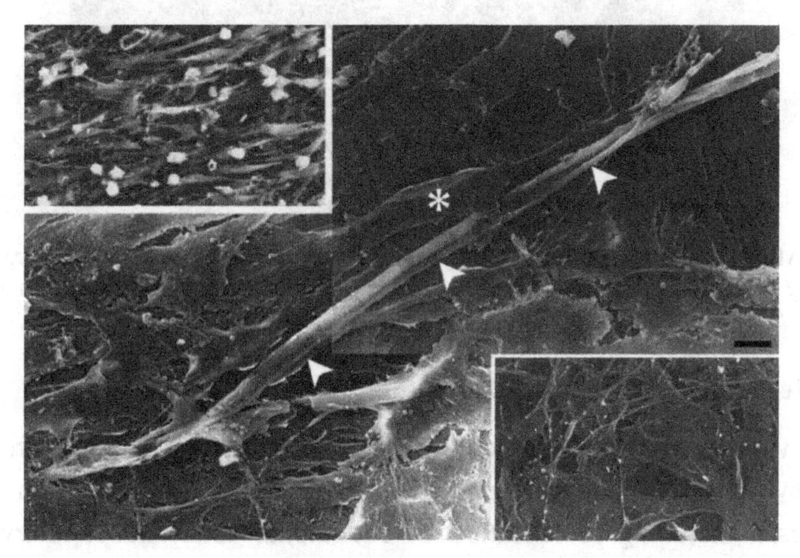

Figure 6: Scanning electron microphotograph after 14 days of culture. Capillary strand (arrowheads) formed in the gel; the vicinity, flat cell (*). Scale bar = 10 μm (See Akita *et al.,* [9] Annals of anatomy, 1997;179:137-47). Inset (upper left): Scanning electron microphotograph of the surface of the culture. Inset (lower right): Scanning electron microphotograph of the bottom of the culture.

IMMUNOHISTOCHEMISTRY OF ANGIOGENIC FACTORS

The *in vivo* mechanism of angiogenesis is very complex, involving not only endothelial cells but also other cellular components such as pericytes, smooth muscle cells, fibroblasts and macrophages. Angiogenic factors, adhesive molecules (integrins) and ECM also play an important role in angiogenesis. Endothelial proliferation is required for the new vessel formation. Angiogenic factors are thought to be inducers of mitogenesis in vascular endothelial cells. Angiogenesis can be induced by a variety of growth factors.

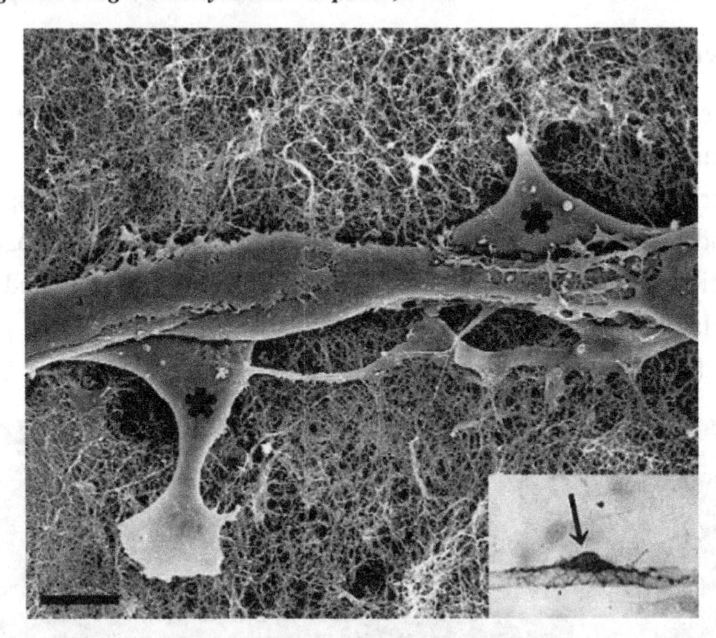

Figure 7: After 14 days of culture: capillary tubes from a mouse aortic explant were observed. Pericyte-like cells (*). Scale bar = 10 μm. Inset. The pericyte-like cell (indicated with arrow) was positive for α smooth muscle actin antibody staining.

Tumor Necrosis Factor Alpha (Tnf-α)

In the previous study [14] using collagen gel culture, capillary tube growth was induced by Tnf-α in a dose-dependent manner. Tnf-α play important roles in angiogenesis, including proliferation of endothelial cells and capillary sprouts [15]. Fig. **8** shows the Tnf-α positive capillary tubes. Tnf-α positive cells were observed in the capillary tube, especially at the leading edge. It is suggested that Tnf-α-positive cells relate to the elongation of capillary tubes.

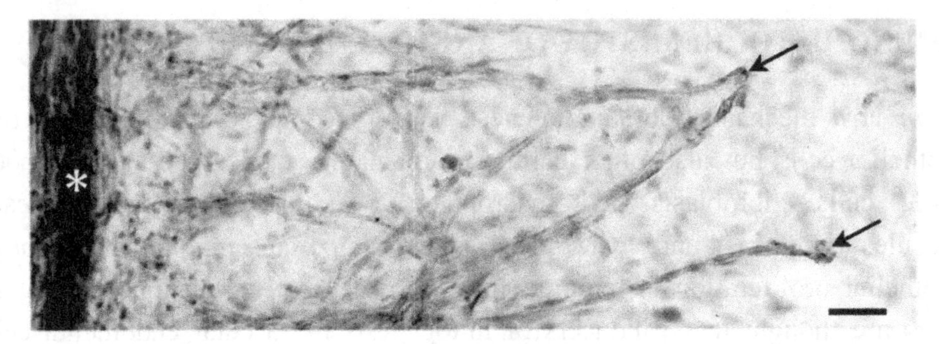

Figure 8: After 14 days of culture, capillary tubes showed Tnf-α positive reaction. Arrows indicated the leading edge of the capillary tube from a mouse aortic explant (*). Scale bar = 100 μm.

Fibroblast Growth Factors (Fgfs)

Fgfs are a family of heparin-binding proteins. Fgfs involved in many biological processes including angiogenesis [16]. The actions of the Fgfs, particularly the basic form (basic Fgf = Fgf-2), were reported on in a large number of cells (for a review see Bikfalvi *et al.,* [17]). Fgf-2 is well recognized as an angiogenic factor. In the collagen gel culture, the expressions of both of Fgf-7 and -9 as well as Fgf-2, is revealed by reverse transcription-polymerase chain reaction (RT-PCR) and immunohistochemistry (Fig. **9**). We proposed that Fgf-7 and -9 also relate to angiogenic events under in the collagen gel culture [18].

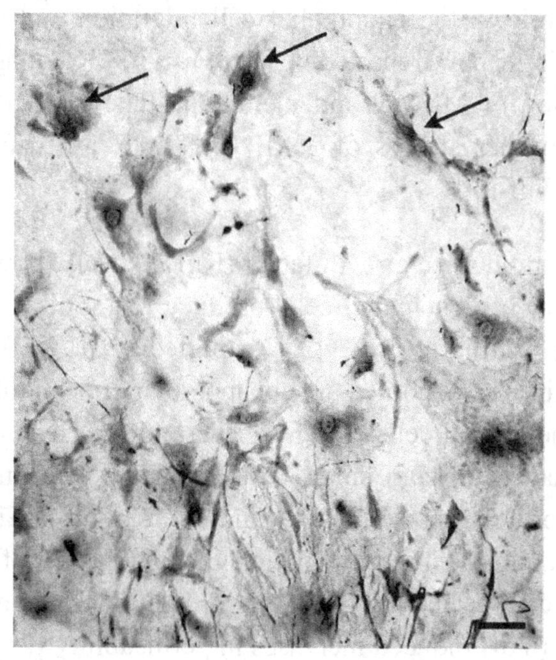

Figure 9: After 14 days of culture, migrating cells from the aortic explant showed Fgf-9 positive reaction. Arrows indicated the cells spreading out in the collagen gel. Scale bar = 20 μm.

Fgf-2, is a potent angiogenic factor *in vivo* and *vitro*, excites endothelial cell growth [19, 20], and smooth muscle cell growth, wound healing, and tissue repair [21-24]. Kuzuya *et al.,* [23] and Satake *et al.,* [24] reported that Fgf-2 plays important roles in angiogenesis. When fetal bovine aortic endothelial cells (BAECs) were embedded in collagen gels, they underwent apoptotic death within 2 days unless the cultures were repeatedly supplied with Fgf-2 [23, 24]. Supplementation with Fgf-2 induced endothelial cell differentiation, resulting in the capillary-like tube formation inside the collagen gels [23, 24]. Endothelial cells derived from BAECs undergo apoptosis in 3D type I collagen gels in the absence of Fgf-2. In the presence of Fgf-2, BAECs survive and form a capillary-like tube structure in 3D culture [23, 24].

Fgf-9 originally cloned as a glial-activatingfactor. Fgf-9 stimulated the cell growth of fibroblasts and glial cells [25, 26]. The biological roles of Fgf-9 are not well known. Pilcher *et al.,* [27] and Miyagi *et al.,* [28] reported that Fgf-9 relates to the expression of matrix metalloproteinases (MMPs) and their inhibitor through the Fgf cell-surface receptor (FgfR). Haas *et al.,* [29] hypothesized that MMPs play an important role in the endothelial cell migration and matrix remodeling during angiogenesis. They reported that 3D type I collagen gels induced coordinate expression of MMP-2 and membrane type I-MMP in microvascular endothelial cells. Further studies of MMP and FgfR are needed to clarify the angiogenic roles of Fgf-9.

Fgf-7, also called keratinocyte growth factor (Kgf), is well known as a growth and differentiation factor. Mesenchymal cells produce Kgf, and Kgf acts specifically on epithelial cells. Gillis *et al.,* [30] reported that Kgf induced *in vivo* neovascularization in the rat cornea at subnanomolar concentrations. Kgf was not effective against endothelial cells cultured from large vessels *in vitro,* but did act directly on those cultured from small vessels, inducing chemotaxis, activating mitogen activated protein kinase (Mapk) and stimulating proliferation [30]. Kgf also helped to maintain the barrier function of monolayers of capillary, but not aortic endothelial cells, protecting against hydrogen peroxide and Vegf /vascular permeability factor (Vpf) induced increases in permeability [29]. Gillis *et al.,* [30] suggested that these newfound abilities of Kgf advanced their repair after major damage.

Vegf and Vascular Endothelial Growth Factor Receptor (VegfR)

In a study usingfibrin gel, Brown *et al.,* [31] showed that both Vegf and Fgf-2 mRNAs were expressed in the cells migratingfrom the plasental blood vessel stump. Gerber *et al.,* [32] suggested that Vegf is a key regulator of angiogenesis during bone development. Vegf is a growth factor that acts specifically active on vascular endothelial cells. According to Roy *et al.,* [33], Vegf and VegfR are considered to be key molecules in the angiogenesis [33]. The Vegf family currently includes Vegf-a, -b, -c, -d, -e, -f and placental growth factor (Plgf) proteins [33]. Vegf family acts as a chemoattractant and directs capillary growth, and Vegf concentration gradients are important for the activation and chemotactic guidance of capillary sprouting [34, 35]. Fig. **10a** shows Vegf positive cells around the capillary tubes. In our previous experiment, Vegf promoted cell outgrowth from aortic explant [36]. We show that VegfR-1 expression is restricted to the pericyte-like cells around the capillary tubes and is not present in the endothelial cells of the capillary tubes. Feng *et al.,* [37] and Witmer *et al.,* [38] reported that VegfR-1 was localized to only the pericytes. VegfR-2 was localized to only the endothelial cells. These findings are consistent with observations of VegfR-1 expression in pericytes *in vitro* [39]. VegfR-1 is mainly a pericyte receptor for Vegf. VegfR-2 is strongly expressed in endothelial tip cells, and the VegfR-2 distributes along the tip-cell filopodia [40, 41]. The data also show that VegfR-2 is expressed in the capillary tube (Fig. **10b**). Hamada *et al.,* [42] noted that both VegfR-2 and VegfR-3 show similar immuno-histochemical staining patterns in vascular endothelial cells of mouse embryos at E9.5. Witmer *et al.,* [38] also observed that VegfR-3 expression in blood vessels always co-localized with VegfR-2 expression. This suggests that there is a synergistic or antagonistic role for different members of the Vegf family at these sites.

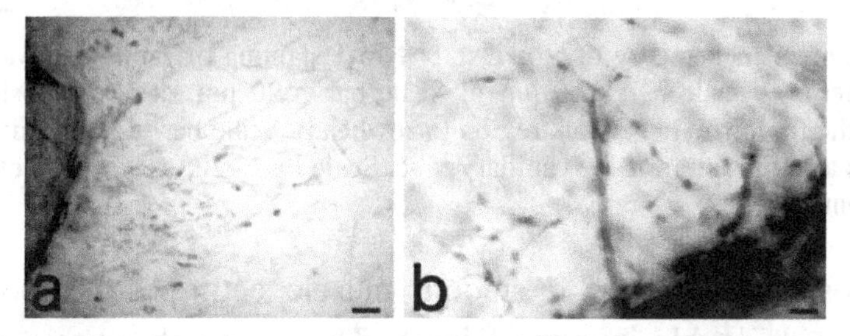

Figure 10: Immunohistochemistry of Vegf and Vegf-2. a: Migrating cells into collagen gels were positive for Vegf. Capillary tubes did not show immunoreactivity of Vegf. Scale bar = 20 μm. b. Capillary tubes showing immunoreactivity of VegfR-2. Scale bar = 20 μm.

Neuropilin (Nrp)

Although the role of Vegf family is well established, the role of VegfR family, especially newly described VegfR-related proteins, Nrp-1 and -2, remain unknown in angiogenesis. We studied the expression of VegfR-related proteins, Nrp-1 and -2, during angiogenesis *in vitro*. Immunohistochemical studies here show that Nrp-1 staining was present in a tube-like pattern (Fig. **11a, b**). Nrp-1 staining appeared to localize to the endothelial cells. Previously, Nrp-1 was thought to be expressed in endothelial cells [43, 44]. The data also support these findings. However, some Nrp-1-positive cells were present around the capillary tube (Fig. **11c, d**).

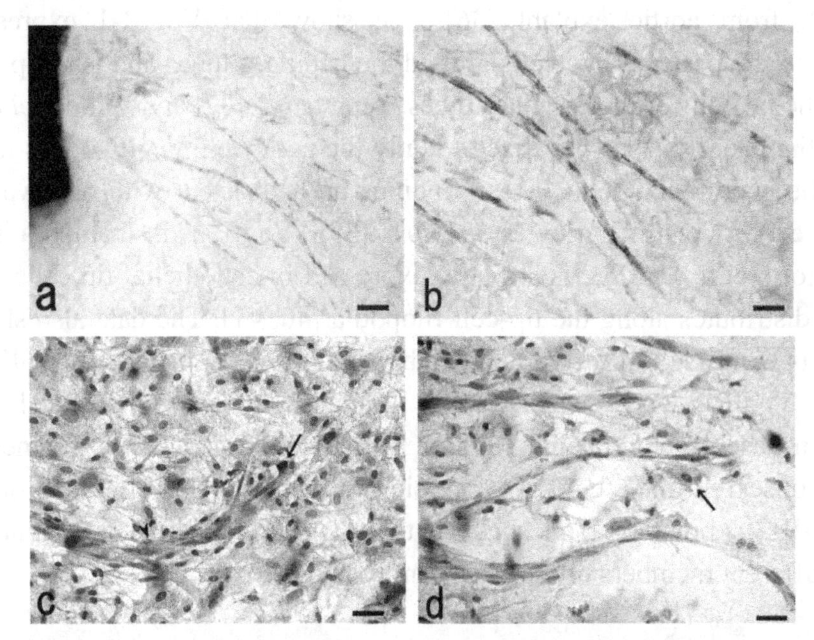

Figure 11: Immunohistochemistry of Nrp-1. a, b: Nrp-1 staining of capillaries was mostly found in tube-like pattern. a: Scale bar = 50 μm b: Scale bar = 20 μm. c: Nrp-1-positive cells were present in both the tip (arrow) and stalk region (arrowhead). Scale bar = 20 μm. d: Nrp-1-positive cell (arrow) was also present near the capillary tube. Scale bar = 20 μm. c, d ; nuclei were stained with Mayer's hematoxylin.

The Nrp-1-positive cells and α smooth muscle actin-positive pericytes were distinguished with double immunostaining. The α smooth muscle actin-positive pericytes closely associated with the Nrp-1-positive endothelial cell strand. Cells co-localized with Nrp-1 and α smooth muscle actin were present (Fig. **12**).

Many investigators demonstrated that Nrp-1 was expressed on nonendothelial cells, including various tumor cells [45-49]. Recently, Liu *et al.,* [50] found that vascular smooth muscle cells also express Nrp-1. Vascular smooth muscle cells and pericytes have the same origin. Pericytes are generally assumed to belong to the same cell lineage as mural cells in the blood vessel wall [50]. Interactions between endothelial cells and mural cells (pericytes and vascular smooth muscle cells) have recently come into focus as central processes in the regulation of vascular formation, stabilization, remodeling, and function [50]. The results from our study suggest that crosstalk between endothelial cells and pericytes plays an important role in angiogenesis.

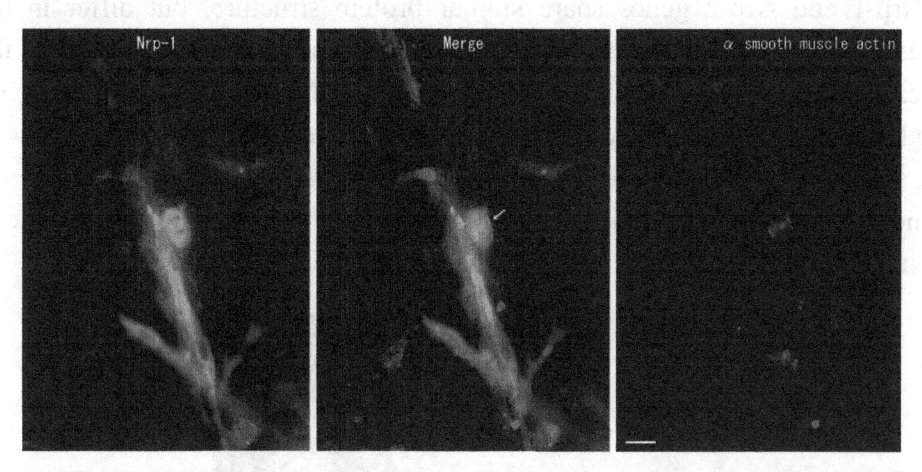

Figure 12: Double immunostaining with Nrp-1 and α smooth muscle actin antibodies. Cell strand of capillary tubes was strongly positive for Nrp-1 staining (green). Alpha smooth muscle actin-positive cells (red) were present around the capillary tubes. Arrow indicated the cell co-localized with Nrp-1 and α smooth muscle actin. Nuclei were stained blue with DAPI (4', 6-diamidino-2-phenylindole). Scale bar = 20 μm.

Nrp-1 was originally characterized as a receptor for class III semaphorin/collapsin family members, and Nrp-1 involved in neuronal guidance in the nervous system [52]. Nrp-1 was subsequently identified as a co-receptor for specific isoforms of Vegf, and Nrp-1 regulated tumor angiogenesis *via* binding the Vegf isoform Vegf 165 [53]. Nrp-1 also bound other Vegf family members such as Vegf-a, Vegf-e, Plgf-2 [54-56]. In endothelial cells, Nrp-1 binds Vegf 165 and enhances its binding to VegfR-2 [53, 57]. As was reviewed by Shibuya [58], Vegf-165 modestly binds to Nrp-1 and heparin, which leads to the formation of a Vegf-165 gradient around Vegf-a-expressing cells. Overexpression of Nrp-1 in chimeric mice leads to excessive capillary and blood vessel formation and to cardiac malformations including haemorrhage [59]. The defective vascularization has been observed in many organs of Nrp-1-deficient mouse embryos [60-62]. These findings are compatible with the hypothesis that Nrp-1 functions in endothelial tip cell guidance [41]. However, our immunohistochemical data here show that Nrp-1 expression was present in both the tip and stalk regions. Thus, it seems unlikely that Nrp-1 plays a role in endothelial tip cell guidance

In chick embryos, endothelial Nrp-1 expression is mostly confined to the arteries, whereas Nrp-2 primarily marks the veins [63]. Bielenberg *et al.,* [64] noted that the Nrp-1 and Nrp-2 genes share similar protein structure, but differ in their expression patterns, regulation, and ligand-binding specificities. Nrps vary in their expression patterns; endothelial cells express both Nrp-1 and Nrp-2, lymphatic endothelial cells mainly express Nrp-2, and epidermal cells mainly express Nrp-1 [64]. In our study, Nrp-2 positive cells were sparse and were observed only around the capillary tubes in our aortic culture. Unlike the Nrp-1-positive cells, the Nrp-2-positive cells were not found to the capillary tubes (Fig. **13**).

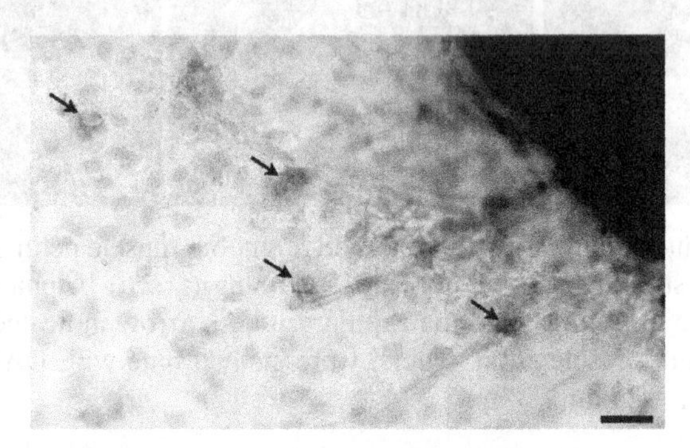

Figure 13: Immunohistochemistry of Nrp-2. Cells around the capillary tubes are Nrp-2 staining positive (arrows), suggesting staining of pericytes. Scale bar = 20 μm.

Thus, it is suggested that Nrp-1 and -2 have different roles in angiogenesis. The presence of Nrp-1 or Nrp-2 enhances VegfR2 signaling [65]. It has been demonstrated that anti-Nrp-1 antibodies reduced angiogenesis and vascular remodeling in reducing tumor growth. Tumor vessels are heterogeneous in their pericyte coverage. Anti-angiogenic therapy directed against the endothelium appears to lead to ablation of the naked endothelial tubes, whereas the tubes covered with pericyte are protected [51, 66]. This observation has led to the idea that combinations of anti-endothelial and anti-pericyte agents might act synergistically in anti-angiogenic therapy. Pan *et al.,* [67] showed that a combination of an anti-Vegf antibody and an antibody targeting Nrp-1 resulted in a greater tumor growth inhibition. Anti-Nrp-1 antibodies reduced angiogenesis and vascular remodeling, and had an additive effect with anti-Vegf therapy (Bevacizumab, Avastin) along with in reducing tumor growth [67]. Vessels from tumors treated with anti-Vegf showed a close association with pericytes, while tumors treated with both anti-Nrp-1 and anti-Vegf lacked this organization [67]. They proposed that blocking Nrp-1 function inhibits vascular remodeling, keeping the vessels in a "Vegf-dependence" state, and renders them more susceptible to anti-Vegf therapy [67]. These results suggest that Nrp-1 is potential targets for anti-angiogenic cancer therapy [68].

ADHESIVE MOLECULES (INTEGRINS)

The interactions between integrins and ECM were identified as important regulators of vascular cell proliferation, invasion and survival during the complex process of blood vessel formation [69]. As pointed out by Eliceiri and Cheresh [70], analysis of the cell signalling pathways downstream of integrins are shedding light on the molecular basis of known anti-angiogenic therapies, as well as the design of novel therapeutic strategies.

The $\alpha_v\beta_3$ integrin heterodimer is believed to play an important role in angiogenesis (see the review of Varner [71]). However, findings on mice lacking either the $\alpha_v\beta_3$ integrin alone or both the $\alpha_v\beta_3$ and $\alpha_v\beta_5$ integrins have called these results into question [72-74]. Moreover, mice lacking both the $\alpha_v\beta_3$ and $\alpha_v\beta_5$ integrins develop more extensive tumors with richer vascular supplies than their littermate controls [74]. Several findings have proven that $\alpha_v\beta_3$ integrin is not indispensable to all vasculogenesis or angiogenesis: mice and humans deficient in β_3 integrin exhibited platelet defects but nonetheless appeared to develop normal vascular beds. β_3 null mice showed normal retinal angiogenesis [73], and α_v knockout mice exhibiting abnormalities in the brain and gut vasculogenesis developed normal vascular trees in other tissues [75]. Whereas $\alpha_v\beta_3$ may play a role in angiogenesis, other mechanisms must be involved or able to recompense for the lack of $\alpha_v\beta_3$ [76]. Gonzalez *et al.,* [77] demonstrated that the G domain of laminin α_4 chain was a specific, high-affinity ligand for the $\alpha_v\beta_3$ and $\alpha_3\beta_1$ integrin heterodimers, and these ligands functioned cooperatively with $\alpha_6\beta_1$ to mediate endothelial cell-α_4 laminin interaction and hence blood vessel development. The study [78] suggests that the α_1 and α_2 integrin subunits relate to the cell-migration and that the α_3 integrin subunit plays some role in the tube formation (Fig. **14**).

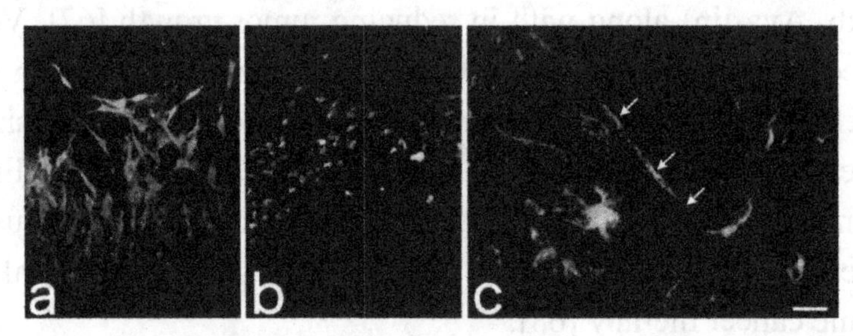

Figure 14: a: Immunofluorescence staining with anti α_1 integrin antibody. After 7 days of culture, the cells located in the outermost periphery were positive for the antibody. The inner cells near the aortic explant were almost negative. b: Immunofluorescence staining with anti α_2 integrin antibody. After 7 days of culture, the cells located at the periphery of the aortic explant were positive for anti α_2 integrin subunit antibody. c: Immunofluorescence staining with anti α_3 integrin antibody. After 7 days of culture, positive stainingfor anti α_3 integrin subunit antibody was clearly observed in the cells located near the aortic explant, and the cord like cell-strand (arrows) was also positively stained. (See Suda *et al.,* [78] Histo Histopathol 2004; 19: 735-42).

These three α subunits have been reported to mediate cell-type I collagen interactions [79]. The α_1 and α_2 integrin subunits are known to act as major collagen receptors in most cell types, whereas the α_3 integrin subunit plays a less established role as a collagen receptor [79]. Both $\alpha_1\beta_1$and $\alpha_2\beta_1$ bind to native collagen, and in some cell types they can also bind to laminin, although with lower affinity (see review of Eble [80]). The binding sites for α_1 and α_2 integrin subunits in type I and type IV collagens are found in the triple helical area [81, 82]. Denatured collagen can be recognized by cells *via* RGD-binding integrin subunits such as α_3 and α_v, whereas collagen binding by α_1 and α_2 integrin subunits require a native conformation [80]. The peptide, KDGEA (Lys-Asp-Gly-Glu-Ala), contains the recognition sequence for $\alpha_2\beta_1$ integrin in type I collagen. Heino [83] reported that integrin $\alpha_2\beta_1$ is required for cell migration through collagenous matrix. In our study [78], the peptide KDGEA decreased the migration of myofibroblasts from the aortic explant, suggesting that the α_2 integrin subunit is involved in the cell migration through the collagenous matrix. After ligand binding, both $\alpha_1\beta_1$ and $\alpha_2\beta_1$ trigger cellular diverse responses such as mechanical contraction of collagen gels [84], gene activation of collagen-degrading MMP-1 and decreased expression of $\alpha 1$ (I) chain of type I-collagen [85, 86]. Riikonen *et al.,* [86] also found that $\alpha_2\beta_1$ integrin mediated the cellular responses to extra-cellular 3D collagen, including the induction of MMP-1 production. Although the native collagen molecules contain the $\alpha_v\beta_3$ integrin ligands (RGD [Arg-Gly-Asp] sequences), these ligands remain masked within the triple helix and hardly bind to integrin $\alpha_v\beta_3$. However, when the collagen molecules are degraded by heat or proteinases, the RGD sites are exposed and can interact with $\alpha_v\beta_3$ integrin [76]. Nicosia and Bonanno [87] found that the addition of a high concentration of GRGDS (300 mg/ml) to the culture medium brought about a marked inhibition of angiogenesis in collagen gel culture, whereas GRGES, a control peptide lacking the RGD sequence, failed to exert such an inhibitory effect. The considerably lower concentration of GRGDS (100 ng/ml) used in the study [78] did not inhibit angiogenesis. As the RGD failed to decrease the cell migration, the α_2 integrin subunit evidently played a more important role in the cell migration than the α_3 and α_v integrin subunits. Although the $\alpha_v\beta_1$ unit has also been described as a receptor for collagen and laminin [88], it binds more readily to laminin-5 than to collagen [89, 90], and it fails to bind to a collagen-matrix [81]. In the immunofluorescence microscope images from the study [78], the α_3 integrin subunit was expressed in the cord-like cell strands. Moreover, immunoelectron microscopy revealed the expression of the α_3 integrin subunit on the cells surrounding the tube-like structure with lumen. The proximity of these cells to the basal lamina suggests that the α_3 integrin subunit is closely related to tube formation. The study also suggested that the α_3 integrin subunit was very important for angiogenesis. However, the α_3 integrin subunit was not demonstrated on the endothelial cells with lumen, but rather on the cells surrounding the tube-like structure [78]. In earlier studies, $\alpha_3\beta_1$ was ascribed a role in cell-cell interaction mediated by heterophilic $\alpha_2\beta_1$-$\alpha_3\beta_1$ interaction [91] and homophilic $\alpha_3\beta_1$-$\alpha_3\beta_1$ interaction [92]. Thus, further investigations will be needed to clarify the interaction of endothelial cells and surrounding pericytes.

CD133 (HEMATOPOIETIC STEM/PROGENITOR CELL MARKER)

Immunohistochemical Detection of CD133

In 1997, Miraglia *et al.,* [93] and Weigmann *et al.,* [94] isolated and cloned CD133, also known as prominin-1 in humans and rodents. Hematopoietic stem and progenitor cells expressed CD133. CD133-positive cell population has hematopoietic potential, and also has the capacity to differentiate into endothelial cells [95]. Sun *et al.,* [96] found a specific binding peptide targeted mouse CD133 using phage-displayed peptide library technology [95]. However, the biological function of CD133 remains largely unknown.

We detected CD133 positive cells during angiogenesis in the collagen gel culture [11]. The samples were fixed in 4% paraformaldehyde/0.1 M phosphate buffer (pH 7.2) and treated with CD133 antibody (rabbit polyclonal; Abcam®, Japan). Alexa Fluor 488- and Nanogold (1.4 nm)-conjugated goat anti-rabbit IgG (Nanoprobes, USA) were used as a secondary antibody. GoldEnhance EM (Nanoprobes, USA) was used as an enhancer of the Nanogold signal for electron and light microscopic observations. In the study [11], CD133-positive cells were observed in the capillary tube (Fig. **15**). The tip of the capillary tube was strongly positive for CD133. We proposed that CD133-positive cells related to the capillary tube elongation and/or branching [11]. Peichev *et al.,* [97] reported that endothelial precursors expressed CD133 and rapidly lost during maturation. The newly formed capillary tubes may immature. Hif-1 is an important enhancer of endothelial cell differentiation [98]. Hypoxic conditions may relate to the CD133-positive capillary tube formation.

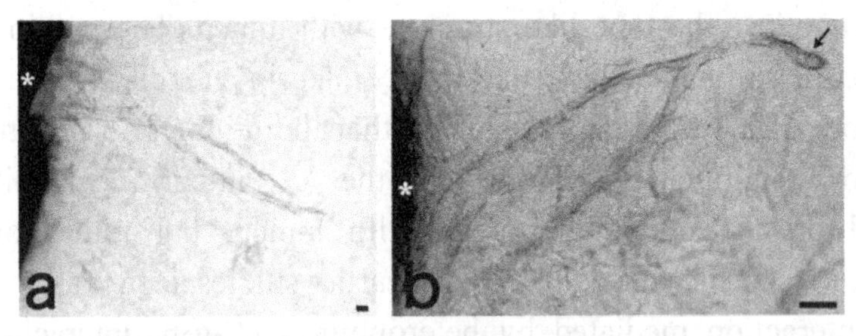

Figure 15: Immunohistochemistry of CD133. a: CD133-positive cells were observed in a tubular pattern. Other capillary tubes were almost negative. Rat aortic explant (*). Scale bar = 10 µm. b: Arrow indicated the tip region of a capillary tube. The stalk region near the aortic explant (*) did not show clearly positive reaction. Scale bar = 20 µm. (See Akita *et al.,* [11] Stem Cells Int, Epub 2013 Jun 23).

CD133 is also known as a marker for cancer stem cells (CSCs) in many different types of solid tumors [99-108]. Although solid tumors recruit new blood vessels from outside the tumor, brain tumors produce endothelial cells within the tumor from tumor stem-like cells [1, 109, 110]. According to Wang *et al.,* [109], Vegf and Notch related to the differentiation of CSCs into endothelial cells. Wang *et al.,* [109] also proposed that Notch regulated the initial differentiation of CSCs to endothelial progenitor cells. Vegf regulated the differentiation of endothelial progenitor cell to tumor-derived endothelial cells. Various tumors suggested that tumor cells were capable of differentiating into endothelial cells [111-115]. To evaluate the possibility of differentiation of CSCs into endothelial cells, we need further investigation.

CD133-positive cells were also presented within the aortic explant [11]. Human fetal aorta contained vascular progenitor cells. These cells are inducible vasculogenesis and angiogenesis [116]. Endothelial precursor and stem cells are present in a zone of the vascular wall, which is located between the smooth muscle and the adventitial layer. Endothelial precursor and stem cells differentiate into mature endothelial cells [117]. These progenitor cells are present as a cluster between the outer smooth muscle layer and inner adventitia [118]. We demonstrated that epithelium scraping from the aorta occurred only spindle-shaped cells migration. However, capillary tube formation did not appear into the collagen gel. It is suggested that the primary source of the capillary tube is the intimal endothelial cells [11]. According to Nicosia [119], completely de-endothelialized rat carotid artery with a balloon catheter failed to an angiogenic response. No-endothelialized artery produced microvessels. Whether the primary source of tube formation is derived from between the smooth muscle and adventitial layer or from the endothelium, further studies are required to clear these issues.

GENE ARRAY ANALYSIS

We examined angiogenesis-related gene expression profiles using collagen gel culture and a DNA chip [120]. At day 5 of the culture, numerous cells migrated from the aortic explant into the collagen gel, but the capillary tubes had not yet formed. The capillary tubes began appearing after 7 days of culture in the collagen gels. Fig. **16** shows phase contrast micrographs before tube formation (day 5) and after tube formation (day 10).

Total RNA was extracted by TRIZOL® (Invitrogen) from each culture plate (day 5 and day 10) without the aortic explant. For microarray analysis, RNA was purified using a RNeasy® MinEluteTM Cleanup Kit (Qiagen). Gene array analysis was conducted as a 'custom order' by Kurabo Industries (Osaka, Japan) with CodeLinkTM Bioarray Mouse Whole Genome Array (Amersham Biosciences). This DNA microarray chip displays probe DNAs for about 35,000 mouse genes and expressed sequence tags [120]. Double-stranded cDNA was prepared from RNA in the culture (day 5 and day 10). cDNA was transcribed with biotin-labeled dUTP as a substrate. For hybridization of the array, the complementary RNA was used as a probe. Indodicarbocyanine-conjugated streptavidin was used to detecte the hybridized RNA. Biochip Reader (Applied Precision) and CodeLink System software (Amersham Biosciences) were applied [120].

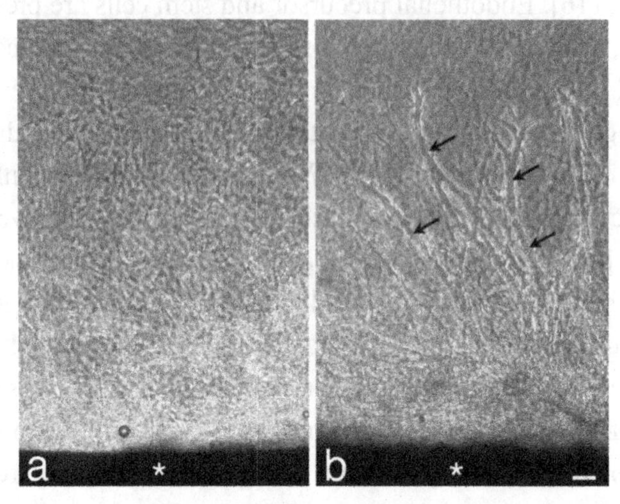

Figure 16: Phase-contrast microphotograph, after 5 days and 10 days of culture. a: After 5 days of culture, cell migration occurred from the end of the aortic explant (asterisk). Capillary-like tubes were not formed at this time. b: After 10 days of culture, many capillary-like tubes (arrows) were seen. Scale bar = 20 µm. (See Akita M, Fujita K. [120] Clinical Medicine: Cardiology 2008; 2: 49-57).

Data Analysis for the CodeLink Arrays

Data from the arrays were exposed to a quantile normalization procedure. Following normalization, *p*-value calculation for individual probes was done. Data with *p*-value <.05 were used to select significant genes for further studies. The data were log transformed and the statistical significance of differences in intensity values was calculated by logfold-change (logFC) analysis (CodeLink Expression Analysis software, v 4.1). Positive logFC (> 2) indicate the logarithmic foldness of up-regulation. Negative logFC (< -2) indicate the logarithmic foldness of down-regulation.

ACCN	DESCRIPTION (OGS)	Normalized_Intensity
NM_010217.1	connective tissue growth factor (Ctgf)	264.7248831
NM_007505.2	annexin A2 (Anxa2)	204.1702274
BC048857.1	integrin alpha V (Ivtgav)	77.6946662
NM_009396.1	tumor necrosis factor, alpha-induced protein 2 (Tnfaip2)	61.64709116
NM_024226.2	reticulon 4 (Rtn4)	52.70839676
NM_178444.3	EGF-like domain 7 (Egfl7)	38.05129471
NM_008486.1	alanyl (membrane) aminopeptidase (Anpep)	34.59025436
NM_013749.1	tumor necrosis factor receptor superfamily, member 12a (Tnfrsf12a)	27.56429037
NM_010216.1	c-fos induced growth factor (Figf)	25.80474246
NM_025630.2	RIKEN cDNA 2010009L17 gene (Aggf1)	22.60618802
NM_009505.2	vascular endothelial growth factor A (Vegfa)	21.34198692
NM_009929.2	procollagen, type XVIII, alpha 1 (Col18a1)	20.12021705
NM_007950.1	epiregulin (Ereg)	19.09632627
NM_010431.1	hypoxia inducible factor 1, alpha subunit (Hif1a)	16.69576181
NM_008737.1	neuropilin 1 (Nrp1)	12.27930901
NM_008827.2	placental growth factor (Pgf)	11.90281775
NM_007554.1	bone morphogenetic protein 4 (Bmp4)	11.74764624
NM_009035.1	recombining binding protein suppressor of hairless (Drosophila) (Rbpsuh)	8.984081808
NM_011697.1	vascular endothelial growth factor B (Vegfb)	8.080468525
NM_007426.2	angiopoietin 2 (Angpt2)	7.619408385
NM_010612.2	kinase insert domain protein receptor (Kdr)	7.134782369
BM118719.2	L0915B11-3 NIA Mouse Newborn Kidney cDNA Library (Long) cDNA clone NIA:L0915B11 IMAGE:30001462 3' (Elk3)	7.13212412
CF617858.1	AGENCOURT_15766118 NIH_MGC_222 cDNA clone IMAGE:30525742 5'(Flt1)	6.373779042
NM_ 013508.1	ELK3, member of ETS oncogene family (Elk3)	6.152306736
NM_009506.1	vascular endothelial growth factor C (Vegfc)	5.723748621
AI645296.1	mq49a04y1 Soares_thymus_2NbMT cDNA clone IMAGE:582030 5'(Vegfb).	5.319707708
BY717943.1	RIKEN adult male thymus cDNA clone 5830454J16 5' (Mapk14)	4.790681592
NM_153423.3	WAS protein family, member 2 (Wasf2)	3.856966912
NM_ 008213.1	heart and neural crest derivatives expressed transcript 1 (Hand1)	3.672770541
NM_019454.1	delta-like 4 (Drosophila) (Dll4)	3.575385362
NM_080463.2	protein O-fucosyltransferase 1 (Pofut1)	3.547570894
BB089170.1	RIKEN 12 days embryo, embryonic body between diaphragm region and neck cDNA clone 9430020N21 3' similar to U58112 Mus musculus Flt4 ligand J7 mRNA (Vegfc)	2.849753874
NM_ 11614.1	tumor necrosis factor (ligand) superfamily, member 12 (Tnfsf12)	2.604519239
NM_021412.1	matrix metalloproteinase 19 (Mmp19)	2.59866663
NM_ 011951.1	mitogen activated protein kinase 14 (Mapk14)	2.544310688
NM_009633.2	adrenergic receptor, alpha 2b (Adra2b)	2.075063515
NM_007448.1	angiogenin, ribonuclease A family, member 3 (Ang3)	1.810189211
AF332049.1	adrenergic receptor alpha 2B [Mus musculus] (Adra2b)	1.611589074
M92416.1	fibroblast growth factor (Fgf6)	1.543600791
NM_028783.2	roundabout homolog 4 (Drosophila) (Robo4)	1.543287759
CD742412.1	UI-M-AO0-cib-b-13-0-UIs1 NIH_BMAP_MPG cDNA clone UI-M-AO0-cib-b-13-0-UI 3'(MAPK14)	1.541944242
BC020092.1	cDNA clone IMAGE:3987110 (Nte)	1.515874115
BB533541.2	RIKEN 0 day neonate lung cDNA clone E030030J17 3' (Epas1)	1.436922533

ACCN;accession nnumber, OGS;official gene symbol

Figure 17: Highly expressed genes associated with angiogenesis after tube formation. (See Akita M, Fujita K. [120] Clinical Medicine: Cardiology 2008; 2: 49-57).

The CodeLink® Mouse Whole Genome array consists of probe sets representing over 35,000 transcripts [120]. The Gene Ontology terms associated with angiogenesis (GO:0001525), and negative regulation of angiogenesis (GO:0016525) were examined. Seventy-three out of over 35,000 transcripts were expressed after tube formation. The expression of 43 transcripts was high while 30 showed lower expression [120]. Fig. **17** showed the highly expressed genes. Comparing day 10 with day 5, Fig. **17** shows a scatter of all the genes of Gene Ontology terms associated with angiogenesis and negative regulation of angiogenesis. The majority of genes examined showed only small differences, with ratios ranging between 2.0 and -2.0 [120]. However, there were 7 up-regulated genes (those with ratios of more than 2.0). These up-regulated genes (Tnfaip2, Vegf-b, Crhr2, Vegf-c, Fgf-6, Pofut1 and Hif1-a) showed good levels of expression (sufficient signal intensity in the microarray system). There were 11 down-regulated genes (those with ratios of less than -2.0). These down-regulated genes were as follows; Rhob, Fgf2, Thbs1, Ang3, Gna13, Tnfrsf12a, Plg, Elk3, Ubp1, Itgav, Mapk14. The up- and down-regulated genes are summarized in Fig. **18**.

A combination of the up- and down-regulated genes identified in this study promotes strategy for angiogenic and anti-angiogenic therapy.

Up-Regulated Genes

Seven up-regulated genes were identified. These genes play an important role for capillary tube formation and growth.

Tnf α-Induced Protein 2 (Tnfαip2)

Tnfαip2 was originally described as a Tnfα-inducible primary response gene in endothelial cells. During vasculogenesis in the mouse embryo, Tnfαip2 was expressed in the myocardium, aortic arch and aortic endothelium [121]. In our previous study using collagen gel culture, capillary tube growth was induced by Tnf-α in a dose-dependent manner [14]. Tnf-α plays an important role in angiogenesis, including proliferation of endothelial cells and capillary sprouts [15]. Tnfip2, as well as Tnf-α, could be involved in the capillary tube formation and growth.

Up-regulated genes

Tnfaip2 ; tumor necrosis factor, alpha-induced protein 2
Vegfb ; mq49a04y1 Soares_thymus_2NbMT cDNA clone IMAGE:582030 5'.
Crhr2 ; corticotropin releasing hormone receptor 2
Vegfc ; vascular endothelial growth factor C
Fgf6 ; fibroblast growth factor (Fgf6) mRNA, 3' end
Pofut1 ; protein O-fucosyltransferase 1
Hif1a ; hypoxia inducible factor 1, alpha subunit

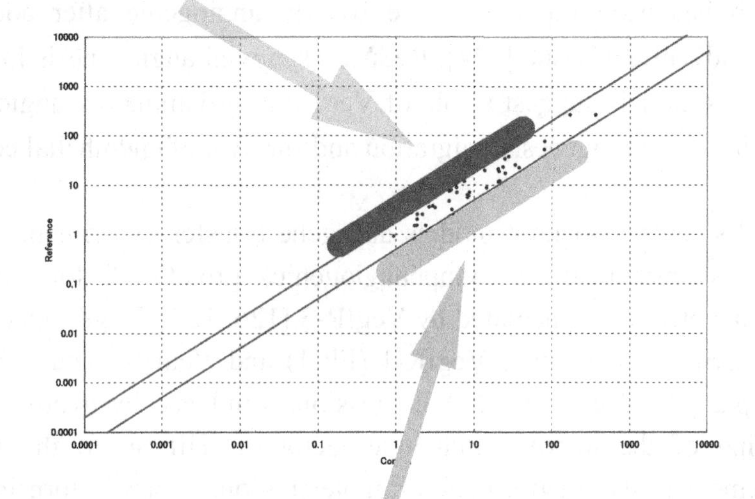

Down-regulated genes

Rhob; ras homolog gene family, member B
Fgf2; fibroblast growth factor 2
Thbs1; thrombospondin 1 mRNA
Ang3; angiogenin, ribonuclease A family, member 3
Gna13; guanine nucleotide binding protein, alpha 13
Tnfrsf12a; tumor necrosis factor receptor superfamily, member 12a
Plg; plasminogen
Elk3;UI-M-BH3-axa-f-03-0-UIs1 NIH_BMAP_M_S4 cDNA clone UI-M-BH3-axa-f-03-0-UI 3'
Ubp1; upstream binding protein 1
Itgav; integrin alpha V
Mapk14; mitogen activated protein kinase 14

Figure 18: Up- and down-regulated genes associated with angiogenesis after tube formation. (See Akita M, Fujita K. [120] Clinical Medicine: Cardiology 2008; 2: 49-57).

Vegf-b, Vegf-c and Hypoxia Inducible Factor 1 Alpha Subunit (Hif-1a)

Vegfs are considered as key molecules in the process of angiogenesis. The Vegf family consists of seven members; Vegf-a, Vegf-b, Vegf-c, Vegf-d, Vegf-e, Vegf-f and Pgf [122]. Vegf-a is a key molecule in induction of angiogenesis and vasculogenesis it causes proliferation, sprouting, migration and tube formation of endothelial cells [122]. Hao *et al.,* [123] reported that gene transfer of plasmid encoding Vegf-a was applicable for therapeutic angiogenesis in a rat myocardial infarction model. In the study [120], Vegf-a, -b, -c and Pgf were highly expressed, especially Vegf-b and -c were up-regulated. The precise role of Vegf-b is not known. Vegf-b has been reported to be weakly angiogenic after adenoviral delivery to periadventitial tissue [124]. Reduced synovial angiogenesis in Vegf-b knockout arthritis models suggest a role of Vegf-b in inflammatory angiogenesis [125]. Vegf-c induces mitogenesis, migration and survival of endothelial cells.

Developmental studies, knockout models and gene transfer experiments suggest that Vegf-c is primarily a lymphangiogenic growth factor, and its lymphangiogenic effects are mediated by VegfR-3 [126, 127]. Vegf-a mediates its responses primarily by activating VegfR-1 (Flt-1) and VegfR-2 and it binds to Nrp-1 and Nrp-2 [43]. Vegf-a mRNA expression is induced by hypoxia, which induced binding of the hypoxia inducible factor-1a (Hif-1a) to the hypoxia responsive element in the Vegf-a gene promoter region, which in turn increases Vegf-a transcription [33]. In our study, Vegf R-1, -2, Nrp-1 and Hif-1a were highly expressed. In this culture, Vegfs and its receptors play an important role for the formation and growth of capillary tubes [18, 120].

Fibroblast Growth Factor-6 (Fgf-6)

Fgfs are a family of heparin-binding proteins involved in many biological processes including angiogenesis [16, 128]. Pizette *et al.,* [129] noted that Fgf-6 displays a strong mitogenic activity on fibroblastic cells, and it is able to morphologically transform. Fgf-6 showed only a limited mitogenic response on adult bovine aortic endothelial cells. It is suggested that Fgf-6 relates mainly to the proliferation of fibroblasts than that of endothelial cells.

Corticotropin-Releasing Hormone Receptor 2 (Crhr2)

According to Kokkotou *et al.,* [130], Crhr2 deficient mice display cardiovascular abnormalities. The significance of Crhr2 up-regulation is obscure.

Protein O-Fucosyltransferase 1 (Pofut1)

Pofut1 has been shown to be essential in Notch signaling, which has an important role in vascular development, differentiation, proliferation, apoptosis and tumorigenesis [131-133], a Notch signaling molecule, was also expressed in this study. Dll4, a membrane-bound ligand for Notch1 and Notch4, is expressed selectively in the developing endothelium, and it is induced by Vegf-a and hypoxia [134]. Up-regulation of Pofut1 may relate to the up-regulation of Hif-1a.

Down-Regulated Genes

Eleven down-regulated genes were identified in the present study. These genes were strongly expressed before tube formation. They act at the early stage of angiogenesis (see Akita and Fujita [120]).

ANTI-ANGIOGENIC DRUGS

Thalidomide

Since thalidomide has an anti-angiogenic property, it is designated as a drugfor multiple myeloma. However, the anti-angiogenic property of thalidomide remains obscure. In the study using 3D collagen gel-cultures, we confirmed the anti-angiogenic potential of thalidomide with cytochrome P-450 [14]. The anti-angiogenic property of thalidomide depended not on the type and concentration of thalidomide. It depended on a species-specific difference in the metabolism of the agent [14, 135]. Suppression of Tnf-α induced the inhibition of angiogenesis [14, 135]. On the basis of these data, we described that apoptosis was an important factor to explain the anti-angiogenic property of thalidomide [135]. Capillary tube formation was significantly inhibited with thalidomide plus cytochrome P-450 (CYP2B4) (Fig. **19**). The capillary tube and migrating cells were positive for active caspase-3 (Fig. **20**). TEM revealed that active caspase-3 positive cells demonstrated apoptotic characteristics (Fig. **21**). This morphological study is the first report to demonstrate the anti-angiogenic effect of thalidomide. Taken together with earlier findings, the new results indicated that the anti-angiogenic property of thalidomide involved apoptosis with the suppression of Tnf-α, and inhibition of cell migration [135].

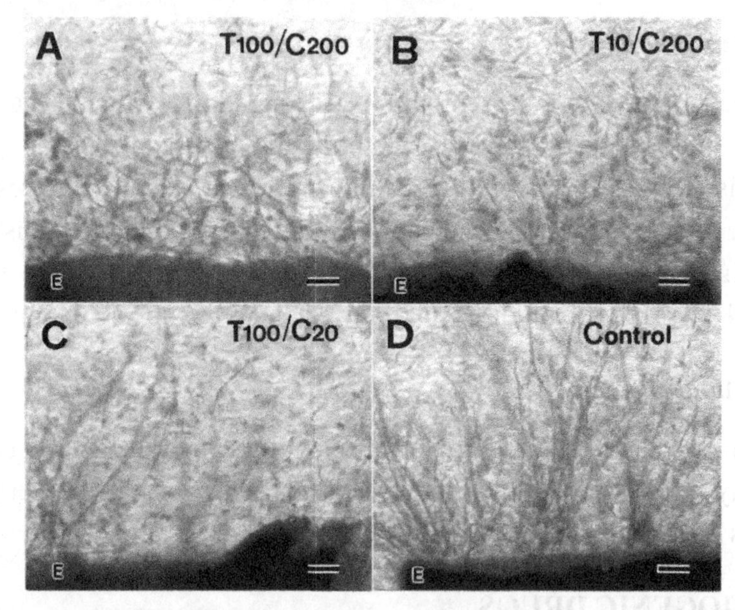

Figure 19: Treatment with thalidomide and cytochrome P-450. (A) 100 mg/ml thalidomide, (B) 10 mg/ml thalidomide, (C) 200 mg/ml cytochrome P-450, (D) 20 mg/ml cytochrome P-450. The number of capillary tubes was decreased (A-C) compared with the control (D). E=explant. Scale bars=50 μm. (See Fujita *et al.,* [135] Histochem Cell Biol 2004; 122: 27-33).

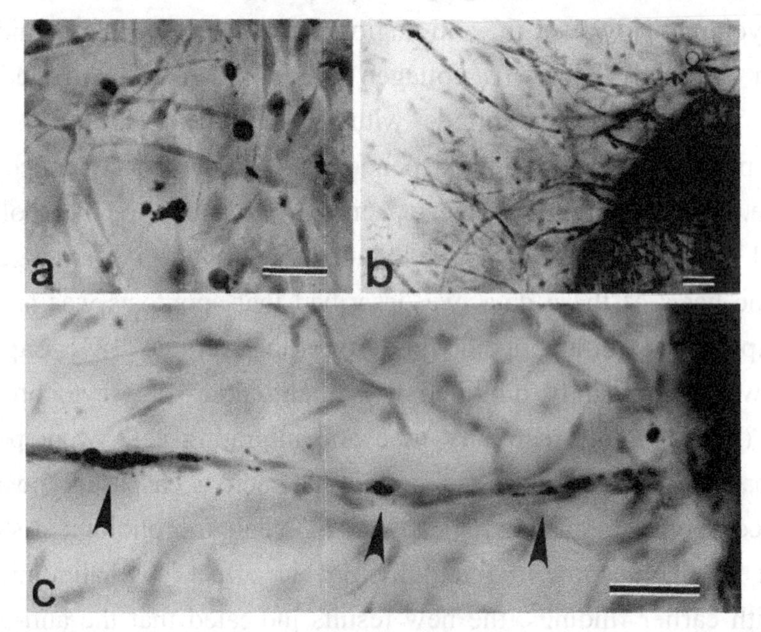

Figure 20: (a) Anti-caspase-3 immunostaining. Intense immunostaining was observed in the some migrating cells. (b, c) Tube-like structures were observed from a mouse aortic explant. E; explant Scale bar = 50 μm (See Fujita *et al.,* [135] Histochem Cell Biol 2004; 122: 27-33).

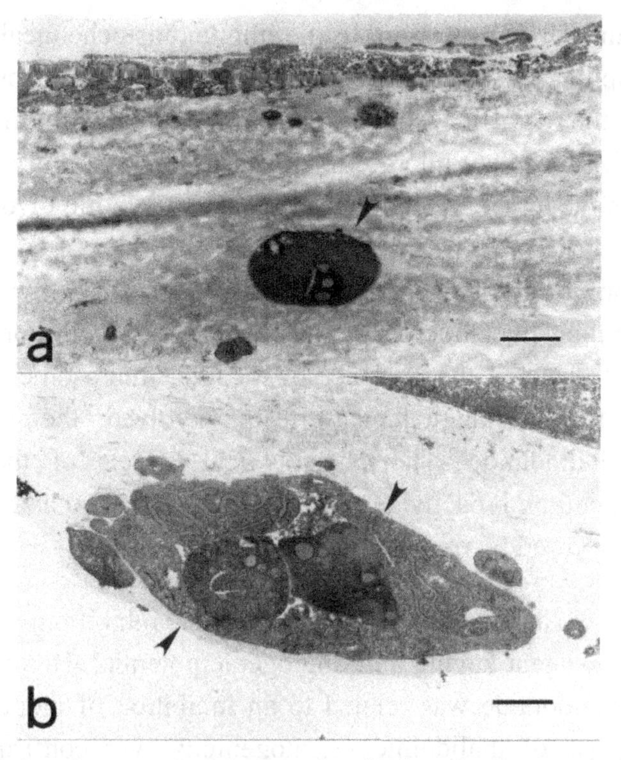

Figure 21: Transmission electron microphotographs of cultured cells treated with thalidomide and cytochrome P-450. After active caspase-3 immunostaining, paraffin sections were re-fixed with OsO_4 and embedded in epoxy resin. Ultrathin sections were examined with an electron microscope. a: In the vicinity of capillary tubes, an apoptotic cell (arrowhead) was detected. b: Arrowheads indicated some apoptotic endothelial cells in the lumen of capillary tube. Scale bars = 1 µm. (See Fujita *et al.,* [135] Histochem Cell Biol 2004; 122: 27-33).

Mitsiades *et al.,* [136] reported that thalidomide and its analogs (IMiDs) had an anti-angiogenic effect inducing apoptosis or growth arrest of multiple myeloma cells. IMiDs induced caspase-8-dependent apoptosis, down-regulated nuclear factor (NF)-κB transcriptional activity, and sensitized multiple myeloma cells to apoptosis-induced Fas cross-linking [136]. Other criteria such as the cellular morphology also have to be thought to assure that the form of cell death is apoptotic [136]. In our study, the cells treated with thalidomide and cytochrome P-450 showed the characteristics of apoptotic cells in the lumen of capillary tubes and migrating myofibroblasts (Fig. **21**). Caspase-3 was identified as a member of the caspase family of cysteine proteases [137]. Initiation of cellular events during the apoptotic process depends on the activation of caspase-3. Since caspase-3 is activated at the early stage of apoptosis, the detection of active caspase-3 is useful for apoptosis than TUNEL reaction. In the immunohistochemical study of active caspase-3, the capillary tubes and migrating cells were positive for active caspase-3 [135]. When the paraffin sections were pretreated with the antibody and examined by TEM after light microscopic observation, apoptotic endothelial cells were identified in the lumen of capillary tubes, and apoptotic myofibroblasts were recognized around the tubes [135]. Bauer *et al.,* [138] reported that thalidomide induced inhibition of microvessel formation in the presence of cytochrome P-450 from human or rabbit liver microsomes. However, the inhibition of microvessel formation did not occur in the presence of rat liver microsomes. These findings suggested that the anti-angiogenic effects involved the metabolism and metabolites of thalidomide. The physiological effects of thalidomide were considered to depend on bioactivation by cytochrome P-450 and prostaglandin H synthase (PHS) followed by oxidizing DNA [139, 140].

A target of novel anticancer strategies is the inhibition of angiogenesis. Thalidomide is one agent having anti-angiogenic potential. D'Amato *et al.,* [141] postulated that thalidomide was related to an inhibition of blood vessel growth. The principal cause of thalidomide teratogenicity was confirmed to be anti-angiogenic activity [142, 143]. Therapeutic effects of thalidomide on multiple myeloma and several other types of cancer increased in recent years. Several studies demonstrated the increased angiogenesis in myeloma [144, 145] and multiple myeloma [146]. No statistical significance was found in the bone marrow microvessel density after treatment between the responder and nonresponder group [147]. They suggested that the lack of a consistent decrease in microvessel density following thalidomide treatment indicated the involvement of other mechanisms [147]. Recent studies proposed that thalidomide teratogenicity was related to the metabolism of thalidomide *in vivo* [148, 149]. Vesela *et al.,* [148] shown the teratogenic thalidomide hydrolysis products in the presence of liver microsome in chick embryo. Bauer *et al.,* [149] demonstrated that microvessel formation was inhibited in rodents as well as in human and rabbit by treatment of thalidomide with liver microsome.

Inhibition of angiogenesis is currently regarded as one of the hopeful approaches in the therapy of cancer. The anti-angiogenic property of thalidomide excited the research on this teratogenic drug. Metabolites of thalidomide have a responsibility for the pharmacological actions. Ng *et al.,* [150] synthesized 118 thalidomide analogues on the basis of the structures of the metabolites. In the aortic ring assay, all 4 analogues in the N-substituted class and 2 of the 3 analogues in the tetrafluorinated class significantly inhibited microvessel outgrowth at 12.5-200 microM, although thalidomide failed to block angiogenesis at similar concentrations [150]. Those analogues also demonstrated anti-proliferative effect in the human umbilical vein endothelial cells, and capillary tube formation was suppressed with treatment of all 7 analogues as well as thalidomide [150].

Liu *et al.,* [151] investigated the changes of genes in the U266 multiple myeloma cell line following culture with s-thalidomide. There were changes in the expression profile of genes involved in angiogenesis and apoptosis. In particular, expression of I-kappaB kinase was decreased by two-fold, which was associated with a four-fold decrease in NF-κB expression. Additionally, the Bax: Bcl-2 ratio was significantly increased. A dramatic decrease in Bcl-2 expression with s-thalidomide suggested a possible enhancement of cytotoxic effect if combined with other cytotoxic agents.

Vacca *et al.,* [152] studied the anti-angiogenic effect of thalidomide. The expression of key angiogenic genes was studied following exposure to therapeutic doses of thalidomide in the endothelial cells (ECs) of bone marrow from patients with active and nonactive multiple myeloma (MM); monoclonal gammopathy unattributed/unassociated [MG(u)]; diffuse large B-cell non-Hodgkin's lymphoma; in a Kaposi's sarcoma (KS) cell line; in healthy human umbilical vein ECs (HUVECs). Thalidomide markedly down-regulated the genes in a dose-dependent manner in active MMECs and KS cell line, but up-regulated them or was ineffective in nonactive MMECs, MG(u)ECs, NHL-ECs, and in HUVECs [152]. Secretion of Vegf, bFgf and hepatocyte growth factor (Hgf) also diminished in the conditioned culture media (CM) of active MMECs and KS in a dose-dependent manner, whereas it did not change in the other CM. They concluded that inhibition by thalidomide was probably confined to the genes of active MMECs and KS [152].

Aerbajinai *et al.* [153] reported that thalidomide induced the expression of γ-globin mRNA in a dose-dependent manner. They also found that treatment with thalidomide for 48 hours increased the intracellular reactive oxygen species (ROS) levels. Thalidomide activated the p38 Mapk signaling pathway in a time- and dose-dependent manner and increased histone H4 acetylation [153]. Aerbajinai *et al.* [153] also noted that the antioxidant enzyme catalase and the hydroxyl scavenger dimethylthiourea (DMTU) suppressed the thalidomide-induced p38 Mapk activation and histone H4 acetylation. They also found that thalidomide-induced γ-globin gene expression was diminished after catalase and DMTU exposure. They concluded that thalidomide increased the expression of the γ-globin mRNA *via* ROS-dependent activation of the p38 Mapk signaling pathway and histone H4 acetylation [153].

De Luisi *et al.,* [154] determined the *in vivo* and *in vitro* anti-angiogenic potential of lenalidomide (a thalidomide analogue). *In vitro*, lenalidomide inhibited endothelial cell migration and angiogenesis in the bone marrow of patients with multiple myeloma. They also demonstrated that changes of Vegf/VegfR2 signaling pathway, several proteins controlling EC motility, cytoskeleton remodeling and energy metabolism pathways were caused by lenalidomide treatment [154].

Lovastatin

Lovastatin inhibited the cell migration from the aortic explant and induced capillary tube degradation (Figs. **22, 23** and **24**) [11].

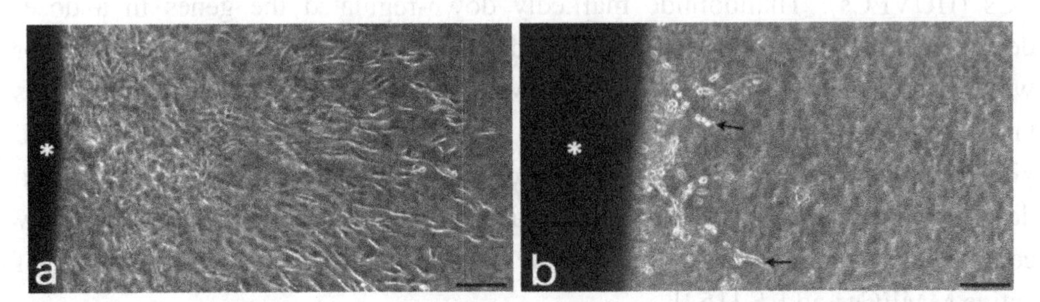

Figure 22: Effect of lovastatin treatment before capillary tube formation. a: Control. Rat aortic explant (*), Scale bar = 100 μm. b: Lovastatin treatment. Lovastatin strongly inhibited the cell migration (arrows) from a rat aortic explant (*), Scale bar = 100 μm. (See Akita *et al.,* [11] Stem Cells Int, Epub 2013 Jun 23).

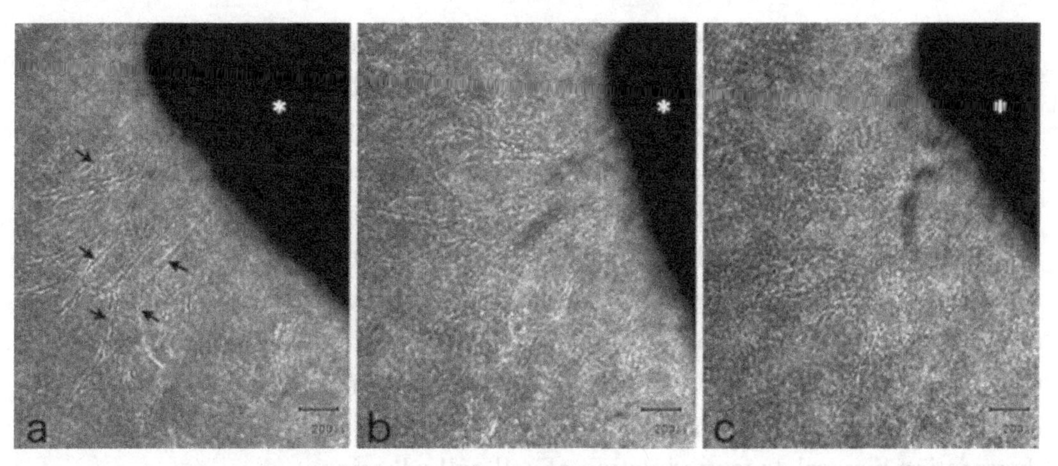

Figure 23: Effect of lovastatin treatment after capillary tube formation. a: Before lovastatin treatment, arrows indicated newly formed capillary tubes. b: After 24 hours. c: After 48 hours; Lovastatin caused the degradation of capillary tubes. Rat aortic explant (∗), Scale bar = 200 μm. (See Akita *et al.,* [11] Stem Cells Int, Epub 2013 Jun 23).

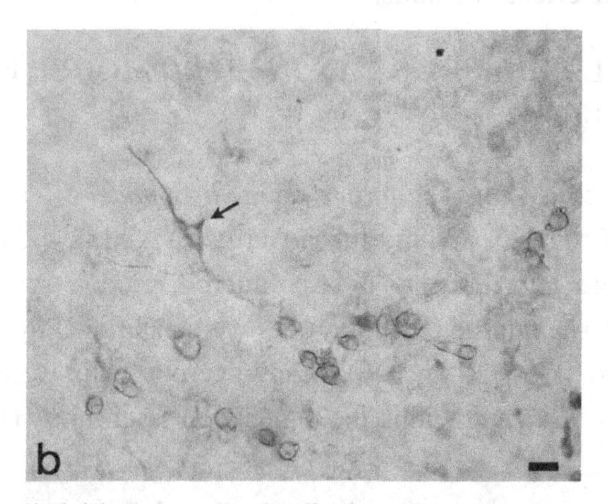

Figure 24: Lovastatin diminished the cell-cell adhesion. Many CD133-positive cells changed the morphology. Degradation of the capillary tubes was observed (arrows). Scale bar = 20 μm. (See Akita *et al.,* [11] Stem Cells Int, Epub 2013 Jun 23).

Lovastatin inhibited hydroxymethyl glutaryl coenzyme A (HMG-CoA) reductase. HMG-CoA reductase was involved in lipid metabolism [155]. The inhibition caused defective the normal membrane function and impaired the adhesive properties of membrane [155, 156]. According to Karthikeyan and Lip [157], statin-treated patients diminished intraplaque angiogenesis. It is suggested that statins have angiostatic effects *in vivo*. Statins have also been reported to reduce the abnormal growth and spread of tumors [158, 159]. Inhibition of angiogenesis may be related to reduce the tumor growth [160]. Our study also demonstrated that lovastatin abrogated the cell-cell adhesion and capillary tube degradation [11]. Khaidakov *et al.,* [160] suggested that statins modulate cell-cell adhesion through VE-cadherin. According to the recent study of Koyama-Nasu *et al.,* [161], CD133 interacts with plakoglobin (desmosomal linker protein). They also noted that CD133 knockdown by RNA interference (RNAi) resulted in the downregulation of desmoglein-2 (desmosomal cadherin). CD133 knockdown diminished cell-cell adhesion and tumorigenic potential of ovary carcinoma [161]. Cholesterol chelating agent, methyl-β-cyclodextrin diminished cell adhesion by decreasing desmosomes and intercellular digitations [162]. The capillary tube degradation may relate to modulation of cell-cell adhesion.

The present study provides the useful data of the biological function of CD133 and anti-angiogenic effect of statins.

COLLAGEN GEL CULTURE AS AN *IN VITRO* MODEL FOR STUDYING VASCULAR INJURY

We applied the 3D collagen gel culture as an *in vitro model* of vascular injury [163]. After 7-10 days of culture, capillary tubes were observed in the collagen gels. The capillary tubes were injured using a laser microdissection system or a scraping method with razor. The injured capillary tubes were examined by phase contrast and electron microscopy (Figs. **25** and **26**). Many necrotic cells were observed around the injured capillary tubes and within the lumen after laser injury (Fig. **27**).

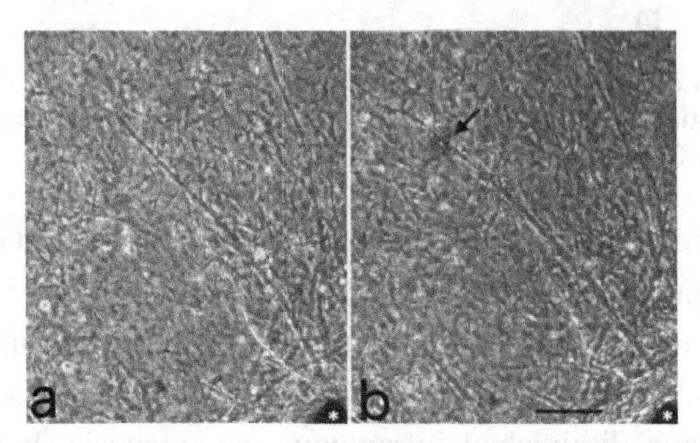

Figure 25: Phase contrast microphotographs of the capillary tubes. Mouse aortic explant (∗). Scale bar=250 μm. a: Before laser injury. b: 30 min after laser injury. Arrow indicated the damaged site. (See Fujita *et al.,* [163] Histochem Cell Biol 2006; 125: 509-14).

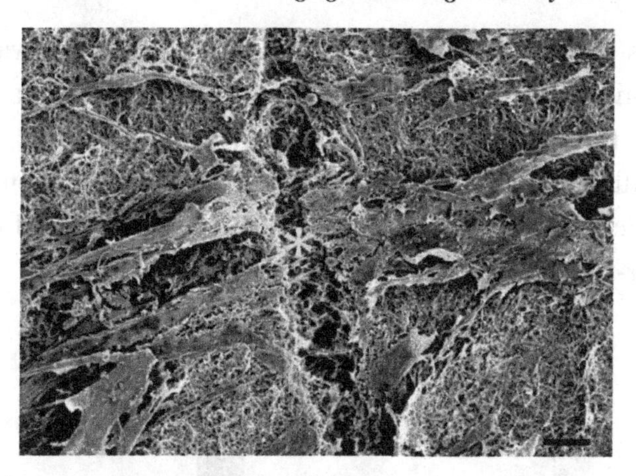

Figure 26: 30 min after laser injury. Asterisk indicated the damaged site. Scale bar = 20 μm.

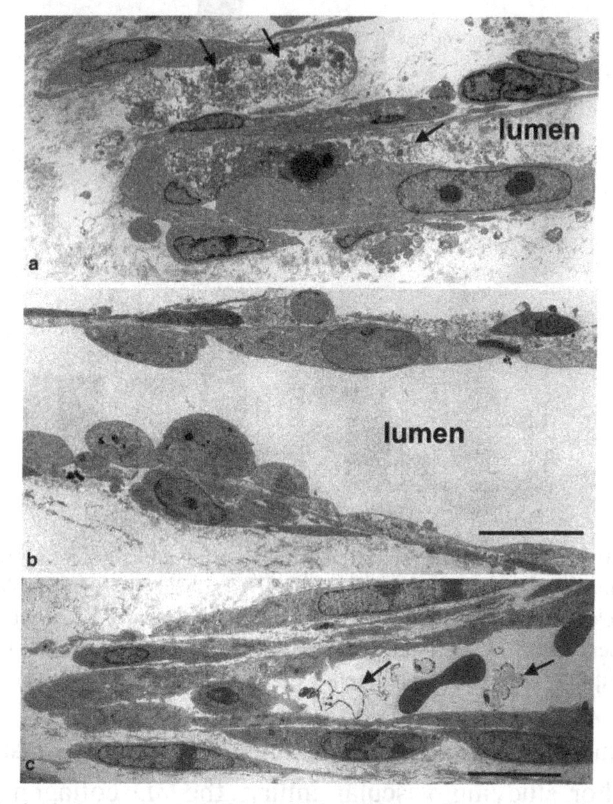

Figure 27: Transmission electron microphotograph of the capillary tubes. Scale bar = 5 μm. a: 30 min after laser injury. Arrows indicated necrotic cells around the damaged end of capillary tube and in the lumen. b: Control; A capillary tube before laser injury did not show necrotic cells. c: 30 min after razor injury. Transmission electron microphotograph showed the damaged end of capillary tube. Necrotic and/or apoptotic cells were observed some in the lumen. (See Fujita *et al.*, [163] Histochem Cell Biol 2006; 125: 509-14).

Total RNA was isolated from the cultures, and cDNA was prepared for quantitative real-time reverse transcription polymerase chain reaction (RT-PCR). Quantitative real time RT-PCR demonstrated the up-regulation of transcription factor early growth response-1 (Egr-1) after injury. Up-regulation of Fgf-2, which is a proangiogenic factor downstream of Egr-1, is accompanied by the increased expression of Egr-1 (Fig. **28**).

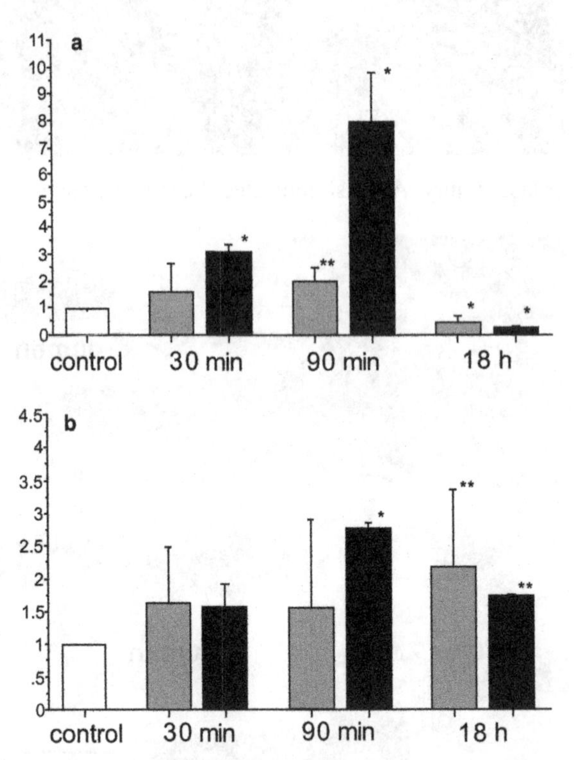

Figure 28: Quantitative RT-PCR analysis after injury. Egr-1 and Fgf2 mRNAs expressions were measured by quantitative RT-PCR. (a) Egr-1 gene, (b) Fgf-2 gene. Gray column indicated after laser injury, and black column indicated after razor injury. Student's t-test was used for statistical analysis. Data represent mean +/- SE (standard error). Statistically significant results with *$p<0.01$ or **$p<0.05$. (See Fujita *et al.,* [163] Histochem Cell Biol 2006; 125: 509-14).

Effective laser energy was concentrated only on the limited small area of the capillary tubes. For studying vascular injury, the 3D collagen gel culture is a useful *in vitro* model [163]. DiI-Ac-LDL (Biomedical Technol. Inc., USA) was added to the growth medium. After incubation for 6 hours, capillary tubes were injured from a laser beam by using a laser microdissection system (337 nm pulsed nitrogen laser; P.A.L.M. Microlaser Technologies A.G., Germany). After laser injury, capillary tubes were observed by fluorescent and phase microscopy, immunohistochemistry and electron microscopy. There are many advantages in the 3D collagen gel culture. As compared with the environment of monolayer culture, the 3D culture is much close to the actual environment *in vivo*. The healing process of the capillary tubes can be observed as a time-lapse image. Simultaneously, angiogenic and transcription factors can be analyzed. Compared to the scrape method with razor, the laser energy is concentrated only on the limited small area of the capillary tubes. The laser method makes it possible to choose a specific capillary tube. Furthermore, it is possible to perform microsurgery on a specific site. The disadvantage of laser method is the necessities for specialized equipment. A locally restricted ablative photodecomposition process without heating of adjacent material is the principle of laser cutting [164]. The excimer wavelength of 193 nm was recognized to be particularly suitable for surgery on the cornea [164]. Since we used 337 nm pulsed nitrogen laser in this study, there is possibility of thermal damage. The differences between laser and razor injury may attribute the thermal effect [163]. However, both laser and razor injury caused the up-regulation of Egr-1 and Fgf-2 genes.

Egr-1 regulated many pathophysiologically related genes involved in the growth, differentiation, immune response, wound healing, and blood clotting of the vasculature [165]. Khachigian *et al.,* [166] reported that Egr-1 was rapidly activated by growth factors, cytokines, hypoxia, fluid shear stresses and vascular injury. Activated Egr-1 regulated the expression of proangiogenic genes such as encoding growth factors, cytokines, receptors, adhesion molecules and proteases [167, 168]. Egr-1 bound and activated many promoter genes, and the products influenced vascular healing as angiogenic factors [169]. Egr-1 stimulated the production of platelet-derived growth factor-ab (Pdgf-ab), Hgf, Vegf, Fgf-2 [170] and transforming growth factor-β1 (Tgf-β1) and its receptor [171-173]. Egr-1 supported Fgf-dependent angiogenesis and the cellular growth and neovascularization of the microvascular endothelium [174]. Cancilla *et al.,* [16] noted that Fgfs, a family of at least 21 heparin-binding proteins, associated with angiogenesis. Fgf-2, a potent angiogenic factor *in vivo* and *in vitro*, stimulated endothelial cell growth [19, 20], smooth muscle cell growth, wound healing and tissue repair [21, 22]. The up-regulation of Fgf-2 following the up-regulation of Egr-1 found after injury in the study [163]. The method using the 3D culture showed the many advantages over the conventional scrape assay method using monolayer culture. It provided a useful tool as an *in vitro* model for studying angiogenesis and vascular healing [163].

COLLAGEN GEL CULTURE OF VARIOUS ORGANS

In this chapter, capillary tube formation from aorta was mainly described. However, the capillaries are the vessels closest to the tissue they supply and their wall is in intimate relation with the tissue. Their structure varies in different locations. Typically a single endothelial cell forms the wall of a capillary so that the junctional complex occurs between extensions of the same cell. The endothelial cells of some capillaries have fenestrations or pores. Fenestrated capillaries occur in renal glomeruli, in the intestinal mucosa and the endocrine and exocrine glands. Capillaries without fenestrations are known, such as those in the brain, striated and smooth muscles, lung and connective tissue, are known as continuous capillaries. As discontinuous capillaries, there are sinusoids. These are capillaries, large and irregular in shape. They have true discontinuities in their wall. Sinusoids occur in large numbers in the liver, spleen, bone marrow and suprarenal medulla [175]. We examined a capillary tube formation from explant of skeletal muscles, brain, kidney, liver and spleen. Capillary tube formation occurred only from the skeletal muscle explant.

Skeletal Muscles

Soleus muscles; After 2 days of culture, migrating cells were observed around the explant of muscle tissues in the collagen gel. Spindle-shaped cells were observed. Their orientation is radial with the long axis of the muscle tissues. After 6 days of culture, the strands of elongated cells were recognizable. Lumen formation could not be recognized in these early stages. After 10 days of culture, the capillary formation with lumen was well recognized (Fig. **29a**). As revealed by cross-section, several endothelial cells were always involved in the luminal wall. Pericyte-like cells were resting around the capillary tubes (Fig. **29b**). Capillaries without fenestrations, such as those in the striated muscles, are known as continuous capillaries [175]. Typical gap junctions and tight junctions were not observed.

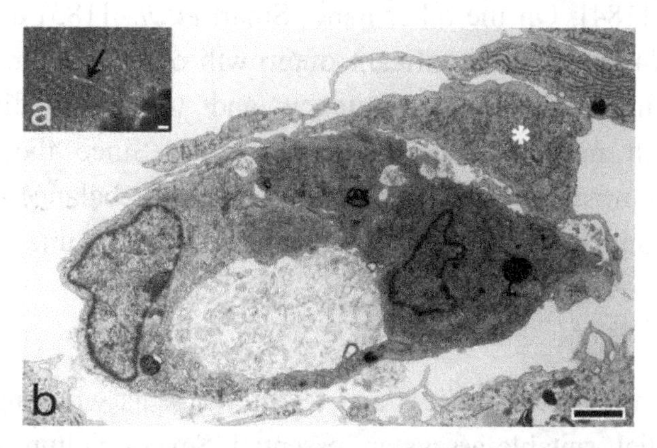

Figure 29: After 10 days cultivation of skeletal muscles. a: Phase-contrast microphotograph showing the tube formation (arrows) from mouse soleus muscles in the collagen gel. Scale bar=20 μm. b: Transmission electron microphotograph of the capillary tube. A cross section showed a capillary tube with lumen. Pericyte like cell (*) was observed around the capillary tube. Caveolae were visible on both the luminal and the abluminal surfaces of the endothelial cell, together with endoplasmic reticulum, microtubules and bundles of microfilaments. The edges of the cell were in contact with each other at the bottom, and they formed specialized junctions. Scale bar=2 μm.

Since endothelial cells are highly heterogeneous in their morphology and functional behavior, *in vitro* studies of endothelial cells are crucial for understanding angiogenic events. We here described a rapid, simplified method for the formation of capillary tube using the 3D collagen gel culture, although only skeletal muscles were successful. The reason why capillary tube formation did not occur from the other organs was unknown. In the present study, we observed pericyte like cells around the capillary tube. Pericytes were recognized to stimulate angiogenesis through secretion of growth factors such as Fgf [176], Vegf and Pgf [177]. In addition, pericytes promoted EC survival [178] and affected EC behaviour such as sprouting [179]. Vegf is a key promoter of angiogenesis. Vegf acts as a chemoattractant and directs capillary growth. Concentration gradient of Vegf is important for activation and chemotactic guidance of capillary sprouting [180, 181]. Zhang *et al.,* [182] reported that myocytes (skeletal muscle fibers) secreted Vegf at different rates. Ji *et al.,* [183] also reported that Vegf was secreted by myocytes, and bound Vegf receptors and Nrp-1 on endothelial cell surface. It may attribute to the density of capillary in the organs, or may relate unknown factors such as thymosin. Recently, thymosin β10 (Tβ10) inhibits cell migration and capillary tube formation of coronary artery endothelial cells [184]. On the other hand, Smart *et al.,* [185] demonstrated that thymosin β4 (Tβ4) induced significant outgrowth of cells from quiescent adult epicardial explant, restoring pluripotency and triggering differentiation of fibroblast, smooth muscles and endothelial cells. Since the most abundant thymosin β in mammalian cells are Tβ4 and Tβ10, the balance of expression of these angiogenic and anti-angiogenic factors may involve in this issue.

CONCLUSION

For studying the angiogenic process and test new agents with angiogenic or anti-angiogenic potential, suitable assays are essential. Several culture techniques using matrix components have been developed for *in vitro* studies of angiogenesis. Collagen gel culture with aortic explant was a useful method for assaying of angiogenesis. Using the 3D collagen gel culture, capillary tube formation from the aortic explant in the collagen gel was observed with electron microscopy. The capillary tubes with pericyte-like cells are observed in the collagen gel. Immunohistochemical studies revealed that Tnf-α and Nrp-1 were positive in the capillary tubes. CD133 (hematopoietic stem and progenitor cell marker) was also positive in the capillary tubes. The explant assays using the 3D collagen gel culture were better mimic *in vivo* niche conditions. Anti-angiogenic drugs (thalidomide and lovastatin) were tested by using the 3D collagen gel culture. Thalidomide induced the suppression of Tnf-α and the inhibition of cell migration from the aortic explant. Thalidomide-induced inhibition of angiogenesis involves apoptosis. Lovastatin inhibited cell migration and caused the degradation of the capillary tubes.

The 3D collagen gel culture was also used for DNA micro-array gene expression profiling and an *in vitro* vascular injury model. The 3D collagen gel culture provides a useful method for assaying angiogenesis.

ACKNOWLEDGEMENTS

The author greatly appreciates the technical assistance of Tanaka K, Murai N, Komatsu N and Matsumoto S.

This work was partly supported by the programs of the Grant-in-Aid for Scientific Research (C) from the Japan Society for the Promotion of Science (JSPS) KAKENHI, Grant Numbers 09670025, 16591796.

CONFLICT OF INTEREST

The author confirms that this chapter contents have no conflict of interest.

DISCLOSURES

This is an updated report of the earlier original articles by M. Akita. "Detection of the hematopoietic stem and progenitor cell marker CD133 during angiogenesis in 3D collagen gel culture". Stem Cells Int. 2013, Article ID 927403, 10 pages http://dx.doi.org/10.1155/2013/927403. "DNA micro-array gene expression profiling of angiogenesis in collagen gel culture". Clinical Medicine: Cardiology, 2008; 2: 49-57. "An *in vitro* model for studying vascular injury after laser microdissection". Histochem. Cell Biol. 2006; 125: 509-514. "Immuno-histochemical expression of alpha1, alpha2 and alpha3 integrin subunits during angiogenesis *in vitro*". Histo. Histopathol. 2004; 19: 735-742. "Anti-angiogenic effects of thalidomide: expression of apoptosis-inducible active-caspase-3 in a 3D collagen gel culture of aorta". Histochem. Cell Biol. 2004; 122: 27-33. "Effects of thalidomide, cytochrome P-450 and TNF-α on angiogenesis in a 3D collagen gel-culture". Okajimas Folia Anat. Jpn. 2002; 79: 101-106. "Morphology of capillary-like structures in a 3D aorta/collagen gel culture". Ann. Anat. 1997; 179: 127-136.

REFERENCES

[1] Bautch VL. Cancer: Tumour stem cells switch sides. Nature 2010; 468 (7325): 770-1.

[2] Deckers M, van der Pluijm G, Dooijewaard S, *et al.* Effect of angiogenic and anti-angiogenic compounds on the outgrowth of capillary structures from fetal mouse bone explants. Lab Invest 2001; 81(1): 5-15.

[3] Jakob W, Jentzsch KD, Mauersberger B, Heder G. The chick embryo choriallantoic membrane as a bioassay for angiogenesis factors: reactions induced by carrier materials. Exp Pathol (Jena) 1978; 15(5): 241-9.

[4] Fournier GA, Lutty GA, Watt S, Fenselau A, Patz A. A corneal micropocket assay for angiogenesis in the rat eye. Invest Ophthalmol Vis Sci 1981; 21(2): 351-4.

[5] Kenyon BM, Voest EE, Chen CC, Flynn E, *et al.* A model of angiogenesis in the mouse cornea. Invest Ophthalmol Vis Sci 1996; 37(8): 1625-32.

[6] Jain RK, Schlenger K, Hockel M, Yuan F. Quantitative angiogenesis assays: progress and problems. Nat Med 1997; 3(11): 1203-8.

[7] Staton CA, Stribbling SM, Tazzyman S, *et al.* Current methods for assaying angiogenesis *in vitro* and *in vivo*. Int. J Exp Path 2004; 85(5): 233-48.

[8] Nicosia RF, Bonanno E, Smith M. Fibronectin promotes the elongation of microvessels during angiogenesis *in vitro*. J Cell Physiol 1993; 154(3): 654-61.

[9] Akita M, Murata E, Merker HJ, Kaneko K. Morphology of capillary-like structures in a 3D aorta/collagen gel culture. Ann Anat 1997; 179(2): 127-36.

[10] Akita M, Murata E, Merker HJ, Kaneko K. Formation of new capillary-like tubes in a 3D *in vitro* model (aorta/collagen gel). Ann Anat 1997; 179(2): 137-47.

[11] Akita M, Tanaka K, Matsumoto S, Komatsu K, Fujita K. Detection of the hematopoietic stem and progenitor cell marker CD133 during angiogenesis in 3D collagen gel culture. Stem Cells Int 2013; Article ID 927403. Available from: http://www.hindawi.com/journals/sci/2013/927403/

[12] Hoffmann S, Spee C, Murata T, *et al.* Rapid isolation of choriocapillary endothelial cells by Lycopersicon esculentum-coated Dynabeads. Graefe's Arch Clin Exp Ophthalmol 1998; 236(10): 779-84.

[13] Akita M, Murai N, Murata E, Kaneko K. Collagen gel culture using aortic explant as an *in vitro* angiogenesis model. J Jpn Coll Angiol 1997; 37(6): 331−6.

[14] Fujita K, Asami Y, Murata E, Akita M, Kaneko K. Effects of thalidomide, cytochrome P-450 and TNF-α on angiogenesis in a 3D collagen gel-culture. Okajimas Folia Anat Jpn 2002; 79(4): 101-6.

[15] Yoshida S, Ono M, Shono T. *et al.* Involvement of interleukin-8, vascular endothelial growth factor and basic fibroblast growth factor in tumor necrosis factor alpha-dependent angiogenesis. Mol Cell Biol 1997; 17(7): 4015-23.

[16] Cancilla B, Davies A, Cauchi JA, Risbridger GP, Bertram JF. Fibroblast growth factor receptors and their ligands in the adult rat kidney. Kidney Int 2001; 60(1): 147-55.

[17] Bikfalvi A, Klein S, Pintucci G, Rifkin DB. Biological Roles of Fibroblast Growth Factor-2. Endocrine Reviews 1997; 18(1): 26-45.

[18] Akita M, Fujita K, Tanaka K, *et al.* Remodeling of small blood vessels after laser injury *in vitro* [Abstract]. Cardiovasc Pathol 13 (Suppl 3): S163.

[19] Moscatelli D, Presta M, Rifkin DB. Purification of a factor from human placenta that stimulates capillary endothelial cell protease production, DNA synthesis, and migration. Proc Natl Acad Sci USA 1986; 83(7): 2091-5.

[20] Presta M, Moscatelli D, Joseph-Silverstein J, Rifkin DB. Purification from a human hepatoma cell line of a basic fibroblast growth factor-like molecule that stimulates capillary endothelial cell plasminogen activator production, DNA synthesis, and migration. Mol Cell Biol 1986; 6(11): 4060-6.

[21] Basilico C, Moscatelli D. The FGF family of growth factors and oncogenes [review]. Adv Cancer Res 1992; 59:115-65.

[22] Schwartz SM, Liaw L. Growth control and morphogenesis in the development and pathology of arteries. J Cardiovasc Pharmacol 1993; 21 (Suppl 1): 31-49.

[23] Kuzuya M, Satake S, Ramos MA, *et al.* Induction of apoptotic cell death in vascular endothelial cells cultured in 3D collagen lattice. Exp Cell Res 1999; 248(2): 498-508.

[24] Satake S, Kuzuya M, Ramos MA, Kanda S, Iguchi A. Angiogenic stimuli are essential for survival of vascular endothelial cells in 3D collagen lattice. Biochem Biophys Res Commun 1998; 244(3): 642-6.

[25] Naruo K, Seko C, Kuroshima K, *et al.* Novel secretory heparin-bindingfactors from human glioma cells (glia-activatingfactors) involved in glial cell growth. Purification and biological properties. J Biol Chem 1993; 268(4): 2857-64.

[26] Miyamoto M, Naruo K, Seko C, *et al.* Molecular cloning of a novel cytokine cDNA encoding the ninth member of the fibroblast growth factor family, which has a unique secretion property. Mol Cell Biol 1993; 13(7): 4251-9.

[27] Pilcher BK, Gaither-Ganim J, Parks WC, Welgus HG. Cell type-specific inhibition of keratinocyte collagenase-1 expression by basic fibroblast growth factor and keratinocyte growth factor. A common receptor pathway. J Biol Chem 1997; 272(29): 18147-54.

[28] Miyagi N, Kato S, Terasaki M, Shigemori M, Morimatsu M. Fibroblast growth factor-2 and 9 regulate proliferation and production of matrix metalloproteinases in human gliomas. Int J Oncol 1998; 12(5): 1085-90.

[29] Haas TL, Madri JA. Extracellular matrix-driven matrix metalloproteinase production in endothelial cells: implications for angiogenesis. Trends Cardiovasc Med 1999; 9(3-4): 70-7.

[30] Gillis P, Savla U, Volpert OV, *et al.* Keratinocyte growth factor induces angiogenesis and protects endothelial barrier function. J Cell Sci 1999; 112(Pt 12): 2049-57.

[31] Brown KJ, Maynes SF, Bezos A, *et al.* A novel *in vitro* assay for human angiogenesis. Lab Invest 1996; 75(4): 539-55.

[32] Gerber HP, Vu TH, Ryan A, *et al.* VEGF couples hypertrophic cartilage remodeling, ossification and angiogenesis during endochondral bone formation. Nature Med 1999; 5(6): 623-8.

[33] Roy H, Bhardwaj S, Ylä-Herttuala S. Biology of vascular endothelial growth factors. FEBS letters. 2006; 580(12): 2879-87.

[34] Gerhardt H, Golding M, Fruttiger M, *et al.* VEGF guides angiogenic sprouting utilizing endothelial tip cell filopodia. J Cell Biol 2003; 161(6): 1163-77.

[35] Helm CL, Fleury ME, Zisch AH, Boschetti F, Swartz MA. Synergy between interstitial flow and VEGF directs capillary morphogenesis *in vitro* through a gradient amplification mechanism. Proc Natl Acad Sci USA 2005; 102(44): 15779-84.

[36] Akita M, Murata, Fujita K, Kaneko K, Merker HJ. Observation of capillary-like tubes newly formed from the aorta specimen in collagen gel and the effects of growth factors [abstract]. Acta Anat Nippon 1996; 71: 426.

[37] Feng D, Nagy JA, Brekken RA, *et al.* Ultrastructural localization of the vascular permeability factor/vascular endothelial growth factor (VPF/VEGF) receptor-2 (FLK-1, KDR) in normal mouse kidney and in the hyperpermeable vessels induced by VPF/VEGF-expressing tumors and adenoviral vectors. J Histochem Cytochem 2000; 48(4): 545-56.

[38] Witmer AN, Blaauwgeers HG, Weich HA, *et al.* Altered expression patterns of VEGF receptors in human diabetic retina and in experimental VEGF-induced retinopathy in monkey. Invest Ophthalmol Vis Sci 2002; 43(3): 849-57.

[39] Takagi H, King GL, Aiello LP. Identification and characterization of vascular endothelial growth factor receptor (Flt) in bovine retinal pericytes. Diabetes 1996; 45(8): 1016-23.

[40] Gerhardt H, Golding M, Fruttiger M, *et al.* VEGF guides angiogenic sprouting utilizing endothelial tip cell filopodia. J Cell Biol 2003; 161(6): 1163-77.

[41] Gerhardt H, Betsholtz C. How do endothelial cells orientate? EXS 2005; (94): 3-15.

[42] Hamada K, Oike Y, Takakura N, *et al.* VEGF-C signaling pathways through VEGFR-2 and VEGF R-3 in vasculoangiogenesis and hematopoiesis. Blood 2000; 96(12): 3793-800.

[43] Klagsbrun M, Takashima S, Mamluk R. The role of neuropilin in vascular and tumor biology. Adv Exp Med Biol 2002; 515: 33-48.

[44] Kurschat P, Bielenberg D, Rossignol-Tallandier M, Stahl A, Klagsbrun M. Neuron restrictive silencer factor NRSF/REST is a transcriptional repressor of neuropilin-1 and diminishes the ability of semaphorin 3A to inhibit keratinocyte migration. J Biol Chem 2006; 281(5): 2721-9.

[45] Miao HQ, Lee P, Lin H, Soker S, Klagsbrun M. Neuropilin-1 expression by tumor cells promotes tumor angiogenesis and progression. FASEB J 2000; 14(15): 2532-9.

[46] Fakhari M, Pullirsch D, Abraham D, *et al.* Selective upregulation of vascular endothelial growth factor receptors neuropilin-1 and -2 in human neuroblastoma. Cancer 2002; 94(1): 258-63.

[47] Latil A, Bieche I, Pesche S, *et al.* VEGF overexpression in clinically localized prostate tumors and neuropilin-1 overexpression in metastatic forms. Int J Cancer 2000; 89(2): 167-71.

[48] Bachelder RE, Crago A, Chung J, *et al.* Vascular endothelial growth factor is an autocrine survival factor for neuropilin-expressing breast carcinoma cells. Cancer Res 2001; 61(15): 5736-40.

[49] Parikh AA, Liu W, Fan F, *et al.* Expression and regulation of the novel vascular endothelial growth factor receptor neuropilin-1 by epidermal growth factor in human pancreaticcarcinoma. Cancer 2003; 98(4): 720-9.

[50] Liu W, Parikh AA, Stoeltzing O, *et al.* Upregulation of neuropilin-1 by basic fibroblast growth factor enhances vascular smooth muscle cell migration in response to VEGF. Cytokine 2005; 32(5): 206-12.

[51] Armulik A, Abramsson A, Betsholtz C. Endothelial/Pericyte Interactions. Circ Res 2005; 97(6): 512-23.

[52] Fujisawa H, Kitsukawa T. Receptors for collapsin/semaphorins. Curre Opin Neurobiol 1998; 8(5): 587-92.

[53] Soker S, Takashima S, Miao HQ, Neufeld G, Klagsbrun M. Neuropilin-1 is expressed by endothelial and tumor cells as an isoform-specific receptor for vascular endothelial growth factor. Cell 1998; 92(6): 735-45.

[54] Makinen T, Olofsson B, Karpanen T, *et al.* Differential binding of vascular endothelial growth factor B splice and proteolytic isoforms to neuropilin-1. J Biol Chem 1998; 274(30): 21217-22.

[55] Migdal M, Huppertz B, Tessler S, *et al.* Neuropilin-1 is a placenta growth factor-2 receptor. J Biol Chem 1998; 273(35): 22272-8.

[56] Wise LM, Veikkola T, Mercer AA, *et al.* Vascular endothelial growth factor (VEGF)-like protein from of virus NZ2 binds to VEGFR2 and neuropilin-1. Proc Natl Acad Sci USA 1999; 96(6): 3071-6.

[57] Fuh G, Garcia KC, de Vos AM. The interaction of neuropilin-1 with vascular endothelial growth factor and its receptor flt-1. J Biol Chem 275(35): 26690-5.

[58] Shibuya M. Vascular endothelial growth factor-dependent and -independent regulation of angiogenesis. BMB Rep 2008; 41(4): 278-86.

[59] Kitsukawa T, Shimono A, Kawakami A, Kondoh H, Fujisawa H. Over expression of a membrane protein, neuropilin, in chimeric mice causes anomalies in the cardiovascular system, nervous system and limbs. Development 1995; 121(12): 4309-18.

[60] Kawasaki T, Kitsukawa T, Bekku Y, *et al.* A requirement for neuropilin-1 in embryonic vessel formation. Development 1999; 126(21): 4895-902.

[61] Takashima S, Kitakaze M, Asakura M, *et al.* Targeting of both mouse neuropilin-1 and neuropilin-2 genes severely impairs developmental yolk sac and embryonic angiogenesis. Proc Natl Acad Sci USA 2002; 99(6): 3657-62.

[62] Gu C, Rodriguez ER, Reimert DV, *et al.* Neuropilin conveys semaphorin and VEGF signaling during neural and cardiovascular development. Dev Cell 2003; 5(1): 45-57.

[63] Herzog B, Pellet-Many C, Britton G, Hartzoulakis B, Zachary IC. VEGF binding to NRP1 is essential for VEGF stimulation of endothelial cell migration, complex formation between NRP1 and VEGFR2, and signaling *via* FAK Tyr407 phosphorylation. Mol Biol Cell 2011; 22(15): 2766-76.

[64] Bielenberg DR, Pettaway CA, Takashima S, Klagsbrun M. Neuropilins in neoplasms: expression, regulation, and function. Exp Cell Res 2006; 10; 312(5): 584-93.

[65] Holderfield MT, Hughes CC. Crosstalk between vascular endothelial growth factor, notch, and transforming growth factor-beta in vascular morphogenesis. Circ Res 2008; 102(6): 637-52.

[66] Baluk P, Hashizume H, McDonald DM. Cellular abnormalities of blood vessels as targets in cancer. Curr Opin Genet Dev 2005; 15(1): 102-11.

[67] Pan Q, Chanthery Y, Liang WC, *et al.* Blocking neuropilin-1 function has an additive effect with anti-VEGF to inhibit tumor growth. Cancer Cell 2007; 11(1): 53-67.

[68] Geretti E, Shimizu A, Klagsbrun M. Neuropilin structure governs VEGF and semaphorin binding and regulates angiogenesis. Angiogenesis 2008; 11(1): 31-3.

[69] Strömblad S, Cheresh DA. Cell adhesion and angiogenesis. Trends Cell Biol 1996; 6(12): 462-8.

[70] Eliceiri BP, Cheresh DA. Adhesion events in angiogenesis. Curr Opin Cell Biol 2001; 13(5): 563-8.

[71] Varner JA. The role of vascular cell integrins αvβ3 and αvβ5 in angiogenesis. In: Goldberg ID and Resen EM (eds) Regulation of Angiogenesis, Birkhuser verlag Basel/Switzland, 1997, pp 361-90.

[72] van der Flier A, Badu-Nkansah K, Whittaker CA, *et al.* Endothelial α5 and αv integrins cooperate in remodeling of the vasculature during development. Development 2010; 137(14), 2439-49.

[73] Hodivala-Dilke KM, McHugh KP, Tsakiris DA, *et al.* Beta3-integrin-deficient mice are a model for Glanzmann thrombasthenia showing placental defects and reduced survival. J Clin Invest 1999; 103(2): 229-38.

[74] Reynolds LE, Wyder L, Lively JC, *et al.* Enhanced pathological angiogenesis in mice lacking beta3 integrin or beta3 and beta5 integrins. Nat Med 2002; 8(1): 27-34.

[75] Bader BL, Rayburn H, Crowley D, Hynes RO. Extensive vasculogenesis, angiogenesis, and organogenesis precede lethality in mice lacking all αv integrins. Cell 1998; 95(4): 507-19.

[76] Kuzuya M. Interaction of vascular endothelial cell with extracellular matrix protein: Implication of atherosclerosis and angiogenesis. Connective Tissue 2002; 34: 309-16.

[77] Gonzalez AM, Gonzales M, Herron GS, *et al.* Complex interactions between the laminin alpha 4 subunit and integrins regulate endothelial cell behavior *in vitro* and angiogenesis *in vivo*. Proc Natl Acad Sci USA 2002; 99(25): 16075-80.

[78] Suda H, Asami Y, Murata E, Fujita K, Akita M. Immuno-histochemical expression of alpha1, alpha2 and alpha3 integrin subunits during angiogenesis *in vitro*. Histo Histopathol 2004; 19(3): 735-742

[79] Broberg A, Heino J. Integrin α2β1-dependent contraction of floating collagen gels and induction of collagenase are inhibited by tyrosine kinase inhibitors. Exp. Cell Res 1996; 228(1): 29-35.

[80] Eble JA. Integrins-A Versatile and Old Family of Cell Adhesion Molecules. In: Chapter 1, Eble JA, Kühn K Eds. Integrin-Ligand Interaction. Texas : R. G. Landes Company 1997; pp 1-40.

[81] Gullberg D, Gehlsen KR, Turner DC, *et al.* Analysis of α1β1, α2β1 and α3β1 integrins in cell-collagen interactions: identification of conformation dependent α1β1 binding sites in collagen type I. EMBO J 1992; 11(11): 3865-73.

[82] Eble JA, Golbik R, Mann K, Kühn K. The alpha 1 beta 1 integrin recognition site of the basement membrane collagen molecule [alpha 1(IV)]2 alpha 2(IV). EMBO J 1993; 12(12): 4795-802.

[83] Heino J. Biology of tumor cell invasion: interplay of cell adhesion and matrix degradation. Int. J Cancer 1996; 65(6): 717-22.

[84] Schiro JA, Chan BM, C, Roswit W, *et al.* Integrin α2β1 1 (VLA-2) mediates reorganization and contraction of collagen matrices by human cells. Cell 1991; 67(2): 403-10.

[85] Langholz O, Rockel D, Mauch C, *et al.* Collagen and collagenase gene expression in 3D collagen lattices are differentially regulated by α1β1 and α2β1 integrins. J. Cell Biol 1995; 131: 1903-15.

[86] Riikonen T, Westermarck J, Koivisto L, *et al.* Integrin α2β1 is a positive regulator of collagenase (MMP-1) and collagen α1(I) gene expression. J. Biol. Chem. 1995; 270(6 Pt 2): 13548-52.

[87] Nicosia RF, Bonanno E. Inhibition of angiogenesis *in vitro* by Arg-Gly-Asp-containing synthetic peptide. Am J Pathol 1991; 138(4): 829-33.

[88] Elices MJ, Urry LA, Hemler ME. Receptor functions for the integrin VLA-3: fibronectin, collagen, and laminin binding are differentially influenced by Arg-Gly-Asp peptide and by divalent cations. J Cell Biol 1991; 112(1): 169-81.

[89] Carter WG, Ryan MC, Gahr PJ. Epiligrin, a new cell adhesion ligand for integrin alpha 3 beta 1 in epithelial basement membranes. Cell 1991; 65(4): 599-610.

[90] Delwel GO, de Melker AA, Hogervorst F, *et al.* Distinct and overlapping ligand specificities of the α3Aβ1 and α6Aβ1 integrins: recognition of laminin isoforms. Mol Biol Cell 1994; 5(2): 203-15.

[91] Symington BE, Takada Y, Carter WG. Interaction of integrins α3β1 and α2β1: potential role in keratinocyte intercellular adhesion. J Cell Biol 1993; 120(2): 523-35.

[92] Sriramarao P, Steffner P, Gehlsen KR. Biochemical evidence for the homophilic interaction of the α3β1 integrin. J. Biol Chem 1993; 268(29): 22036-41.

[93] Miraglia S, Godfrey W, Yin AH, *et al.* A novel five-transmembrane hematopoietic stem cell antigen: isolation, characterization, and molecular cloning. Blood 1997; 90(12): 5013-21.

[94] Weigmann A, Corbeil D, Hellwig A, Huttner WB. Prominin, a novel microvilli-specific polytopic membrane protein of the apical surface of epithelial cells, is targeted to plasmalemmal protrusions of non-epithelial cells. Proc Natl Acad Sci USA 1997; 94(23): 12425-30.

[95] Gehling UM, Ergün S, Schumacher U, *et al. In vitro* differentiation of endothelial cells from AC133-positive progenitor cell. Blood 2000; 95(10): 3106-12.

[96] Sun J, Zhang C, Liu G, *et al.* A novel mouse CD133 binding-peptide screened by phage displayinhibits cancer cell motility *in vitro.* Clin Exp Metastasis 2012; 29(3): 185-96.

[97] Peichev M, Naiyer AJ, Pereira D, *et al.* Expression of VEGFR-2 and AC133 by circulating human CD34(+) cells identifies a population of functional endothelial precursors Blood 2000; 95(3): 952-8.

[98] Soda Y, Marumoto T, Friedmann-Morvinski D, *et al.* Transdifferentiation of glioblastoma cells into vascular endothelial cells. Proc Natl Acad Sci USA 2011; 108(11): 4274-80.

[99] O'Brien CA, Pollett A, Gallinger S, Dick JE. A human colon cancer cell capable of initiating tumour growth in immunodeficient mice. Nature 2007; 445; 106-10.

[100] Ricci-Vitiani L, Lombardi DG, Pilozzi EM, *et al.* Identification and expansion of human colon-cancer-initiating cells. Nature 2007; 445(7123): 111-5.

[101] Singh SK, Hawkins C, Clarke ID, *et al.* Identification of human brain tumour initiating cells. Nature 2004; 432(7015): 396-401.

[102] Liu G, Yuan X, Zeng Z, *et al.* Analysis of gene expression and chemoresistance of CD133+ cancer stem cells in glioblastoma. Mol Cancer 2006; 5: 67.

[103] Monzani E, Facchetti F, Galmozzi E, *et al.* Melanoma contains CD133 and ABCG2 positive cells with enhanced tumourigenic potential. Eur J Cancer 2007; 43(5): 935-46.

[104] Olempska M, Eisenach PA, Ammerpohl O, *et al.* Detection of tumor stem cell markers in pancreatic carcinoma cell lines. Hepatobiliary Pancreat Dis Int 2007; 6(1): 92-7.

[105] Suetsugu A, Nagaki M, Aoki H, *et al.* Characterization of CD133+ hepatocellular carcinoma cells as cancer stem/progenitor cells. Biochem Biophys Res Commun 2006; 351(4): 820-4.

[106] Yin S, Li J, Hu C, *et al.* CD133 positive hepatocellular carcinoma cells possess high capacity for tumorigenicity. Int J Cancer 2007; 120(7): 1444-50.

[107] Hayashi S, Fujita K, Matsumoto S, Akita M, Satomi A. Isolation and identification of cancer stem cells from a side population of a human hepatoblastoma cell line, HuH-6 clone-5. Pediatr Surg Int 2011; 27(1): 9-16.

[108] Collins AT, Berry PA, Hyde C, Stower MJ, Maitland NJ. Prospective identification of tumorigenic prostate cancer stem cells. Cancer Res 2005; 65(23): 10946-51.

[109] Wang R, Chadalavada K, Wilshire J, *et al.* Glioblastoma stem-like cells give rise to tumour endothelium. Nature 2010; 468(7325): 829-33.

[110] Ricci-Vitiani L, Pallini R, Biffoni M, *et al.* Tumour vascularization *via* endothelial differentiation of glioblastoma stem-like cells. Nature 2010; 468(7325): 824-8.

[111] Gunsilius E, Duba HC, Petzer AL, *et al.* Evidence from a leukaemia model for maintenance of vascular endothelium by bone-marrow-derived endothelial cells. Lancet 2000; 355(9216): 1688-91.

[112] Streubel B, Chott A, Huber D, *et al.* Lymphoma-specific genetic aberrations in microvascular endothelial cells in B-cell lymphomas. N Engl J Med 2004; 351(3): 250-9.

[113] Rigolin GM, Fraulini C, Ciccone ME, *et al.* Neoplastic circulating endothelial cells in multiple myelomawith 13q14 deletion. Blood 2006; 107(6): 2531-5.

[114] Pezzolo A, Parodi F, Corrias MV, *et al.* Tumor origin of endothelial cells in human neuroblastoma. J Clin Oncol 2007; 25(4): 376-83.

[115] Bussolati B, Grange C, Sapino A, Camussi G. Endothelial cell differentiation of human breast tumour stem/progenitor cells. J Cell Mol Med 2009; 13(2): 309-19.

[116] Invernici G, Emanueli C, Madeddu P, *et al.* Human fetal aorta contains vascular progenitor cells capable of inducing vasculogenesis, angiogenesis, and myogenesis *in vitro* and in a murine model of peripheral ischemia. Am J Pathol 2007; 170(6): 1879-92.

[117] Zengin E, Chalajour F, Gehling UM, *et al.* Vascular wall resident progenitor cells: a source for postnatalvasculogenesis. Development 2006; 133(8): 1543-51.

[118] Majesky MW, Dong XR, Hoglund V, Daum G, Mahoney WM Jr. The adventitia: a progenitor cell niche for the vessel wall. Cells Tissues Organs 2012; 195(1-2): 73-81.

[119] Nicosia RF. The aortic ring model of angiogenesis: a quarter century of search and discovery. J Cell Mol Med 2009: 13(10): 4113-36.

[120] Akita M, Fujita K. DNA micro-array gene expression profiling of angiogenesis in collagen gel culture. Clinical Medicine: Cardiology 2008; 2: 49-57. Available from: www.la-press.com/redirect_file.php?fileId=580...Fujita...

[121] Wolf FW, Sarma V, Seldin M, *et al.* B.94, a primary response gene inducible by tumor necrosis factor-alpha, is expressed in developing hematopoietic tissues and the sperm acrosome. J Biol Chem 1994; 269(5): 3633-40.

[122] Ferrara N, Gerber HP, LeCouter J. The biology of VEGF and its receptors. Nat Med 2003; 9(6): 669-76.

[123] Hao X, Mansson-Broberg A, Grinnemo KH, *et al.* Myocardial angiogenesis after plasmid or adenoviral VEGF-A(165) gene transfer in rat myocardial infarction model. Cardiovasc Res 2007; 73(3): 481-7.

[124] Bhardwaj S, Roy H, Gruchala M, *et al.* Angiogenic responses of vascular endothelial growth factors in periadventitial tissue. Hum Gene Ther 2003; 14(15): 1451-62.

[125] Mould AW, Tonks ID, Cahill MM, *et al.* Vegfb gene knockout mice display reduced pathology and synovial angiogenesis in both antigen-induced and collagen-induced models of arthritis. Arthritis Rheum 2003; 48(9): 2660-9.

[126] Enholm B, Karpanen T, Jeltsch M, *et al.* Adenoviral expression of vascular endothelial growth factor-C induces lymphangiogenesis in the skin. Circ Res 2001; 88(6): 623-9. Erratum in Circ Res 2001; 89(1): E15.

[127] Karkkainen MJ, Haiko P, Sainio K, *et al.* Vascular endothelial growth factor C is required for sprouting of the first lymphatic vessels from embryonic veins. Nat Immunol 2004; 5(1): 74-80.

[128] Antoine M, Wirz W, Tag CG, *et al.* Fibroblast growth factor 16 and 18 are expressed in human cardiovascular tissues and induce on endothelial cells migration but not proliferation. Biochem Biophys Res Commun 2006; 346(1): 224-33.

[129] Pizette S, Batoz M, Prats H, *et al.* Production and functional characterization of human recombinant FGF-6 protein. Cell Growth Differ 1991; 2(11): 561-6.

[130] Kokkotou E, Torres D, Moss AC, *et al.* Corticotropin-releasing hormone receptor 2-defi cient mice have reduced intestinal infl ammatory responses. J Immunol 2006; 177(5): 3355-61.

[131] Luo Y, Nita-Lazar A, Haltiwanger RS. Two distinct pathways for O- fucosylation of epidermal growth factor-like or thrombospondin type 1 repeats. J Biol Chem 2006; 281(14): 9385-92.

[132] Shi W, Harris AL. Notch signaling in breast cancer and tumor angiogenesis: cross-talk and therapeutic potentials. J Mammary Gland Biol Neoplasia 2006; 11(1): 41-52.

[133] Rehman AO, Wang CY. Notch signaling in the regulation of tumor angiogenesis. Trends Cell Biol 2006; 16(6): 293-300.

[134] Williams CK, Li JL, Murga M, Harris Al, Tosato G. Up-regulation of the Notch ligand Delta-like 4 inhibits VEGF -induced endothelial cell function. Blood 2006; 107(3): 931-9.

[135] Fujita K, Asami Y, Tanaka K, Akita M, Merker HJ. Anti-angiogenic effects of thalidomide: expression of apoptosis-inducible active-caspase-3 in a 3D collagen gel culture of aorta. Histochem Cell Biol 2004; 122(1): 27-33

[136] Mitsiades N, Mitsiades CS, Poulaki V, *et al.* Apoptotic signaling induced by immunomodulatory thalidomide analogs in human multiple myeloma cells: therapeutic implications. Blood 2002; 99(12): 4525-30.

[137] Alnemri ES, Livingston DJ, Nicholson DW, *et al.* Human ICE/CED-3 protease nomenclature [letter]. Cell 1996; 87(2): 171.

[138] Bauer KS, Dixon SC, Figg WD. Inhibition of angiogenesis by thalidomide requires metabolic activation, which is speciesdependent. Biochem Pharmacol 1998; 55(11): 1827-34.

[139] Wells PG, Kim PM, Laposa RR, *et al.* Oxidative damage in chemical teratogenesis. Mutat Res 1997; 396(1-2): 65-78.

[140] Parman T, Wiley MJ, Wells PGFree radical-medicated oxidative DNA damage in the mechanism of thalidomide teratogenicity. Nat Med 1999; 5(5): 582-5.

[141] D'Amato RJ, Loughnan MS, Flynn E, Folkman J. Thalidomide is an inhibitor of angiogenesis. Proc Natl Acad Sci USA 1994; 91(9): 4082-5.

[142] Neubert R, Merker HJ, Neubert D. Developmental model for thalidomide action. Nature 1995; 400(6743): 1500-2.

[143] Diggle GE. Thalidomide: 40 years on. Int J Clin Pract 2001; 55(9): 627-31.

[144] Vacca A, Ribatti D, Roncali L, *et al.* Bone marrow angiogenesis and progression in multiple myeloma. Br J Haematol 1994; 87(3): 503-8.

[145] Rajkumar SV, Leong T, Roche PC, *et al.* Prognostic value of bone marrow angiogenesis in multiple myeloma. Clin Cancer Res 2000; 6(8): 3111-6.

[146] Singhal S, Mehta J, Desikan R, *et al.* Antitumor activity of thalidomide in refractory multiple myeloma. N Engl J Med 1999; 341(21): 1565-71.

[147] Rajkumar SV. Thalidomide in the treatment of multiple myeloma. Expert Rev Anticancer Ther 2001; 1(1): 2-28.

[148] Vesela D, Vesely D, Jelinek R. Embryotoxicity in chick embryo of thalidomide hydrolysis products following metabolic activation by rat liver homogenate. Funct Dev Morphol 1994; 4(4): 313-6.

[149] Bauer KS, Dixon SC, Figg WD. Inhibition of angiogenesis by thalidomide requires metabolic activation, which is species-dependent. Biochem Pharmacol 1998; 55(11): 1827-34.

[150] Ng SS, Gütschow, M, Weiss M, *et al.* Anti-angiogenic activity of N-substituted and tetrafluorinated thalidomide analogues. Cancer Res 2003; 63(12): 3189-94.

[151] Liu WM, Strauss SJ, Chaplin T, *et al.* s-thalidomide has a greater effect on apoptosis than angiogenesis in a multiple myeloma cell line. Hematol J 2004; 5(3): 247-54.

[152] Vacca A, Scavelli C, Montefusco V, *et al.* Thalidomide downregulates angiogenic genes in bone marrow endothelial cells of patients with active multiple myeloma. J Clin Oncol 2005; 23(23): 5334-46.

[153] Aerbajinai W, Zhu J, Gao, Z, *et al.* Thalidomide induces gamma-globin gene expression through increased reactive oxygen species-mediated p38 MAPK signaling and histone H4 acetylation in adult erythropoiesis. Blood 2007; 110(8): 2864-71.

[154] De Luisi A, Ferrucci A, Coluccia AM, *et al.* Lenalidomide restrains motility and overangiogenic potential of bone marrow endothelial cells in patients with active multiple myeloma. Clin Cancer Res 2011; 17(7): 1935-46.

[155] Singer II, Scott S, Kazazis DM, Huff JW. Lovastatin, an inhibitor of cholesterol synthesis, induces hydroxymethylglutaryl-coenzyme A reductase directly on membranes of expanded smooth endoplasmic reticulum in rat hepatocytes. Proc Natl Acad Sci USA 1988; 85(14): 5264-8.

[156] Depasquale I, Wheatley DN. Action of Lovastatin (Mevinolin) on an *in vitro* model of angiogenesis and its co-culture with malignant melanoma cell lines. Cancer Cell Int 2006; 6: 9. Available from: http://www.cancerci.com/content/6/1/9

[157] Karthikeyan VJ, Lip GY. Statins and intra-plaque angiogenesis in carotid artery disease. Atherosclerosis 2007; 192(2): 455-6.

[158] Poynter JN, Gruber SB, Higgins PD, *et al.* Statins and the risk of colorectal cancer. New Engl J Med 2005; 352(21): 2184-92.

[159] Friis S, Poulsen AH, Johnsen SP, *et al.* Cancer risk among statin users: a population-based cohort study. Int J Cancer 2005; 114(4): 643-7.

[160] Khaidakov M, Wang W, Khan JA, *et al.* Statins and angiogenesis: Is it about connections? Biochem Biophys Res Commun 2009; 387(3): 543-7.

[161] Koyama-Nasu R, Takahashi R, Yanagida S, *et al.* The cancer stem cell marker CD133 interacts with plakoglobin and controls desmoglein-2 protein levels. PLoS One 2013; 8(3): e53710.

[162] Akita M, Tanaka K, Murai N, *et al.* Detection of CD133 (prominin-1) in a human hepatoblastoma cell line (HuH-6 clone 5). Microsc Res Tech 2013; 76(8): 844-52.

[163] Fujita K, Komatsu K, Tanaka K, *et al.* An *in vitro* model for studying vascular injury after laser microdissection. Histochem Cell Biol 2006; 125(5): 509-14.

[164] Srinivasan R. Ablation of polymers and biological tissue by ultraviolet lasers. Science 1986; 234(4776): 559-65.

[165] Pendurthi UR, Rao LV. Suppression of transcription factor Egr-1 by curcumin. Thromb Res 2000; 15; 97(4): 179-89.

[166] Khachigian LM, Lindner V, Williams AJ, Collins T. Egr-1-induced endothelial gene expression: a common theme in vascular injury. Science 1996; 271(5254): 1427-31.

[167] Gashler A, Sukhatme VP. Early growth response protein 1 (Egr-1): prototype of a zinc-finger family of transcription factors. Prog Nucleic Acid Res Mol Biol 1995; 50: 191-224.

[168] Khachigian LM, Collins T. Inducible expression of Egr-1-dependent genes. A paradigm of transcriptional activation in vascular endothelium. Circ Res 1997; 81(4): 457-61.

[169] Fahmy RG, Dass CR, Sun LQ, Chesterman CN, Khachigian LM. Transcription factor Egr-1supports FGF-dependent angiogenesis during neovascularization and tumor growth. Nat Med 2003; 9(8): 1026-32.

[170] Biesiada E, Razandi M, Levin ER. Egr-1 activates basic fibroblast growth factor transcription. Mechanistic implications for astrocyte proliferation. J Biol Chem 1996; 271(31): 18576-81.

[171] Liu C, Yao J, de Belle I, *et al.* The transcription factor EGR-1 suppresses transformation of human fibrosarcoma HT1080 cells by coordinated induction of transforming growth factor-beta1, fibronectin, and plasminogen activator inhibitor-1. J Biol Chem 1999; 274(7): 4400-11.

[172] Du B, Fu C, Kent KC, *et al.* Elevated Egr-1 in human atherosclerotic cells transcriptionally represses the transforming growth factor-beta type II receptor. J Biol Chem 2000; 275(50): 39039-47.

[173] Houston P, Campbell CJ, Svaren J, Milbrandt J, Braddock M. The transcriptional corepressor NAB2 blocks Egr-1-mediated growth factor activation and angiogenesis. Biochem Biophys Res Commun 2001; 283(2): 480-6.

[174] Fahmy RG, Dass CR, Sun LQ, Chesterman CN, Khachigian LM. Transcription factor Egr-1supports FGF-dependent angiogenesis during neovascularization and tumor growth. Nat Med 2003; 9(8): 1026-32.

[175] Standring S. Smooth muscle and the cardiovascular and lymphatic systems. In: Chapter 7, Standring S Ed. Gray'Anatomy, 39th ed. London: Elsevier Churchill Livingstone 2005, pp. 140-4.

[176] Watanabe S, Morisaki N, Tezuka M, *et al.* Cultured retinal pericytes stimulate *in vitro* angiogenesis of endothelial cells through secretion of a fibroblast growth factor-like molecule. Atherosclerosis 1997; 130(1-2): 101-7.

[177] Yonekura H, Sakurai S, Liu X, *et al.* Placenta growth factor and vascular endothelial growth factor B and C expression in microvascular endothelial cells and pericytes. Implication in autocrine and paracrine regulation of angiogenesis. J Biol Chem 1999; 274(49): 35172-8.

[178] Benjamin LE, Hemo I, Keshet E. A plasticity window for blood vessel remodeling is defined by pericyte coverage of the preformed endothelial network and is regulated by PDGF-B and VEGF. Development 1998; 125(9): 1591-8.

[179] Nehls V, Denzer K, Drenckhahn D. Pericyte involvement in capillary sprouting during angiogenesis in situ. Cell Tissue Res 1992; 270(3): 469-74.

[180] Gerhardt H, Golding M, Fruttiger M, *et al.* VEGF guides angiogenic sprouting utilizing endothelial tip cell filopodia. J Cell Biol 2003; 161(6): 1163-77.

[181] Helm CL, Fleury ME, Zisch AH, Boschetti F, Swartz MA Synergy between interstitial flow and VEGF directs capillary morphogenesis *in vitro* through a gradient amplification mechanism. Proc Natl Acad Sci USA 2005; 102(44): 15779-84.

[182] Zhang QX, Magovern CJ, Mack CA, *et al.* Vascular endothelial growth factor is the major angiogenic factor in omentum: mechanism of the omentum-mediated angiogenesis. J Surg Res 1997; 67(2): 147-54.

[183] Ji JW, Mac Gabhann F, Popel AS. Skeletal Muscle VEGF Gradients in Peripheral Arterial Disease: Simulations of Rest and Exercise. Am J Physiol Heart Circ Physiol. 2007; 293(6): H3740-9.

[184] Mu H, Ohashi R, Yang H, *et al.* Thymosin beta10 inhibits cell migration and capillary-like tube formation of human coronary artery endothelial cells. Cell Motil Cytoskeleton 2006; 63(4): 222-30.

[185] Smart N, Risebro CA, Melville AA, *et al.* Thymosin beta4 induces adult epicardial progenitor mobilization and neovascularization. Nature 2007; 445(7124): 177-82.

CHAPTER 3

Clear Cell Renal Cell Carcinoma as a Model of Pathological Angiogenesis: Which Actors to Target for Treatment?

Caroline Hilmi and Gilles Pagès*

University of Nice Sophia Antipolis, Institute for Research on Cancer and Ageing of Nice, UMR CNRS 7284, INSERM U1081, France

Abstract: Angiogenesis is a physiological phenomenon that establishes the vascular tree during development and maintains the supply of oxygen and nutrients to organs during adult life. A tight balance exists between several pro- or anti-angiogenic actors. The level of these molecules determines *de novo* angiogenesis or a highly controlled steady-state. Hence, a slight modification of this equilibrium inducing an increase of angiogenesis may provoke pathological angiogenesis in particular favouring aggressiveness of cancers by contributing to tumour growth and metastasis. Disruption of the angiogenic balance is a result of over production of chemokines, in particular the Vascular Endothelial Growth Factor (VEGF) or members of the ELR+CXCL family (ELR for glutamic acid (E), leucine (L) and arginin (R)), which are major pro-angiogenic factors. Their receptors (VEGFR or CXCR) present at the surface of endothelial and tumour cells are also essential to drive angiogenesis through activation of signalling pathways (RAS/RAF/MEK/ERK and PI3 Kinase/AKT/mTOR). Both signalling pathways drive cell proliferation, survival, and production of angiogenic and inflammatory cytokines. In this context, clear cell renal cell carcinomas (ccRCC) represent a paradigm of deregulated angiogenesis. Thus, ccRCC have been widely used to better understand pathological angiogenesis leading to over vascularization and it will be discussed in this revue the different approaches used by companies to study the implication of angiogenesis in cancers and to develop antibodies or pharmacological inhibitors targeting major receptors or cytokines. Although some therapeutic compounds are used in the clinic, they have given disappointing results particularly on the improvement of global survival. Hence the current challenge is to improve the existing therapeutics or to stratify the patients that will really benefit from them in order to tend to a more personalized therapy.

Keywords: Angiogenesis, clear cell renal cell carcinoma, ELR+CXCL, immunotherapy, tyrosine kinase receptors,VEGF.

***Corresponding author Gilles Pagès:** Centre Antoine Lacassagne, 33 Avenue de Valombrose 06189 Nice France; Tel: +33 4 92 03 12 31; Fax: +33 4 92 03 12 35; E-mail: gpages@unice.fr

Atta-ur-Rahman and Muhammad Iqbal Choudhary (Eds)
Copyright © 2014 Bentham Science Publishers Ltd. Published by Elsevier Inc. All rights reserved.
10.1016/B978-0-12-803963-2.50003-X

INTRODUCTION

Pathological angiogenesis is widely triggered by over-expression of the Vascular Endothelial Growth Factor (VEGF) main sub-family VEGF-A, and other factors. In the 1970's Folkman proposed to target angiogenesis as a useful way to kill tumours [1]. So far VEGF and their receptors VEGFRs have been targeted. It is now well described that solid tumours become hypoxic due to rapid cell proliferation and overgrowth of the tumour mass. This results in an over-expression of VEGF due to stabilization of its major transcription factor Hypoxia-Inducible Factor (HIF). Furthermore, ccRCC often possess inactivating mutations of the von Hippel-Lindau gene that normally contributes to HIF recycling [2]. These inactivating mutations prevent HIF from being degraded and therefore HIF continuously acts on its target genes, in particular VEGF. Secreted cytokines are the major factors controlling angiogenesis. A tight balance of their expression is needed in order to maintain enough *de novo* vessels when necessary or to stop angiogenesis if a steady state is reached. Secreted cytokines exist as pro-angiogenic forms such as VEGF, ELR+CXCL [3], and angiopoietin and also as anti-angiogenic forms, which include forms of VEGF called VEGFxxxb (xxx number of amino acids of the protein) [4], and ELR-CXCL [3]. Deregulation of expression of one or several of these factors can lead to pathological angiogenesis, which is the case in many cancers and some retinopathies in particular (Fig. **1**). Above all, an increase of VEGF pro-angiogenic isoforms leads to tumour growth [5]. All of these cytokines bind their specific tyrosine kinase receptors and induce downstream signalling pathways leading to angiogenesis, cell proliferation and tumour progression. These cytokines play a major role in tumour progression of many cancers and metastatic ccRCC (mccRCC) in particular [6-11]. Since mccRCC are poorly sensitive to chemotherapy they have received the most attention for targeted drug development in renal cancer.

TARGETED THERAPIES IN USE

So far pathological angiogenesis has been targeted to shrink the tumour volume by reducing the nutrient and oxygen supply in an attempt to cure patients. Different strategies have been considered to reach the desired effect.

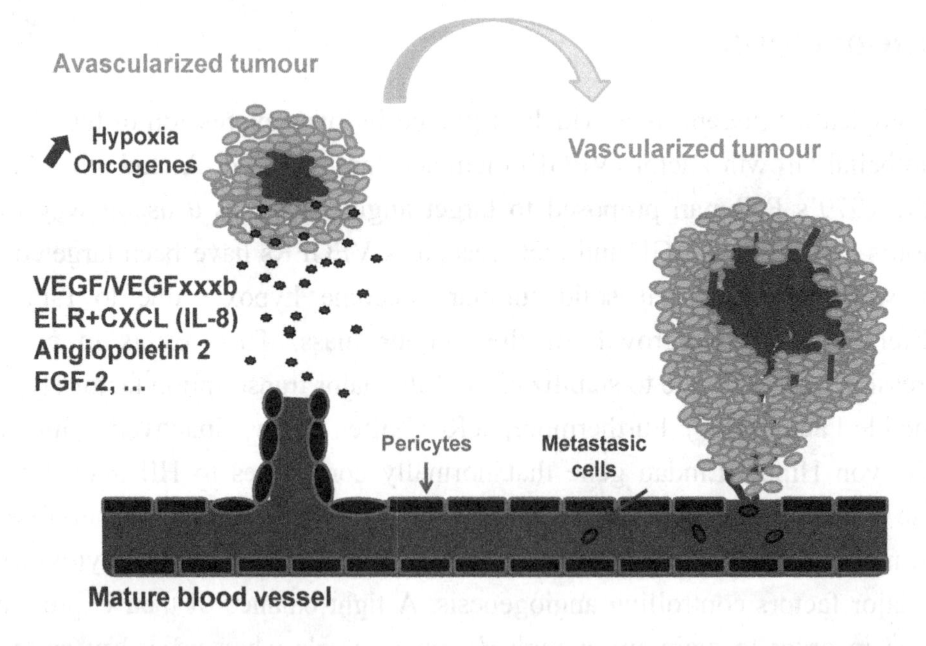

Figure 1: A tumour cannot grow beyond a few millimeters if it is not vascularized. Hypoxia or oncogenes triggers the production of growth factors by tumour cells (VEGF, ELR+CXCL, angiopoietin 2, FGF-2, …). These factors stimulate endothelial cell proliferation which results in tumour neo-vascularization. Tumour vessels are not correctly maturated (devoid of smooth muscle cells also called pericytes) which favors the entry of tumour cells into the circulation. This mechanism promotes metastasis.

Cytokines as Targets for Angiogenesis

One of the most important cytokines involved in promoting angiogenesis is VEGF. Thus, the first targeted anti-angiogenic treatment was aimed at VEGF. VEGF has been targeted in diseases such as retinopathies and also in different cancers in association with standard chemotherapy (Folinic acid, irinotecan, 5 Fluoro uracil oxaliplatin for colon cancers, platinum salts for lung cancers, taxanes for breast cancers and interferon alpha for kidney cancers). Bevacizumab (BVZ), a human neutralizing monoclonal antibody targeting VEGF has obtained food and drug administration (FDA) approval in combination with chemotherapy for colon, breast, lung and kidney cancers [7, 12-14]. Although BVZ treatment improved progression-free survival, it did not increase overall survival. After a decrease in tumour size, a relapse with a particularly aggressive form of the disease has been described [15]. These results were a great disappointment as BVZ had been expected to benefit patients expressing VEGF by decreasing tumour associated capillaries. Furthermore, the use of anti-VEGF treatment in retinopathies (Ranibizumab, RNZ [16]) has been questioned as the side effects could by themselves impair the treatment efficacy. RNZ induces in particular a high level of inflammation when injected in the eye [17]. Moreover, systemic neutralization of VEGF with RNZ in mice leads to the death of a significant number of photoreceptors and reduces retinal function [18]. Therefore treatments targeting VEGF need to be properly justified considering the available negative clinical data.

Tyrosine Kinase Receptors as Targets for Anti-Angiogenic Therapy

Following the disappointment of BVZ trials in different diseases, VEGF receptors and other cytokines have been considered instead to block these signalling pathways and thereby block angiogenesis and proliferation of tumour cells.

Sunitinib (Sutent)

Sunitinib is a multi-target tyrosine kinase inhibitor (TKI) that inhibits VEGF receptor 1, 2, 3 (VEGFRs), Platelet-derived growth factor receptor (PDGFR), Colony stimulating factor receptor (CSFR) and the stem cell factor receptor c-KIT at nanomolar concentrations [19-21]. Sutent is the first line treatment for mccRCC disease and for patients who relapsed with BVZ treatment. When patient tumours progress with Sutent treatment, they are treated with another multi-kinase inhibitor, sorafenib (Nexavar), which shares the same target receptors as sunitinib but also inhibits RAF. For this reason sorafenib was used for the treatment of melanoma [22]. After relapse with sorafenib treatment, sunitinib can be given again as the resistance is not acquired [23].

Axitinib (Inlyta)

Axitinib is also a multi-target tyrosine kinase receptor inhibitor, which can bind and block the signalling pathway of VEGFRs at nanomolar concentrations. In a phase III clinical trial, Axitinib increased progression-free survival (PFS) compared to the use of sorafenib in a second line treatment in mccRCC [24] after a failure of sunitinib, BVZ, temsirolimus or cytokines. Axitinib has recently been licensed by the FDA for the treatment of mccRCC.

Strategies targeting the immune system in combination are also being developed in order to help the tumour shrinkage. By developing other molecules the hope is to either identify more efficient treatments or at least to have the possibility to switch treatments when resistance is unfortunately acquired.

Kinases as Targets for Anti-Angiogenic Therapy

Mammalian target of Rapamycin (mTOR) is widely targeted after mccRCC patients have become refractory to anti-VEGF/VEGFRs therapy. Temsirolimus and everolimus have been approved as a second line treatment of RCC but the safety of these molecules is still under investigation [25]. Furthermore, a decrease in brain metastasis for mccRCC patients has been described proving that temsirolimus is effective against the primary tumour as well as the spreading of tumour cells [26].

EMERGENT MOLECULES IN PHASE III CLINICAL TRIALS

Tyrosine Kinase Receptors as Targets for Anti-Angiogenic Therapy

Tivozanib (AV-951) inhibits VEGFRs and other tyrosine kinase receptors [27]. The phase III clinical trial (NCT01030783) comparing tivozanib and sorafenib in people who have never received VEGF targeted therapy is ongoing.

Dovitinib (TKI258) targets VEGFRs as well as Fibroblast Growth Factor Receptor (FGFR)-1, -2 and -3. The phase III clinical trial (NCT01223027) comparing dovitinib with sorafenib in patients who did not respond to VEGF or mTOR targeted therapies is ongoing.

Targeting the Immune System for Anti-Angiogenic Therapy

The first therapeutic vaccine developed for mccRCC is now under a phase III clinical trial NCT01265901. **IMA901** is a RCC-associated peptide that activates T cells. Furthermore, APOA1 and CCL17 have been identified as two predictive biomarkers. APOA1 is often down-regulated in cancers while CCL17 is a chemo-attractant for CD4+ and CD8+ T cells in particular. Thus, high levels of APOA1 as well as CCL17 may indicate a good response to the vaccination [28].

AGS-003 is an autologous immunotherapy that induces T-cell proliferation to specifically target metastases of mRCC patients following nephrectomy [29]. This molecule will be used in a phase III clinical trial in combination with sunitinib as a first line treatment for mccRCC.

EMERGENT MOLECULES IN PHASE II CLINICAL TRIALS

Tyrosine Kinase Receptors as Targets for Anti-Angiogenic Therapy

Novel molecules have recently passed phase I studies with encouraging results. **Cediranib** (AZD2171) use has demonstrated benefit to patients in two trials [30, 31]. Cediranib is an inhibitor of VEGFRs and is currently used in a phase II clinical trial (NCT00303862) on patients with refractory mRCC who have not received anti VEGF therapy.

BIBF 1120 inhibits VEGFRs, FGFRs and PDGFR. The phase II clinical trial (NCT01024920) is based on first line treatment for RCC comparing BIBF1120 with sunitinib.

Lenvatinib E7080 is a multi-targeted TKI inhibiting VEGFRs, FGFRs, PDGFR and c-KIT. In the phase II clinical trial (NCT01136733) it is used either alone or in combination with everolimus on patient with refractor mccRCC who have not received anti VEGF therapy.

Lenvatinib (E7080) is also a multi-target TKI triggering down-regulation of PGFR, VEGFRs, FGFRs and c-KIT signalling pathways. The preliminary studies showed the safe use of Lenvatinib as well as encouraging tumour regression for both melanoma and mccRCC [32]. The phase II clinical trial (NCT01136733) is studying the effect of Lenvatinib alone or in combination with everolimus on patients with refractory mccRCC to anti VEGF therapies.

Regorafenib (BAY 73-4506) has received approval for treatment of mccRCC. It inhibits VEGFRs, PDGFR, and other tyrosine kinase receptors such as c-KIT, "rearranged during transfection" (RET), FGFRs as well. It also inhibits "tyrosine kinase with immunoglobulin and EGF homology domains" TIE2, which is involved in tumour escape mechanisms and the RAF/BRAF, stress activated protein kinase 2 (SAPK2), Abelson (ABL) pathway. Although the preliminary studies showed certain benefit and that drug safety needs close monitoring [33, 34], a phase II trial (NCT 00664326) as a first line treatment on mccRCC has been achieved in 2011.

Other molecules have been developed to down regulate pathological angiogenesis including monoclonal antibodies such as **Ramucirumab (IMC1121-B)**, which target the extracellular ligand-binding domain of VEGFR2, avoiding the binding of VEGF [35].

Cytokines as Targets for Anti-Angiogenic Therapy

Aflibercept (VEGF Trap) is a decoy receptor made with part of VEGFR-1 and -2 fused with the Fc fragment of human IgG. This molecule can thus bind to VEGF avoiding the binding and the signalling of their receptors [36]. However, Aflibercept is not so well investigated for mccRCC compared to colorectal cancers [37], lung cancers [38], and eye disease [39]. Thus far Aflibercept is under a phase II clinical trial (NCT00357760) for patients with a refractory mccRCC to TKI treatment.

Other Targets Investigated to Target Pathological Angiogenesis

Another molecule developed is **MDX-1106**, a monoclonal antibody that targets the programmed cell death protein 1 (PD-1) receptor expressed on activated T cells membrane surface [40]. This receptor and its ligand negatively regulate the immune response. Patients with a high proportion of PD-1 positive cells have been evaluated with a poor outcome [41]. Thus the use of MDX-1106 in these patients is predicted to boost the immune system and enable anti-tumour immunity. MDX-1106 is now under an ongoing phase II clinical trial (NCT01354431) in patients who received an anti-angiogenic therapy and a phase III study is planned for the future.

Targeting angiogenic receptors and cytokines is a well-developed strategy but targeting kinases that are downstream of these signalling pathways is also considered. In this strategy, **MK-2206** is an allosteric inhibitor of the kinase Akt. Akt is widely involved in angiogenesis, survival and tumour growth. A phase II clinical study (NCT01239342) is comparing the efficacy of MK-2206 with another inhibitor of the pathway, everolimus that targets mTOR [42]. This study is being performed on patients with refractory mccRCC to anti VEGF therapy [42].

AMG 386 is a fusion-Fc peptide that neutralizes angiopoietin -1 and -2 and prevents their binding to TIE2. It is currently in use in a phase II clinical trial (NCT00853372) combined with sunitinib as first line treatment or for cytokine refractory mccRCC patients [43].

ABX-IL-8 is a humanized monoclonal antibody targeting interleukin-8 (IL-8). IL-8 binds to CXCR1 and CXCR2 with high affinity [3] and is involved in inflammatory process, tumour growth and angiogenesis. This induction of tumour growth and angiogenesis in particular has been described in melanoma [44, 45]. The inhibition of IL-8 decreases tumour growth and invasion process *via* inhibition of metalloproteinases in bladder cancer [46] and metastatic properties of melanoma cells [47]. Thus, also targeting the CXCR1/2 receptors is a strategy developed with molecules such as **SCH-479833** and **SCH-527123,** which are giving encouraging results [48].

IDENTIFICATION OF PREDICTIVE BIOMARKERS OF THE EFFICIENCY OF ANTI-ANGIOGENESIS TREATMENTS

The poor benefit on overall survival of the different treatments described above and their elevated costs put in evidence the urgency to identify predictive markers of efficiency in order to tend towards personalized therapies. Some markers have been described in small cohorts but need confirmation studies on a larger number of patients. Moreover these predictive markers may not be relevant for all types of tumours. The following paragraph describes relevant predictive markers of anti-angiogenesis treatment success/failure in mccRCC but also in other cancers.

Predictive Markers of the Efficiency of BVZ

Some relevant markers were identified following clinical trials.

Colon Cancers

BVZ was first approved by the FDA for the treatment of metastatic colon cancers in association with chemotherapy [12]. The presence of anti-angiogenic forms of

VEGF appears as a predictive marker of BVZ treatment inefficacy for colorectal cancers [49]. Hypertension was suggested as a marker of BVZ efficacy for colorectal cancer but only one on seven clinical trials correlated increased blood pressure with progression free and overall survival [50-51]. Pharmacogenetic profiling (tumour levels and polymorphisms) has identified CD133, a surface protein used to isolate stem cells, as a predictive marker of progression free and overall survival [52]. Genomic profiling also identified the VEGFR-1 319 C/A single nucleotide polymorphism as a relevant marker, patients with the A allele appearing to have increased response rates (Nordic ACT trial) [53]. The univariate and multivariate analyses demonstrated that the serine hydroxymethyltransferase 1 1420T allele was associated with better response, longer progression-free survival (PFS) and overall survival (OS) [54]. A significant correlation between circulating endothelial cells and endothelial progenitors levels and BVZ efficacy was also shown on a small cohort of patients [55].

Lung Cancers

The pivotal study of Sandler and collaborators resulted in the approval of BVZ in association with platinum salts for the treatment of non-small-cell lung cancers [13]. A baseline systemic inflammatory status was suggested as a marker of resistance to BVZ treatment in advanced non-squamous non-small-cell lung cancer patients [56]. As for colon cancers, conflicting results on lung cancers and hypertension were also reported [57].

Breast Cancers

The association of BVZ with taxanes was approved after the work of Miller and co-workers [14]. High baseline plasma VEGF concentrations were associated with greater BVZ benefit (not statistically significant) for patients with Human Epidermal Growth Factor Receptor 2 (HER2)-negative breast cancer (AVEREL clinical trial) [58].

mccRCC

The association of BVZ plus interferon alpha was approved in 2007 following the AVOREN clinical trial [7]. A single nucleotide polymorphism of the VEGFR1 gene is predictive for progression free survival (AVOREN clinical trial) [59-60]

Pancreatic Cancers

A single nucleotide polymorphism of the VEGFR1 gene is predictive for progression free and overall survival. This polymorphism was identified in the AVITA clinical trial [59]. Lactate deshydrohenase serum levels correlated with poor prognosis and BVZ treatment improved outcome in this group of patients [61].

Gastric Cancers

Plasma VEGF and tumour neuropilin are biomarkers of clinical outcome in patients with advanced gastric cancer treated with BVZ (AVAGAST clinical trial) [62].

Glioblastoma

Phospho lipid metabolites were recently suggested as predictive biomarkers of BVZ efficacy in glioblastomas [63].

Melanoma

Whereas contradictory results were reported for other cancers, hypertension is more promising to predict BVZ activity in melanoma [64].

Predictive Markers of the Efficiency of Sunitinib

Breast Cancers

Changes in soluble c-KIT and VEGF may be predictive of clinical outcome in metastatic breast cancers treated with sunitinib [65].

mccRCC

VEGF and VEGFR polymorphisms were indicative of outcome in patient receiving sunitinib [60]. Polymorphisms in VEGFR3 and CYP3A5*1 were also shown to define a subset of patients with decreased sunitinib response and tolerability [66]. The same group identified hypoxia-related proteins as markers for the outcome of treatment with sunitinib (HIF2α, PDGFR and VEGFR3) [67]. A subset of plasmatic miRNA was also shown as good indicators of sunitinib response [68]. A delayed response to sunitinib was also related to the presence of circulating tumour and endothelial cells [69]. C-reactive protein is also an independent predictive marker of increased progression free survival of patient treated with sunitinib [70]. High CXCR4 expression on the primary tumour correlates with sunitinib poor response [71]. Whereas some controversies were described for BVZ, hypertension seems predictive of sunitinib efficiency [72]. Serum levels of VEGF, neutrophil gelatinase-associated lipocalin, metallo proteinase 9 and tumour necrosis factor alpha are significant predictors of progression-free survival in patients treated with sunitinib [73-75]. An increase in dentritic cell is also indicative of sunitinib efficiency [76].

Ovarian Cancers

Angiopoietin-2 could define a patient population with a better response to sunitinib [77]. Skin toxicity was related to tumour control in patient treated with sorafenib [78].

Predictive Markers of the Efficiency of Sorafenib

mccRCC

No surrogate markers have been identified to date in mccRCC. None of the usual markers tested in a large clinical trial (900 patients) including VEGF, soluble VEGFR2, carbonic anhydrase 9, tissue inhibitor of metalloproteinase 1 and Ras p21 showed predictive value on the sorafenib treatment [79]. Dynamic contrast magnetic resonance imaging cannot be considered either as a predictive marker of efficacy [80].

Hepatocellular Carcinomas

Activation of extracellular signal regulated kinase (ERK), c-JUN N-terminal kinase (JNK), high expression level of CD133, high level of serum cytokines and des-γ-carboxyprothrombin (DCP) predict a poor response to sorafenib in hepatocellular carcinoma [81-83].

Predictive Markers of Everolimus Efficiency

mccRCC

Serum lactate deshydrogenase was identified as a predictive marker of the survival benefit [84].

Neuroendocrine Tumours

A single nucleotide polymorphism in the fibroblast growth factor receptor 4 was shown as a surrogate marker of treatment failure [85].

Bladder Cancers

Full exome sequencing of the DNA of a tumour from a patient with a durable remission of metastatic bladder cancer has identified mutations in the tuberous sclerosis complex 1 gene (TSC1), a negative regulator of mTOR pathway activation. The relevance of this mutation was confirmed on a cohort of 109 patients, 8% of which were highly responders [86].

DISCUSSIONS

Renal cell carcinoma is a cancer for which the treatments evolve rapidly. One of the reasons is that unfortunately patients often relapse even when they respond to the initial treatment. We postulate that the most important tool we still lack is a predictive marker of reference or a group of markers although some have been described in small cohorts of patients. However, this signature may differ from a cancer to another. Aflibercept will potentially provide this where the proportion of PD-1 + cells determines the likely-hood of treatment efficacy. We can explain the slow development of predictive markers by the fact that RCC is a very heterogeneous disease. Thus, we would need to combine different molecule to make sure we target all the different responses within the same tumour and therefore increase the likely-hood to avoid relapse and thus increase overall survival. As described in this review, we can predict that combining VEGF therapy/TKI with immune system or kinases targeted therapies to cover a wide range of cancer weapons will optimize the best-personalized therapy for every patient. The targets used so far to decrease pathological angiogenesis are represented in Fig. **2**. Furthermore therapies targeting other angiogenic cytokines receptors such as CXCR are developing as these pathways are commonly used by RCC in particular to promote tumour growth. These novel forms of giving therapy are essential first of all for the patients and to benefit health costs as more efficient treatments will be given to every patient.

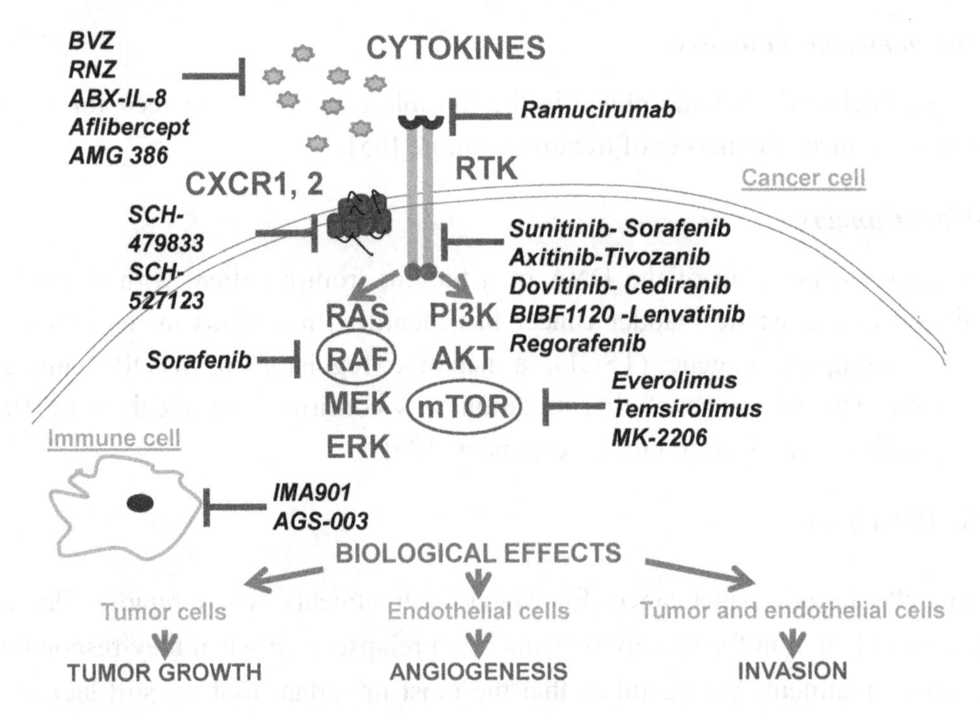

Figure 2: Targets involved in drug development to trigger down pathological angiogenesis. On cancer cells, compounds can act at several levels: targeting soluble cytokines preventing their bonding, blocking the binding site on the receptors on these cytokines; blocking the auto-phosphorylation of the receptors, or blocking directly mTOR, a kinase involved in numerous of biological effects such as proliferation, angiogenesis, and protein synthesis in particular. On immune cells, IMA901 and AGS-003 can activate RCC associated T cells.

ACKNOWLEDGEMENTS

This work was supported by the Institut National pour la Santé et la Recherche Médicale (INSERM) and the Centre National de la Recherche Scientifique (CNRS). This work was supported by the National Institute of Cancer (INCA), the Association for Cancer Research (ARC), the Fondation de France, the Fondation pour la Recherche Médicale (FRM), and The « Association pour la Recherche sur les Tumeurs du Rein (ARTuR) ». We gratefully thank Dr Scott Parks for editorial corrections.

CONFLICT OF INTEREST

The authors confirm that this chapter contents have no conflict of interest.

DISCLOSURE

Declared none.

REFERENCES

[1] Folkman J. Tumor angiogenesis: therapeutic implications. N Engl J Med. 1971 Nov 18;285(21):1182-6.

[2] Seizinger BR, Rouleau GA, Ozelius LJ, Lane AH, Farmer GE, Lamiell JM, *et al.* Von Hippel-Lindau disease maps to the region of chromosome 3 associated with renal cell carcinoma. Nature. 1988 Mar 17;332(6161):268-9.

[3] Vandercappellen J, Van Damme J, Struyf S. The role of CXC chemokines and their receptors in cancer. Cancer Lett. 2008 Aug 28;267(2):226-44.

[4] Harper SJ, Bates DO. VEGF-A splicing: the key to anti-angiogenic therapeutics? Nat Rev Cancer. 2008 Nov;8(11):880-7.

[5] Folkman J, Merler E, Abernathy C, William G. Isolation of a tumor factor responsible for angiogenesis. J Exp Med. 1971;133:275-88.

[6] Motzer RJ, Hutson TE, Tomczak P, Michaelson MD, Bukowski RM, Rixe O, *et al.* Sunitinib versus interferon alfa in metastatic renal-cell carcinoma. N Engl J Med. 2007 Jan 11;356(2):115-24.

[7] Escudier B, Pluzanska A, Koralewski P, Ravaud A, Bracarda S, Szczylik C, *et al.* Bevacizumab plus interferon alfa-2a for treatment of metastatic renal cell carcinoma: a randomised, double-blind phase III trial. Lancet. 2007 Dec 22;370(9605):2103-11.

[8] Escudier B, Eisen T, Stadler WM, Szczylik C, Oudard S, Siebels M, *et al.* Sorafenib in advanced clear-cell renal-cell carcinoma. N Engl J Med. 2007 Jan 11;356(2):125-34.

[9] Rini BI, Halabi S, Rosenberg JE, Stadler WM, Vaena DA, Ou SS, *et al.* Bevacizumab plus interferon alfa compared with interferon alfa monotherapy in patients with metastatic renal cell carcinoma: CALGB 90206. J Clin Oncol. 2008 Nov 20;26(33):5422-8.

[10] Rini BI, Michaelson MD, Rosenberg JE, Bukowski RM, Sosman JA, Stadler WM, *et al.* Antitumor activity and biomarker analysis of sunitinib in patients with bevacizumab-refractory metastatic renal cell carcinoma. J Clin Oncol. 2008 Aug 1;26(22):3743-8.

[11] Sternberg CN, Davis ID, Mardiak J, Szczylik C, Lee E, Wagstaff J, *et al.* Pazopanib in locally advanced or metastatic renal cell carcinoma: results of a randomized phase III trial. J Clin Oncol. 2010 Feb 20;28(6):1061-8.

[12] Hurwitz H, Fehrenbacher L, Novotny W, Cartwright T, Hainsworth J, Heim W, *et al.* Bevacizumab plus irinotecan, fluorouracil, and leucovorin for metastatic colorectal cancer. N Engl J Med. 2004 Jun 3;350(23):2335-42.

[13] Sandler A, Gray R, Perry MC, Brahmer J, Schiller JH, Dowlati A, *et al.* Paclitaxel-carboplatin alone or with bevacizumab for non-small-cell lung cancer. N Engl J Med. 2006 Dec 14;355(24):2542-50.

[14] Miller K, Wang M, Gralow J, Dickler M, Cobleigh M, Perez EA, *et al.* Paclitaxel plus bevacizumab versus paclitaxel alone for metastatic breast cancer. N Engl J Med. 2007 Dec 27;357(26):2666-76.

[15] Escudier B, Bellmunt J, Negrier S, Bajetta E, Melichar B, Bracarda S, *et al.* Phase III trial of bevacizumab plus interferon alfa-2a in patients with metastatic renal cell carcinoma (AVOREN): final analysis of overall survival. J Clin Oncol. 2010 May 1;28(13):2144-50.

[16] Ciulla TA, Rosenfeld PJ. Antivascular endothelial growth factor therapy for neovascular age-related macular degeneration. Curr Opin Ophthalmol. 2009 May;20(3):158-65.

[17] Sato T, Emi K, Ikeda T, Bando H, Sato S, Morita S, *et al.* Severe intraocular inflammation after intravitreal injection of bevacizumab. Ophthalmology. 2010 Mar;117(3):512-6, 6 e1-2.

[18] Saint-Geniez M, Maharaj AS, Walshe TE, Tucker BA, Sekiyama E, Kurihara T, *et al.* Endogenous VEGF is required for visual function: evidence for a survival role on muller cells and photoreceptors. PLoS ONE. 2008;3(11):e3554.

[19] Eisen T, Sternberg CN, Robert C, Mulders P, Pyle L, Zbinden S, *et al.* Targeted therapies for renal cell carcinoma: review of adverse event management strategies. J Natl Cancer Inst. 2012 Jan 18;104(2):93-113.

[20] Escudier B, Szczylik C, Porta C, Gore M. Treatment selection in metastatic renal cell carcinoma: expert consensus. Nat Rev Clin Oncol. 2012 Jun;9(6):327-37.

[21] Negrier S, Raymond E. Antiangiogenic treatments and mechanisms of action in renal cell carcinoma. Invest New Drugs. 2012 Aug;30(4):1791-801.

[22] Hauschild A, Agarwala SS, Trefzer U, Hogg D, Robert C, Hersey P, *et al.* Results of a phase III, randomized, placebo-controlled study of sorafenib in combination with carboplatin and paclitaxel as second-line treatment in patients with unresectable stage III or stage IV melanoma. J Clin Oncol. 2009 Jun 10;27(17):2823-30.

[23] Hammers HJ, Verheul HM, Salumbides B, Sharma R, Rudek M, Jaspers J, *et al.* Reversible epithelial to mesenchymal transition and acquired resistance to sunitinib in patients with renal cell carcinoma: evidence from a xenograft study. Mol Cancer Ther. 2010 Jun;9(6):1525-35.

[24] Rini BI, Escudier B, Tomczak P, Kaprin A, Szczylik C, Hutson TE, *et al.* Comparative effectiveness of axitinib versus sorafenib in advanced renal cell carcinoma (AXIS): a randomised phase 3 trial. Lancet. 2011 Dec 3;378(9807):1931-9.

[25] van den Eertwegh AJ, Karakiewicz P, Bavbek S, Rha SY, Bracarda S, Bahl A, *et al.* Safety of Everolimus by Treatment Duration in Patients With Advanced Renal Cell Cancer in an Expanded Access Program. Urology. 2013 Jan;81(1):143-9.

[26] Kikuno N, Kennoki T, Fukuda H, Matsumoto Y, Tsunoyama K, Ban S, *et al.* Brain metastasis in a patient with a sarcomatoid variant RCC with well-controlled extracerebral metastases by temsirolimus. Anticancer Res. 2012 Aug;32(8):3443-7.

[27] De Luca A, Normanno N. Tivozanib, a pan-VEGFR tyrosine kinase inhibitor for the potential treatment of solid tumors. IDrugs. 2010 Sep;13(9):636-45.

[28] Walter S, Weinschenk T, Stenzl A, Zdrojowy R, Pluzanska A, Szczylik C, *et al.* Multipeptide immune response to cancer vaccine IMA901 after single-dose cyclophosphamide associates with longer patient survival. Nat Med. 2012 Jul 29.

[29] Figlin R, Sternberg C, Wood CG. Novel agents and approaches for advanced renal cell carcinoma. J Urol. 2012 Sep;188(3):707-15.

[30] Wedge SR, Kendrew J, Hennequin LF, Valentine PJ, Barry ST, Brave SR, *et al.* AZD2171: a highly potent, orally bioavailable, vascular endothelial growth factor receptor-2 tyrosine kinase inhibitor for the treatment of cancer. Cancer Res. 2005 May 15;65(10):4389-400.

[31] Batchelor TT, Sorensen AG, di Tomaso E, Zhang WT, Duda DG, Cohen KS, *et al.* AZD2171, a pan-VEGF receptor tyrosine kinase inhibitor, normalizes tumor vasculature and alleviates edema in glioblastoma patients. Cancer Cell. 2007 Jan;11(1):83-95.

[32] Boss DS, Glen H, Beijnen JH, Keesen M, Morrison R, Tait B, *et al.* A phase I study of E7080, a multitargeted tyrosine kinase inhibitor, in patients with advanced solid tumours. Br J Cancer. 2012 May 8;106(10):1598-604.

[33] Eisen T, Joensuu H, Nathan PD, Harper PG, Wojtukiewicz MZ, Nicholson S, *et al.* Regorafenib for patients with previously untreated metastatic or unresectable renal-cell carcinoma: a single-group phase 2 trial. Lancet Oncol. 2012 Oct;13(10):1055-62.

[34] Demetri GD, Reichardt P, Kang YK, Blay JY, Rutkowski P, Gelderblom H, *et al.* Efficacy and safety of regorafenib for advanced gastrointestinal stromal tumours after failure of imatinib and sunitinib (GRID): an international, multicentre, randomised, placebo-controlled, phase 3 trial. Lancet. 2012 Nov 21.

[35] Spratlin JL, Cohen RB, Eadens M, Gore L, Camidge DR, Diab S, *et al.* Phase I pharmacologic and biologic study of ramucirumab (IMC-1121B), a fully human immunoglobulin G1 monoclonal antibody targeting the vascular endothelial growth factor receptor-2. J Clin Oncol. 2010 Feb 10;28(5):780-7.

[36] Verheul HM, Hammers H, van Erp K, Wei Y, Sanni T, Salumbides B, *et al.* Vascular endothelial growth factor trap blocks tumor growth, metastasis formation, and vascular leakage in an orthotopic murine renal cell cancer model. Clin Cancer Res. 2007 Jul 15;13(14):4201-8.

[37] Van Cutsem E, Tabernero J, Lakomy R, Prenen H, Prausova J, Macarulla T, *et al.* Addition of aflibercept to fluorouracil, leucovorin, and irinotecan improves survival in a phase III randomized trial in patients with metastatic colorectal cancer previously treated with an oxaliplatin-based regimen. J Clin Oncol. 2012 Oct 1;30(28):3499-506.

[38] Ramlau R, Gorbunova V, Ciuleanu TE, Novello S, Ozguroglu M, Goksel T, *et al.* Aflibercept and Docetaxel versus Docetaxel alone after platinum failure in patients with advanced or metastatic non-small-cell lung cancer: a randomized, controlled phase III trial. J Clin Oncol. 2012 Oct 10;30(29):3640-7.

[39] Stewart MW, Grippon S, Kirkpatrick P. Aflibercept. Nat Rev Drug Discov. 2012 Apr;11(4):269-70.

[40] Brahmer JR, Drake CG, Wollner I, Powderly JD, Picus J, Sharfman WH, *et al.* Phase I study of single-agent anti-programmed death-1 (MDX-1106) in refractory solid tumors: safety, clinical activity, pharmacodynamics, and immunologic correlates. J Clin Oncol. 2010 Jul 1;28(19):3167-75.

[41] Thompson RH, Dong H, Lohse CM, Leibovich BC, Blute ML, Cheville JC, *et al.* PD-1 is expressed by tumor-infiltrating immune cells and is associated with poor outcome for patients with renal cell carcinoma. Clin Cancer Res. 2007 Mar 15;13(6):1757-61.

[42] Grabinski N, Ewald F, Hofmann BT, Staufer K, Schumacher U, Nashan B, *et al.* Combined targeting of AKT and mTOR synergistically inhibits proliferation of hepatocellular carcinoma cells. Mol Cancer. 2012 Nov 20;11(1):85.

[43] Rini B, Szczylik C, Tannir NM, Koralewski P, Tomczak P, Deptala A, *et al.* AMG 386 in combination with sorafenib in patients with metastatic clear cell carcinoma of the kidney : A randomized, double-blind, placebo-controlled, phase 2 study. Cancer. 2012 Dec 15;118(24):6152-61.

[44] Xie K. Interleukin-8 and human cancer biology. Cytokine Growth Factor Rev. 2001 Dec;12(4):375-91.

[45] Bar-Eli M. Role of interleukin-8 in tumor growth and metastasis of human melanoma. Pathobiology. 1999;67(1):12-8.

[46] Mian BM, Dinney CP, Bermejo CE, Sweeney P, Tellez C, Yang XD, *et al.* Fully human anti-interleukin 8 antibody inhibits tumor growth in orthotopic bladder cancer xenografts via down-regulation of matrix metalloproteases and nuclear factor-kappaB. Clin Cancer Res. 2003 Aug 1;9(8):3167-75.

[47] Huang S, Mills L, Mian B, Tellez C, McCarty M, Yang XD, *et al.* Fully humanized neutralizing antibodies to interleukin-8 (ABX-IL8) inhibit angiogenesis, tumor growth, and metastasis of human melanoma. Am J Pathol. 2002 Jul;161(1):125-34.

[48] Singh S, Sadanandam A, Nannuru KC, Varney ML, Mayer-Ezell R, Bond R, *et al.* Small-molecule antagonists for CXCR2 and CXCR1 inhibit human melanoma growth by decreasing tumor cell proliferation, survival, and angiogenesis. Clin Cancer Res. 2009 Apr 1;15(7):2380-6.

[49] Bates DO, Catalano PJ, Symonds KE, Varey AH, Ramani P, O'Dwyer PJ, *et al.* Association between VEGF splice isoforms and progression-free survival in metastatic colorectal cancer patients treated with bevacizumab. Clin Cancer Res. 2012 Nov 15;18(22):6384-91.

[50] Hurwitz HI, Douglas PS, Middleton JP, Sledge GW, Johnson DH, Reardon DA, *et al.* Analysis of early hypertension and clinical outcome with bevacizumab: results from seven phase III studies. Oncologist. 2013;18(3):273-80.

[51] Tahover E, Uziely B, Salah A, Temper M, Peretz T, Hubert A. Hypertension as a predictive biomarker in bevacizumab treatment for colorectal cancer patients. Med Oncol. 2013 Mar;30(1):327.

[52] Pohl A, El-Khoueiry A, Yang D, Zhang W, Lurje G, Ning Y, *et al.* Pharmacogenetic profiling of CD133 is associated with response rate (RR) and progression-free survival (PFS) in patients with metastatic colorectal cancer (mCRC), treated with bevacizumab-based chemotherapy. Pharmacogenomics J. 2013 Apr;13(2):173-80.

[53] Hansen TF, Christensen R, Andersen RF, Garm Spindler KL, Johnsson A, Jakobsen A. The predictive value of single nucleotide polymorphisms in the VEGF system to the efficacy of first-line treatment with bevacizumab plus chemotherapy in patients with metastatic colorectal cancer: results from the Nordic ACT trial. Int J Colorectal Dis. 2012 Jun;27(6):715-20.

[54] Budai B, Komlosi V, Adleff V, Pap E, Reti A, Nagy T, *et al.* Impact of SHMT1 polymorphism on the clinical outcome of patients with metastatic colorectal cancer treated with first-line FOLFIRI+bevacizumab. Pharmacogenet Genomics. 2012 Jan;22(1):69-72.

[55] Ronzoni M, Manzoni M, Mariucci S, Loupakis F, Brugnatelli S, Bencardino K, *et al.* Circulating endothelial cells and endothelial progenitors as predictive markers of clinical response to bevacizumab-based first-line treatment in advanced colorectal cancer patients. Ann Oncol. 2010 Dec;21(12):2382-9.

[56] Botta C, Barbieri V, Ciliberto D, Rossi A, Rocco D, Addeo R, *et al.* Systemic inflammatory status at baseline predicts bevacizumab benefit in advanced non-small cell lung cancer patients. Cancer Biol Ther. 2013 Jun 1;14(6):469-75.

[57] Evans T. Utility of hypertension as a surrogate marker for efficacy of antiangiogenic therapy in NSCLC. Anticancer Res. 2012 Nov;32(11):4629-38.

[58] Gianni L, Romieu GH, Lichinitser M, Serrano SV, Mansutti M, Pivot X, *et al.* AVEREL: a randomized phase III Trial evaluating bevacizumab in combination with docetaxel and trastuzumab as first-line therapy for HER2-positive locally recurrent/metastatic breast cancer. J Clin Oncol. 2013 May 10;31(14):1719-25.

[59] Lambrechts D, Claes B, Delmar P, Reumers J, Mazzone M, Yesilyurt BT, *et al.* VEGF pathway genetic variants as biomarkers of treatment outcome with bevacizumab: an analysis of data from the AViTA and AVOREN randomised trials. Lancet Oncol. 2012 Jul;13(7):724-33.

[60] Scartozzi M, Bianconi M, Faloppi L, Loretelli C, Bittoni A, Del Prete M, *et al.* VEGF and VEGFR polymorphisms affect clinical outcome in advanced renal cell carcinoma patients receiving first-line sunitinib. Br J Cancer. 2013 Mar 19;108(5):1126-32.

[61] Scartozzi M, Giampieri R, Maccaroni E, Del Prete M, Faloppi L, Bianconi M, *et al.* Pre-treatment lactate dehydrogenase levels as predictor of efficacy of first-line bevacizumab-based therapy in metastatic colorectal cancer patients. Br J Cancer. 2012 Feb 28;106(5):799-804.

[62] Van Cutsem E, de Haas S, Kang YK, Ohtsu A, Tebbutt NC, Ming Xu J, *et al.* Bevacizumab in combination with chemotherapy as first-line therapy in advanced gastric cancer: a biomarker evaluation from the AVAGAST randomized phase III trial. J Clin Oncol. 2012 Jun 10;30(17):2119-27.

[63] Hattingen E, Bahr O, Rieger J, Blasel S, Steinbach J, Pilatus U. Phospholipid metabolites in recurrent glioblastoma: in vivo markers detect different tumor phenotypes before and under antiangiogenic therapy. PLoS ONE. 2013;8(3):e56439.

[64] Schuster C, Eikesdal HP, Puntervoll H, Geisler J, Geisler S, Heinrich D, *et al.* Clinical efficacy and safety of bevacizumab monotherapy in patients with metastatic melanoma: predictive importance of induced early hypertension. PLoS ONE. 2012;7(6):e38364.

[65] Keyvanjah K, DePrimo SE, Harmon CS, Huang X, Kern KA, Carley W. Soluble KIT correlates with clinical outcome in patients with metastatic breast cancer treated with sunitinib. J Transl Med. 2012;10:165.

[66] Garcia-Donas J, Esteban E, Leandro-Garcia LJ, Castellano DE, del Alba AG, Climent MA, *et al.* Single nucleotide polymorphism associations with response and toxic effects in patients with advanced renal-cell carcinoma treated with first-line sunitinib: a multicentre, observational, prospective study. Lancet Oncol. 2011 Nov;12(12):1143-50.

[67] Garcia-Donas J, Leandro-Garcia LJ, Gonzalez Del Alba A, Morente M, Alemany I, Esteban E, *et al.* Prospective study assessing hypoxia-related proteins as markers for the outcome of treatment with sunitinib in advanced clear-cell renal cell carcinoma. Ann Oncol. 2013 Jun 20.

[68] Gamez-Pozo A, Anton-Aparicio LM, Bayona C, Borrega P, Gallegos Sancho MI, Garcia-Dominguez R, *et al.* MicroRNA expression profiling of peripheral blood samples predicts resistance to first-line sunitinib in advanced renal cell carcinoma patients. Neoplasia. 2012 Dec;14(12):1144-52.

[69] Rossi E, Fassan M, Aieta M, Zilio F, Celadin R, Borin M, *et al.* Dynamic changes of live/apoptotic circulating tumour cells as predictive marker of response to sunitinib in metastatic renal cancer. Br J Cancer. 2012 Oct 9;107(8):1286-94.

[70] Fujita T, Iwamura M, Ishii D, Tabata K, Matsumoto K, Yoshida K, *et al.* C-reactive protein as a prognostic marker for advanced renal cell carcinoma treated with sunitinib. Int J Urol. 2012 Oct;19(10):908-13.

[71] C DA, Portella L, Ottaiano A, Rizzo M, Carteni G, Pignata S, *et al.* High CXCR4 expression correlates with sunitinib poor response in metastatic renal cancer. Curr Cancer Drug Targets. 2012 Jul;12(6):693-702.

[72] Bono P, Rautiola J, Utriainen T, Joensuu H. Hypertension as predictor of sunitinib treatment outcome in metastatic renal cell carcinoma. Acta Oncol. 2011 May;50(4):569-73.

[73] Porta C, Paglino C, De Amici M, Quaglini S, Sacchi L, Imarisio I, *et al.* Predictive value of baseline serum vascular endothelial growth factor and neutrophil gelatinase-associated lipocalin in advanced kidney cancer patients receiving sunitinib. Kidney Int. 2010 May;77(9):809-15.

[74] Perez-Gracia JL, Prior C, Guillen-Grima F, Segura V, Gonzalez A, Panizo A, *et al.* Identification of TNF-alpha and MMP-9 as potential baseline predictive serum markers of sunitinib activity in patients with renal cell carcinoma using a human cytokine array. Br J Cancer. 2009 Dec 1;101(11):1876-83.

[75] Kontovinis LF, Papazisis KT, Touplikioti P, Andreadis C, Mouratidou D, Kortsaris AH. Sunitinib treatment for patients with clear-cell metastatic renal cell carcinoma: clinical outcomes and plasma angiogenesis markers. BMC Cancer. 2009;9:82.

[76] van Cruijsen H, van der Veldt AA, Vroling L, Oosterhoff D, Broxterman HJ, Scheper RJ, *et al.* Sunitinib-induced myeloid lineage redistribution in renal cell cancer patients: CD1c+ dendritic cell frequency predicts progression-free survival. Clin Cancer Res. 2008 Sep 15;14(18):5884-92.

[77] Bauerschlag DO, Hilpert F, Meier W, Rau J, Meinhold-Heerlein I, Maass N, *et al.* Evaluation of potentially predictive markers for anti-angiogenic therapy with sunitinib in recurrent ovarian cancer patients. Transl Oncol. 2013 Jun;6(3):305-10.

[78] Vincenzi B, Santini D, Russo A, Addeo R, Giuliani F, Montella L, *et al.* Early skin toxicity as a predictive factor for tumor control in hepatocellular carcinoma patients treated with sorafenib. Oncologist. 2010;15(1):85-92.

[79] Pena C, Lathia C, Shan M, Escudier B, Bukowski RM. Biomarkers predicting outcome in patients with advanced renal cell carcinoma: Results from sorafenib phase III Treatment Approaches in Renal Cancer Global Evaluation Trial. Clin Cancer Res. 2010 Oct 1;16(19):4853-63.

[80] Hahn OM, Yang C, Medved M, Karczmar G, Kistner E, Karrison T, *et al.* Dynamic contrast-enhanced magnetic resonance imaging pharmacodynamic biomarker study of sorafenib in metastatic renal carcinoma. J Clin Oncol. 2008 Oct 1;26(28):4572-8.

[81] Hagiwara S, Kudo M, Nagai T, Inoue T, Ueshima K, Nishida N, *et al.* Activation of JNK and high expression level of CD133 predict a poor response to sorafenib in hepatocellular carcinoma. Br J Cancer. 2012 Jun 5;106(12):1997-2003.

[82] Miyahara K, Nouso K, Tomoda T, Kobayashi S, Hagihara H, Kuwaki K, *et al.* Predicting the treatment effect of sorafenib using serum angiogenesis markers in patients with hepatocellular carcinoma. J Gastroenterol Hepatol. 2011 Nov;26(11):1604-11.

[83] Caraglia M, Giuberti G, Marra M, Addeo R, Montella L, Murolo M, *et al.* Oxidative stress and ERK1/2 phosphorylation as predictors of outcome in hepatocellular carcinoma patients treated with sorafenib plus octreotide LAR. Cell Death Dis. 2011;2:e150.

[84] Armstrong AJ, George DJ, Halabi S. Serum lactate dehydrogenase predicts for overall survival benefit in patients with metastatic renal cell carcinoma treated with inhibition of mammalian target of rapamycin. J Clin Oncol. 2012 Sep 20;30(27):3402-7.

[85] Serra S, Zheng L, Hassan M, Phan AT, Woodhouse LJ, Yao JC, *et al.* The FGFR4-G388R single-nucleotide polymorphism alters pancreatic neuroendocrine tumor progression and response to mTOR inhibition therapy. Cancer Res. 2012 Nov 15;72(22):5683-91.

[86] Iyer G, Hanrahan AJ, Milowsky MI, Al-Ahmadie H, Scott SN, Janakiraman M, *et al.* Genome sequencing identifies a basis for everolimus sensitivity. Science. 2012 Oct 12;338(6104):221.

CHAPTER 4

Inhibition of Angiogenesis in Cancer Management by Antioxidants: Ascorbate and *P. leucotomos*

Neena Philips, Halyna Siomyk, Hui Jia and Harit Parakandi

Professor of Biology, School of Natural Sciences, Fairleigh Dickinson University, Teaneck, NJ 07666, USA

Abstract: The hallmarks of cancer include cell growth and metastasis, facilitated by angiogenesis and the remodeling of the extracellular matrix (ECM) by vascular endothelial growth factor (VEGF), interleukin-8 (IL-8), transforming growth factor (TGF-β) and matrixmetalloproteinases (MMPs), which are the predominant factors. These factors are secreted by tumors or the stromal cells in the tumor niche. Oxidative stress and inflammation are the primary causes of the pro-angiogenic factors, including VEGF, MMPs, TGF-β, and IL-8 that collectively activate several signal transduction pathways such as MAP kinase and NF-kB to accentuate ECM remodeling, angiogenesis and cancer metastasis.

Ascorbate (Vitamin C) is a major regulator of the ECM and regulates cancer biology. It inhibits the invasiveness of several cancers such as gastric, oral, pulmonary, fibrosarcoma and melanoma. We have reported ascorbate's dose-dependent inverse effects on cancer cell growth and the expression MMPs and TGF-β. An extract from *P. leucotomos* (a fern) in combination with ascorbate simultaneously reduces cancer cell growth as well as the expression of MMP-1 and TGF-β. Further, ascorbate and *P. leucotomos,* independently and in combination, inhibit the expression of VEGF in a dose dependent manner. A combination of ascorbate and *P. leucotomos* would benefit as preventive measure; and well as a supplemental regimen for cancer patients.

Keywords: Ascorbate, interleukin-8 (IL-8), matrixmetalloproteinases (MMPs), *P. leucotomos,* transforming growth factor (TGF-β), Vascular endothelial growth factor (VEGF).

INTRODUCTION

Angiogenesis is integral in cancer pathogenesis; facilitated by increased

***Corresponding author Neena Philips:** Professor of Biology, School of Natural Sciences, Fairleigh Dickinson University Teaneck, NJ 07666, USA; Tel: 201 692 6494; Fax: 201 692 7349;
E-mails: nphilips@fdu.edu; neenaphilips@optonline.net

Atta-ur-Rahman and Muhammad Iqbal Choudhary (Eds)
Copyright © 2014 Bentham Science Publishers Ltd. Published by Elsevier Inc. All rights reserved.
10.1016/B978-0-12-803963-2.50004-1

expression of matrixmetalloproteinases (MMPs), which remodels the extracellular matrix (ECM), and growth factors including vascular endothelial growth factor (VEGF) and transforming growth factor-β (TGF-β). The MMPs, TGF-β, VEGF as well as other cytokines such as interleukin-8 (IL-8) contribute to ECM remodeling, angiogenesis, and tumoroigenesis [1-3]. These tumor promoting factors are at a higher level in cancer tissue than normal tissue, and mediate tumor growth and metastasis [4]. Angiogenesis is associated with oxidative stress and inflammation in tumors [5, 6]. Antioxidative and anti-inflammatory supplements, such as ascorbate and *P. leucotomos*, may alleviate the pathogenesis of cancer.

Oxidative Stress and Inflammation

Oxidative stress and inflammation are the result of the accumulation reactive oxygen species (ROS) due to mitochondrial metabolism, tissue infiltrated leukocytes or NADPH oxidase, which are activated with aging, genetic factors, and exposure to environmental pollutants or ultraviolet (UV) radiation. The major sources of ROS are the mitochondria and NADPH oxidase (Nox family) [7]. The Nox enzymes are associated with ROS production and tumorigenicity in diverse cancer cells [7]. UV radiation and environmental pollutants damage mitochondrial DNA, which leads to senescence and increased accumulation of ROS [8].

The primary cellular ROS include superoxide, hydroxyl radicals, and hydrogen peroxide, The ROS cause direct harm to DNA (8-oxo-dG, pyrimidine dimmers), proteins (carbonyl amino acid derivatives) and lipids (lipid peroxidation, malonaldehyde, lipid inflammatory mediators) [9-13]. ROS induces inflammatory mediators, which are released from leukocytes, damaged tissue, and endothelial lining of blood vessels. The inflammatory mediators include the plasma mediators, bradykinin, plasmin, fibrin; lipid mediators, prostaglandins, leukotrienes and platelet activating factor; and the inflammatory cytokines, interleukin-1 (IL-1), interleukin-6 (IL-6), and tumor necrosis factor- α (TNF-α) [14].

ROS and inflammatory mediators primarily activate mitogen activated protein kinase (MAPK) and the nuclear factor-kappa B (NF-kB) pathways, which activate

AP-1 and NF-kB transcription factors that in turn activate the expression of MMPs, TGF-β and VEGF genes [15, 16]. In addition, ROS and inflammatory mediators stimulate the Akt/Protein kinase B (PKB), and the JAK/STAT (Signal Transduction and Activation of Transcription) pathways that allow for cell proliferation and the suppression of apoptosis [15]. Summatively, the DNA mutations and damage to biomolecules can initiate carcinogenesis; the oxidative stress and activation of signaling pathways/transcription factors such as AP-1 and NF-kB increase cell proliferation and decrease apoptosis to promote carcinogenesis; and the associated ECM remodeling, angiogenesis and metastasis facilitate cancer progression [15, 17] (Fig. **1**).

Angiogenic Factors

The MMPs are also called matrixins or collagenases; and are central in the remodeling of the ECM. The ECM is composed predominantly of fibrous collagen and elastin, linking proteins laminin and fibronectin and space-filling molecules such as glycosaminoglycans. MMPs have been targeted by natural agents and synthetic drugs in curbing cancer angiogenesis and metastasis.

There are three predominant groups of MMPs based on substrate specificity; collagenases (MMP-1) that cleave interstitial (structural) collagens, gelatinases, such as MMP-2 and MMP-9, which degrade basement membrane collagens, and the stromelysins (MMP-3, 10) that degrade basement membrane collagens as well as proteoglycans and matrix glycoproteins; while the other MMP classes that activate pro-MMPs or degrade the basement membrane or elastin include the membrane-type MMPs (MT-MMP), matrilysin, and elastases [18-23]. MMPs have also been classified on the basis of their promoters and the predominant ones contain activator protein-1 (AP-1 site)The AP-1 transcription factor is activated by the mitogen activated protein (MAP) kinase pathway, through the receptor tyrosine kinases [10, 24]. In addition to AP-1, MMP-9 promoter is activated by NF-kB [25]. The active NF-kB also activates nitric oxide synthase, cycloxygenase, and histone acetylase, which aid in ECM remodeling [15, 26-30].

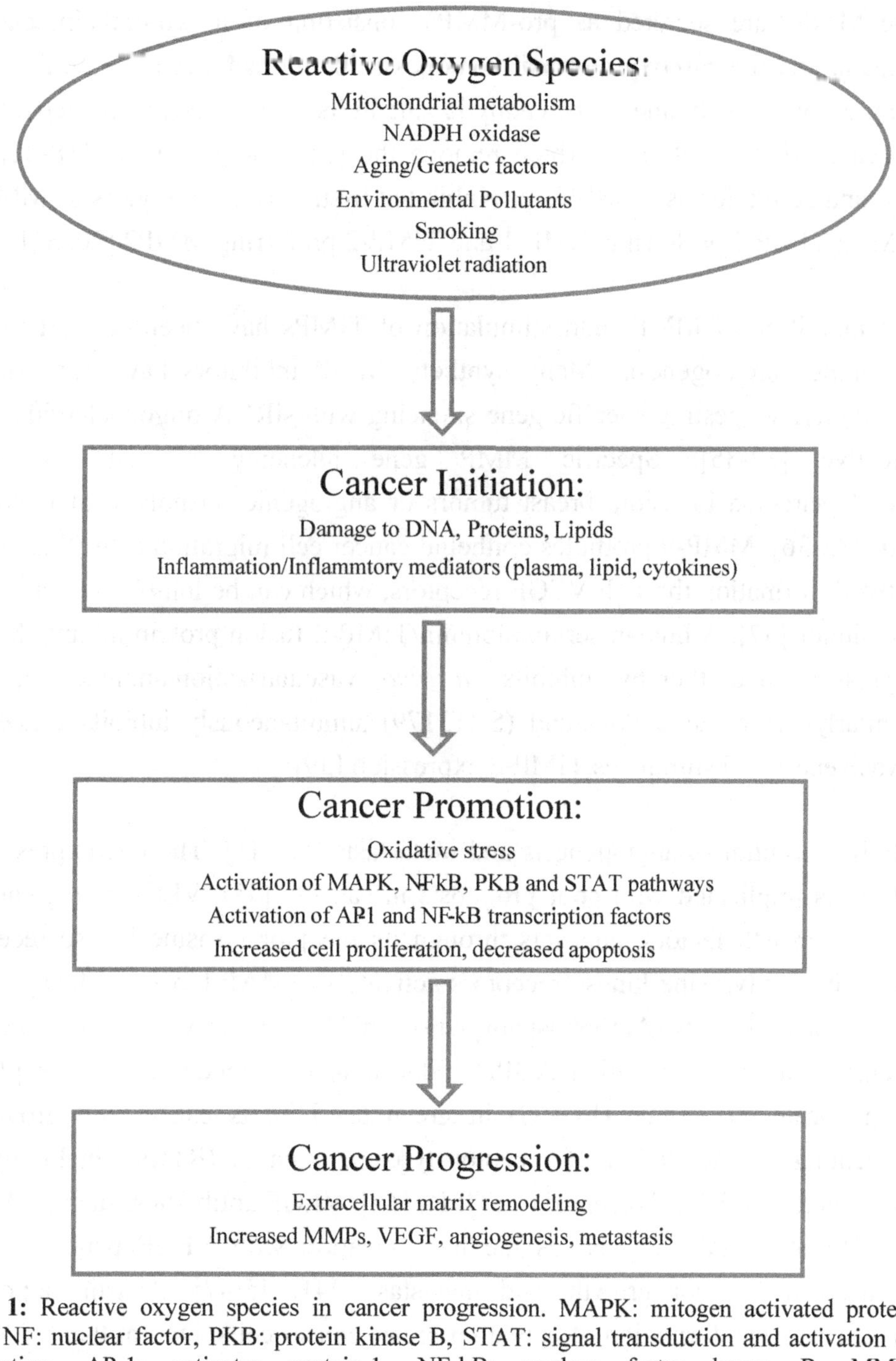

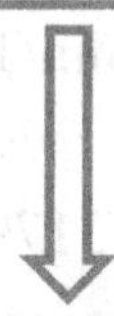

Figure 1: Reactive oxygen species in cancer progression. MAPK: mitogen activated protein kinase, NF: nuclear factor, PKB: protein kinase B, STAT: signal transduction and activation of transcription, AP-1: activator protein-1, NF-kB: nuclear factor kappa B, MMP: matrixmetalloproteinase, VEGF: vascular endothelial growth factor.

The MMPs are secreted as pro-MMPs consisting of a pro-domain, catalytic domain, and the carboxy terminal hemopexin like domain. A cysteine residue in the pro-domain is bound to the catalytic zinc in its inactive form, and activated by proteases such as plasmin which remove the pro-peptide domain [18-21]. The pro- and active forms of MMPs are inhibited by the tissue inhibitors of MMPs or TIMPs; TIMP-1 preferring MMP-1 and TIMP-2 preferring MMP2 [20, 31].

The inhibition of MMPs and stimulation of TIMPs have been active targets in managing carcinogenesis. Many synthetic MMP inhibitors have not exhibited specificity, suggesting specific gene silencing with siRNA oligonucleotides to be effective [32-35]. Specific MMP gene silencing by siRNAs inhibit chondrosarcoma invasion, breast tumors or angiogenic phenotype of melanoma cells [32-36]. MMP-9 promotes epithelial cancer cell migration as well as tubular network formation through VEGF receptors, which can be inhibited by an MMP-9 inhibitor [37]. A human serum albumin/TIMP-2 fusion protein inhibits MMP-2 expression and thereby inhibits *in vivo* vascularization/angiogenesis [38]. Similarly, a synthetic flavanoid (SR13179) simultaneously inhibits cancer cell invasiveness and stimulates TIMP-2 expression [39].

VEGF is central to angiogenesis and metastasis [40, 41]. The overexpression of VEGF is implicated with poor prognosis in cancers [42]. VEGF is regulated by various growth factors, and acts through its receptor tyrosine kinase receptors. The receptor tyrosine kinase receptors activate the MAP kinase pathway, which also induces MMPs. Autophosphorylation of VEGFR as well as angiogeneisis occurs in association with NADPH induced ROS in endothelial cells [7, 43]. Epigallocatechin gallate (EGCG) in green tea inhibits cancer cell growth by preventing the activation of receptor tyrosine kinases (RTKs), including that associated with VEGF receptor (VEGFR) [44]. VEGF antibodies, such as Avastin and BevacizumAb, as well as the immunization with VEGF peptides inhibit angiogenesis, tumor growth, and metastasis [41, 45-47]. Avastin suppresses tubular network formation of epithelial breast cancer cells, but not their migration [37].

TGF-β is associated with the remodeling of the ECM, and thereby angiogenesis. TGF-β stimulates the expression of MMPs in epithelial cells [48-50]. It is expressed in high levels in native tumor cells, and in tumor infiltrating immune cells; and regulates the tumor environment, epithelial-mesenchymal transition, angiogenesis, and tumor invasion [51, 52]

IL-8 is a key angiogenic factor and regulator of tumor immune response [53-55]. Copper enhances the release of IL-8 from endothelial cells, with IL-8 associated with CXCR2 receptor and MAP kinase pathway in the mediation of angiogenesis, and cancer cell survival [38, 56-58]. The transcription factors AP-1 and NF-kB, formed as a result of inflammatory cytokines and oxidative stress, also stimulate IL-8 expression [59, 60]. The chelation of copper reduces expression of IL-8 and inhibits cancer metastasis [53, 55]

Antioxidants: Ascorbate and *P. Leucotomos*

The oxidative stress mediated carcinogenesis can be inhibited by cellular enzymatic and non-enzymatic antioxidants. The primary enzymatic antioxidants are superoxide dismutase (SOD), catalase and glutathione peroxidase [17]. The SOD converts superoxides to hydrogen peroxide, which is converted by catalase to water. The three forms of SOD are cytosolic Cu-SOD, mitochondrial Mn-SOD and extracellular (EC)-SOD [17]. A meta-analysis of SOD and angiogenesis, with regard to psoriasis, indicates the inhibition of inflammation by EC-SOD for its mechanism in inhibiting angiogenesis [61]. The glutathione peroxidase removes peroxides; and inhibits tumor angiogenesis by inhibiting lipooxygenase activity [63]. The non-enzymatic antioxidants include glutathione, ascorbate, α-tocopherol, carotene, and polyphenols such in *P. leucotomos*. Glutathione increases p53 expression and inhibits angiogenesis and oral carcinogenesis [63]. Ascorbate and *P. leucotomos* have shown to have beneficial effects in combination, in our laboratory.

Ascorbate (Vitamin C) has antioxidant and anti-carcinogenic properties. It is chemically synthesized as L-ascorbate through recombinant DNA technology

[64]. Ascorbate scavenges ROS intracellularly, and regenerates antioxidants such as glutathione, vitamin E and β-carotene [65-68].

Ascorbate is essential to the ECM integrity, in particular the synthesis of collagen and elastin. Mutations in glucose transporters, that transport ascorbate into the cell, result in defective collagen and elastin, from the lack of intracellular ascorbate [69]. Ascorbate induces the expression of collagen; and is a co-factor for the hydroxylation of collagen that is essential for collagen stability [70-73]. Increased deposition of the extracellular matrix could encapsulate tumors [74, 75].

The antioxidant activity of ascorbate retards the growth and invasiveness of several cancers [76-81]. It also acts in combination with vitamin K in the inhibition of tumors [82]. Ascorbate reduces inflammation in cancer patients [83]. The lower concentrations of ascorbate significantly inhibit cell viability in several cancers, and induce apoptosis in cancer cells [50, 84]. Further, the growth inhibitory ascorbate doses stimulate the expression of MMPs and TGF-β in cancer cells [50].

P. leucotomos is a tropical fern plant; its extract is rich in polyphenols. *P. leucotomos* exhibits antioxidant properties and inflammatory cytokines; and inhibits direct or cellular damage, inflammatory cytokines, DNA damage, loss of langerhans cells, and angiogenesis [49, 87-89]. *P. leucotomos* directly inhibits the activities of MMPs indicating inhibition of ECM proteolysis [89]. It also inhibits MMPs expression and stimulates TIMPs in fibroblasts, keratinocytes or melanoma cells, suggesting inhibition of skin aging or cancer metastasis [85, 86]. In melanoma cells, *P. leucotomos* inhibits TGF-β expression and thereby MMP-1 [85, 86]. *P. leucotomos* is marketed as Difur® for psoriasis, and as Fernblock® for photoprotection [87, 88].

The combination of *P. leucotomos* a growth inhibitory ascorbate concentration, which stimulates MMPs and TGF-β, simultaneously inhibits cancer cell growth and the expression of MMPs and TGF-β [85, 87, 89].Further, the combination of ascorbate and *P. leucotomos* inhibits the expression of VEGF in melanoma cells (Fig. **2**).

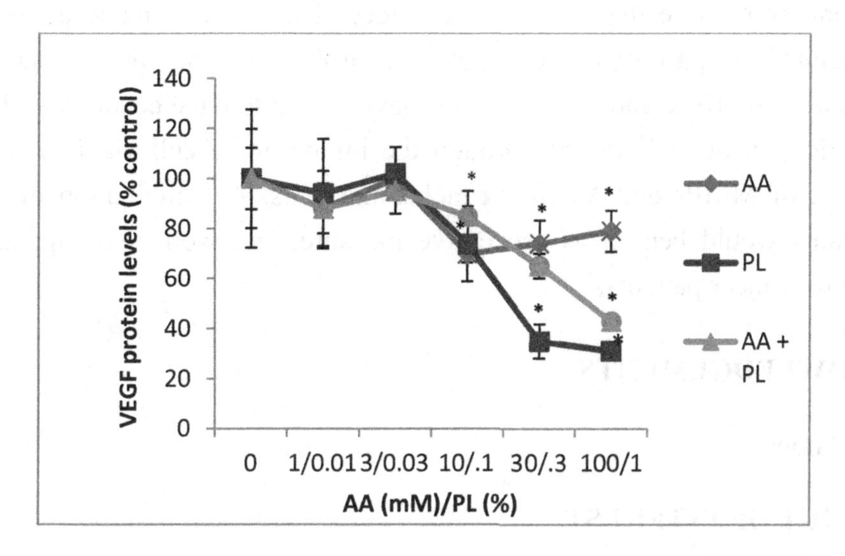

Figure 2: Inhibition of vascular endothelial growth factor (VEGF) expression by ascorbate (AA), *P. leucotomos* (PL), and the combination of ascorbate and *P. leucotomos* (AA+PL) in melanoma cells. Melanoma cells were dosed with 0-100mM ascorbate, in 3-fold dilutions; 0-1% PL, in 3 fold dilutions; or a combination of these concentrations of AA and PL and examined for VEGF expression by ELISA. n:3, error bars:std deviation, *:$p<0.05$ from respective controls.

CONCLUSION

Oxidative stress and inflammation are the result of the accumulation reactive oxygen species (ROS), such as superoxide, hydroxyl radicals, hydrogen peroxide, due to mitochondrial metabolism, NADPH oxidase, aging/genetic factors, environmental pollutants, smoking or ultraviolet (UV) radiation. The ROS support cancer initiation, promotion and progression by causing damage to DNA /biomolecules, and activating MAPK, NF-kB, Akt/Protein kinase B (PKB), and the JAK/STAT pathways that allow for cell proliferation, suppression of apoptosis ECM remodeling, angiogenesis and metastasis. The ECM remodeling and angiogenesis is primarily through the transcription factors activated by ROS mediated signaling, including AP-1 and NF-kB, that increase the expression of MMPs,VEGF, TGF-β, and IL-8. The ROS mediated carcinogenesis can be inhibited by cellular enzymatic and non-enzymatic antioxidants. The primary enzymatic antioxidants are superoxide dismutases, which convert superoxides to hydrogen peroxide, catalase, which convert hydrogen peroxide to water, and glutathione peroxidase that removes peroxides. The non-enzymatic antioxidants include glutathione, ascorbate, α-tocopherol, carotene, and polyphenols such in *P. leucotomos*. Ascorbate and *P. leucotomos* have shown to have beneficial effects in combination, in our laboratory, through the inhibition of cell viability, and the expression of MMPs and VEGF in melanoma cells. A combination of natural antioxidants would benefit as preventive measure; and well as a supplemental regimen for cancer patients.

ACKNOWLEDGEMENTS

Declared none.

CONFLICT OF INTEREST

The authors confirm that this chapter contents have no conflict of interest.

DISCLOSURE

Declared none.

REFERENCES

[1] Liotta LA, Stetler-Stevenson W. Metalloproteinases and malignant conversion: Does correlation imply casuality? Journal of National Cancer Institute 1989; 81:556- 557.

[2] Noël A, Jost M, Maquoi E. Matrixmetalloproteinases at cancer tumor–host interface. Semin. Cell. Dev. Biol 2007; 19:52-60.

[3] Rundhaug JE. Matrix metalloproteinases Angiogenesis and Cancer. Clinical Cancer Research 2003; 9:551- 554.

[4] Razmkhah M, Jaberipour M, Hosseini A, Safaei A, Khalatbari B, Ghaderi A. Expression profile of IL-8 and growth factors in breast cancer cells and adipose-derived stem cells (ASCs) isolated from breast carcinoma. Cellular Immunology 2010; 265:80-85.

[5] Dewing D, Emmett M, Pritchard JR. The Roles of Angiogenesis in Malignant Melanoma: Trends in Basic Science Research over the Last 100 Years. Oncol 2012; PMC3376762.

[6] Philips N, Keller T, Hendrix C, Hamilton S, Arena R, Tuason M, Gonzalez S. Regulation of the extracellular matrix remodeling by lutein in dermal fibroblasts melanoma cells and ultraviolet radiation exposed fibroblasts. Arch Dermatol Res 2007a, 299:373-379.

[7] Ushio-Fukai M, Nakamura Y. Reactive oxygen species and angiogenesis: NADPH oxidase as target for cancer therapy. Cancer letters 2008; 266: 37-52.

[8] Krutmann J. The role of UVA rays in skin aging. Eur J Dermatol 2001; 11:170-171.

[9] Briganti S, Picardo M. Antioxidant activity lipid peroxidation and skin diseases. What's new. J Eur Acad Dermatol Venereol 2003; 17:663-9.

[10] Callaghan TM, Wilhelm KP. A review of ageing and an examination of clinical methods in the assessment of ageing skin. Part I: Cellular and molecular perspectives of skin ageing. Int J Cosmet Sci 2008; 30:313-22.

[11] Melnikova VO, Ananthaswamy HN. Cellular and molecular events leading to the development of skin cancer. Mutation Research 2005; 571:91-106.

[12] Nichols JA, Katiyar SK. Skin photoprotection by natural polyphenols: anti-inflammatory antioxidant and DNA repair mechanisms. Arch Dermatol Res 2010; 302:71-83.

[13] Surjana D, Halliday GM, Damian DL. Role of nicotinamide in DNA damage mutagenesis and DNA repair. J Nucleic Acids 2010; 9:360-365.

[14] Kindt T. J. Goldsby R. A. Osborne B. A. In *Kuby Immunology*. 6th Ed. W. H. Freeman and Company. 2007.

[15] Lodish H, Berk A, Kaiser CA, Krieger M, Scott MP, Bretscher A, Ploegh H, Matsudaira P. In Molecular cell biology. 6th Ed. W.H Freeman and Company. 2008.

[16] Maulik N, Das DK. Redox signaling in vascular angiogenesis. Free Radic Biol Med. *2002;* 33:1047–1060

[17] Valko M, Rhodes CJ, Moncola J, Izakovic M, Mazura M. Free radicals, metals and antioxidants in oxidative-stress induced cancer. Chemico-Biol Interac 2006; 160:1-40.

[18] Bode W, Gomis-Rüth FX, Stöckler W. Astacins serralysins snake venom and matrix metalloproteinases exhibit identical zinc-binding environments (HEXXHXXGXXH and Met-turn) and topologies and should be grouped into a common family the 'metzincins'. FEBS Lett 1993; 331:134-40.

[19] Gomis-Rüth FX, Gohlke U,Betz M, Knäuper V, Murphy G, López-Otín C, Bode W. The helping hand of collagenase-3 (MMP-13): 2.7 A crystal structure of its C-terminal haemopexin-like domain. J Mol Biol 1996; 264:556-66.

[20] Herouy Y. Matrixmetalloproteinases in skin pathology. Int J Mol Med 2001; 7:3-12.

[21] Nagase H, Woessner JF. Matrix Metalloproteinases. J Biol Chem 1999; 274:21491-94.

[22] Van Wart HE, Birkedal-Hansen H. The cysteine switch: a principle of regulation of metalloproteinase activity with potential applicability to the entire matrix metalloproteinase gene family. Proc Natl Acad Sci 1990; 87:5578-82.

[23] Westermarck J, Kahari VM. Regulation of matrix metalloproteinase expression in tumor invasion. FASEB J 1999; 12:781-792.

[24] Bachelor MA, Bowden GT. UVA mediated activation of signaling pathways involved in skin tumor promotion and progression. Semin Cancer Biol 2004; 14:131-8.

[25] Farina AR, Cappabianca L, DeSantis G, Ianni ND, Ruggeri P, Ragone M, Merolle S, Tonissen KF, Gulino A, Mackay AR. Thioredoxin stimulates MMP-9 expression de-regulates the MMP-9/TIMP-1 equilibrium and promotes MMP-9 dependent invasion in human MDA-MB-231 breast cancer cells. FEBS Letters 2011; 585:3328-3336.

[26] Adcock IM, Ito K, Barnes PJ. Histone deacetylation: an important mechanism in inflammatory lung diseases. COPD 2005; 2:445-55.

[27] Fisher GJ, Datta SC, Talwar HS, Wang ZQ, Varani J, Kang S Voorhees JJ. Molecular basis of sun-induced premature skin ageing and retinoid antagonism. Nature 1996; 379:335-9.

[28] Li T, Huang H, Huang B,Huang B, Lu J. Histone acetyltransferase p300 regulates the expression of human pituitary tumor transforming gene (hPTTG). J Genet Genomics 2009; 36:335-42.

[29] Meeran SM, Katiyar SK. Proanthocyanidins inhibit mitogenic and survival-signaling *in vitro* and tumor growth *in vivo*. Front Biosci 2008; 13:887–897.

[30] Mroz RM, Noparlik J, Chyczewska E, Braszko JJ, Holownia A. Molecular basis of chronic inflammation in lung diseases: new therapeutic approach. J Physiol Pharmacol 2007; 58:453-60.

[31] Verstappen J, Von den Hoff JW. Tissue Inhibitors of Metalloproteinases (TIMPs): Their Biological Functions and Involvement in Oral Diseases. Crit. Rev. Oral Biol Med J Dent Res 2006; 85:1074-1084.

[32] Yuan J, Charyl MD, Scully SP. RNAi mediated MMP-1 silencing inhibits human chondrosarcome invasion. Journal of Orthopaedic Research 2005; 23:1467-1474.

[33] Xiaoling J, Dutton CM, Wen-Ning Q, Block JA, Garamszegi N, Scully SP. siRNA Mediated Inhibition of MMP-1 Reduces Invasive Potential of a Human Chondrosoma Cell Line. Journal of Cellular Physiology 2005; 202:723-730

[34] Sateesh K, Lakka SS, Gondi CS, Estes N, Rao JS. RNAi-mediated downregulation of urokinase plasminogen activator receptor and matrix metalloproteinase-9 in human breast cancer cells results in decreased tumor invasion angiogenesis and growth. International Journal of Cancer 2007; 121:2307-2316.

[35] El Mourabit H, Fahem A, Hornebeck W, Bernard P, Bellon G. Silencing of Membrane-Type 1 Metalloprotease (MT1-MMP) by Specific siRNAs Supresses the Angiogenic Phenotype of Dermal Endothelial Cells and Inhibits Type 1 Collagen Invasion by Melanoma Cells. J. Invest. Dermatol 2004; 123:1523-1747.

[36] Overall CM, Kleifeld O. Validating matrix metalloproteinases as drug targets and anti-targets for cancer therapy. Nature Rev. Cancer 2006; 6:227-239.

[37] Karroum A, Mirshahi P, Faussat A, Therwath A, Mirshahi M, Hatmi M. Tubular network formation by adriamycin-resistant MCF-7 breast cancer cells is closely linked to MMP-9 and VEGFR-2/VEGFR-3 over-expressions. European Journal of Pharmacology 2012; 685:1-7.

[38] Lee MS, Jung JI, Kwon SH, Lee SM, Morita K. TIMP-2 Fusion Protein with Human Serum Albumin Potentiates Anti-Angiogenesis-Mediated Inhibition of Tumor Growth by Suppressing MMP-2 Expression. 2012; 7:e35710.

[39] Waleh NS, Murphy BJ, Zaveri NT. Increase in tissue inhibitor of metalloproteinase-2 (TIMP-2) levels and inhibition of MMP-2 activity in a metastatic breast cancer cell line by an anti-invasive small molecule SR13179. Cancer Letters 2010; 289:111-118.

[40] Ali EM, Sheta M, El Mohsen MA. Elevated serum and tissue VEGF associated with poor outcome in breast cancer patients. Alexandria Journal of Medicine 2011; 47:217-224.

[41] Sandler A, Hirsh V, Reck M, von Pawel J, Akerley W, Johnson DH. An evidence-based review of the incidence of CNS bleeding with anti-VEGF therapy in non-small cell lung cancer patients with brain metastases. Lung Cancer 2012; 78:1-7.

[42] Yang L, You S, Kumar V, Zhang C, Cao Y. *In vitro* the behaviors of metastasis with suppression of VEGF in human bone metastatic LNCaP-derivative C4-2B prostate cancer cell line. J Exp Clin Cancer Res 2012; 31:40.

[43] Ushio-Fukai M. Redox signaling in angiogenesis: Role of NADPH oxidase. Cardiovascular research 2006; 71:226-235.

[44] Shimizu M, Shirakami Y, Sakai H, Yasuda Y, Kubota M, Adachi S, Tsurumi H, Hara Y, Moriwaki H. (-)-Epigallocatechin gallate inhibits growth and activation of the VEGF/VEGFR axis in human colorectal cancer cells. Chemico-Biological Interactions 2010; 185:247-252.

[45] Kargi A, Yalcin AD, Erin N, Savas B, Polat HH, Gorczynski RM. IL8 and serum soluble TRAIL levels following anti-VEGF monoclonal antibody treatment in patients with metastatic colon cancer. Clin Lab 2012; 58:501-5.

[46] Morera Y, Bequet-Romero M, Ayala M, Velazco JC, Pérez PP, Alba JS, Ancizar J, Rodríguez M, Cosme K, Gavilondo JV. Immunogenicity and some safety features of a VEGF-based cancer therapeutic vaccine in rats rabbits and non-human primates. Vaccine 2010; 28:3453-3461.

[47] Sandler A, Hirsh V, Reck M, von Pawel J, Akerley W, Johnson DH. An evidence-based review of the incidence of CNS bleeding with anti-VEGF therapy in non-small cell lung cancer patients with brain metastases. Lung Cancer 2012; 78:1-7.

[48] Merryman JI, Neilsen N, Staton DD. Transforming growth factor-beta enhances the ultraviolet-mediated stress response in p53-/- keratinocytes. Int. J. Oncol 1998; 13:781-789.

[49] Philips N. An anti TGF-β increased the expression of transforming growth factor-β matrix metallproteinase-1 and elastin and its effects were antagonized by ultraviolet radiation in epidermal keratinocytes. J. Dermatol. Sci 2003a; 33:177-179.

[50] Philips N, Keller T, Holmes C. Reciprocal effects of ascorbate on cancer cell growth and the expression of matrix metalloproteinases and transforming growth factor-beta. Cancer Letters 2007b; 256:49-55.

[51] Fuxe J, Karlsson MCI. TGF-β-induced epithelial-mesenchymal transition: A link between cancer and inflammation. Seminars in Cancer Biology 2012; 22:455-461.

[52] Juárez P, Guise TA. TGF-β in cancer and bone: Implications for treatment of bone metastases. Bone 2011; 48:23-29.

[53] Brewer GJ, Dick RD, Grover DK, Le Claire V, Tseng M, Wicha M, Pienta K, Redman B.G, Jahan T, Sondak VK, Strawderman M, Le Carpentier G, Merjver SD. Treatment of metastatic cancer with tetrathiomolybdate an anticopper antiangiogenic agent: Phase 1 study. Clin. Cancer Res 2000; 6:1-10.

[54] Gao L, Pan X, Jia J, Liang W, Rao L, Xue H, Zhu Y, Li S, Lv M, Deng W, Chen T, Wei Y, Zhang L. IL-8 –251A/T polymorphism is associated with decreased cancer risk among population-based studies: Evidence from a meta-analysis. European Journal of Cancer 2010; 46:1333-1343.

[55] Pan Q, Kleer CG, van Golen KL, Irani J, Bottema KM, Bias C, De Carvalho M, Mesri EA, Robins DM, Dick RD, Brewer GJ, Merajver SD. Copper deficiency induced by tetrathiomolybdate suppresses tumor growth and angiogenesis. Cancer Res 2002; 62:4854-4859.

[56] Bar-Or D, Thomas GW, Yukl R.L. Copper stimulates the synthesis and release of interleukin-8 in human endothelial cells: a possible early role in systemic inflammatory responses. Shock 2003; 20:154–158.

[57] Hu G.Copper stimulates proliferation of human endothelial cells under culture. J. Biochem 1998; 69:326-335.

[58] Serio D, Doria CL, Pellerito LS, Prudovsky I, Micucci I, Massi D, Landriscina M, Marchionni N, Giulio M, Tarantini F. The release of fibroblast growth factor-1 from melanoma cells requires copper ions and is mediated by phosphatidylinositol 3-kinase/Akt intracellular signaling pathway. Cancer Letters 2008; 267:67-74.

[59] Rahman I. Antioxidant therapeutic advances in COPD. Ther Adv Respir Dis 2008; 2:351-74.

[60] Rahman I, Gilmour PS, Jimenez LA, MacNee W. Oxidative stress and TNF-alpha induce histone acetylation and NF-kappaB/AP-1 activation in alveolar epithelial cells: potential mechanism in gene transcription in lung inflammation. Mol Cell Biochem 2002; 234-235:239-48.

[61] Lee YS, Cheon IS, Kim BH, Kwon MJ, Lee HW, Kim TY. Meta analysis of superoxide dismutase and angiogenesis. J Invest Dermatol. 2013;133:732-41.

[62] Schneider M, Wortmann M, Mandal PK, Arpornchayanon W, Jannasch K, Alves F, Strieth S, Conrad M, Beck H. Absence of glutathione peroxidase 4 affects tumor angiogenesis through increased 12/15-lipoxygenase activity. Neoplasia. 2010; 12:254-263.

[63] Schwartz JL, Shklar G. Glutathione inhibits experimental oral carcinogenesis, p53 expression, and angiogenesis. Nutr Cancer. 1996; 26:229-236.

[64] Glick R, Pasternack JJ. In Molecular biotechnology: Principles and applications of recombinant DNA.3rd Ed. ASM Press. 2003.

[65] Aguirre R, May JM. Inflammation in the vascular bed: importance of vitamin C.Pharmacol Ther, 2008; 119:96-103.

[66] Darr D, Combs S, Dunston S, Manning T, Pinell S. Topical Vitamin C protects porcine skin from ultraviolet radiation-induced damage. British Journal of Dermatology 1992; 127: 247-253.

[67] Harrison FE, May JM. Vitamin C function in the brain: vital role of the ascorbate transporter SVCT2. Free Radic Biol Med 2009; 46:719-30.

[68] Naidu A. Vitamin C in human health and disease is still a mystery? An overview. Nutr J 2003; 2:7-11.

[69] Segade F. Glucose transporter 10 and arterial tortuosity syndrome: the vitamin C connection. FEBS Lett 2010; 584:2990-4.

[70] Booth B, Uitto J. The effects of ascorbic acid on procollagen production and prolyl-hydorxylase activity. Biochem. Biophysics 1981; 675:117-122.

[71] Davidson J, LuValle P, Zoia O, Quaglino D, Giro M. Ascorbate differentially regulates elastin and collagen biosynthesis in vascular smooth muschle cells and skin fibroblasts by pretranslational mechanisms. Amer Soc Biochem and Mol Biol 1997; 272:345-352.

[72] Greesin J, Murad S, Pinnell S. Ascorbic acid stimulates collagen production without altering intacellular degradation in cultured human skin fibroblasts. Biochem and Biophysics 1985; 866:272-274.

[73] Soucy PA, Werbin J, Heinz W, Hoh JH, Romer LH. Microelastic properties of lung cell-derived extracellular matrix. Acta Biomater 2011; 7:96-105.

[74] Frazier K, Williams S, Kothapalli D, Klapper H, Grotendorst GR. Stimulation of fibroblast cell growth matrix production granulation tissue formation by connective tissue growth factor. J. Invest. Dermatol 1996; 107:404-411.

[75] Head KA. Ascorbic acid in the prevention and treatment of cancer. Altern. Med. Rev 1998; 3:174-186.

[76] Liehr JG, Wheeler WJ. Inhibition of estrogen-induced renal carcinoma in Syrian hamsters by vitamin C. Cancer Res 1983; 43:4638-4642.

[77] Liu J, Nagao N, Kageyama K, Miwa N. Antimetastatic and anti-invasive ability of phospho-ascorbyl palmitated through intracellular ascorbate enrichment and the resultant antioxidant action. Oncol Res 1999; 11:479-487.

[78] Lutsenko SV, Feldman NB, Severin SE. Cytotoxic and antitumor activities of doxorubicin conjugates with the epidermal growth factor and its receptor-binding fragment. J Drug Target 2002; 10:567-71.

[79] Oliveira CP, Kassab P, Lopasso FP, Souza HP, Janiszewski M, Laurindo FR, K. Iriya, Laudanna AA. Protective effect of ascorbic acid in experimental gastric cancer: reduction of oxidative stress. World J. Gasteroenterol 2003; 9:446-448.

[80] Potdar P, Kandarkar S, Sirsat S. Modulation by Vitamin C of tumor incidence and inhibition in oral carcinogenesis. Funct. Dev. Morphol 1992; 3:167-172.

[81] Tsao CS. Inhibiting effect of ascorbic acid on the growth of human mammary tumor xenografts. Am J Clin Nutr 1991; 45:1274S-2780S.

[82] De Loecker W, Janssens J, Bonte J, Taper H.Effects of Sodium Ascorbate Vitamin C and 2-methyl-14-naphthoquinone (vitamin k) treatment on human tumor cell growth in-vivo. II. Synergism with combined chemotherapy action. Anticancer Res 1993; 13:103-106.

[83] Mikirova N, Casciari J, Taylor P, Rogers A. Effect of high-dose intravenous vitamin C on inflammation in cancer patients.J Transl Med 2012; 10:189.

[84] Lin SY, Lai WW, Chou CC, Kuo HM, Li TM, Chung GJ, Yang JH. Sodium ascorbate inhibits growth *via* the induction of cell cycle arrest and apoptosis in human malignant melanoma A375.S2 cells. Melanoma Res 2006; 16:509-519.

[85] Philips N, Dulaj L, Upadhya T. Growth inhibitory mechanism of ascorbate and counteraction of its matrix metalloproteinases-1 and transforming growth factor-beta stimulation by gene silencing or P. leucotomos. AntiCancer Res 2009a; 29:3233-8.

[86] Philips N, Conte J, Chen Y, Natrajan P, Taw M, Keller T, Givant J, Tuason M, Dulaj L, Leonardi D, Gonzalez S. Beneficial regulation of matrixmetalloproteinases and its inhibitors fibrillar collagens and transforming growth factor-β by p. leucotomos directly or in dermal fibroblasts ultraviolet radiated fibroblasts and melanoma cells. Arc. Derm. Res 2009b; 301:487-495.

[87] Gonzalez S, Alcaraz MV, Cuevas J, Perez M, Jaen P, Alvarex M., Villarrubia V. An extract of the fern polypodium leucotomos (Difur®) modulates Th1/Th2 cytokines balance *in vitro* and appears to exhibit anti-angiogenic activities *in vivo*: pathogenic relationships and therapeutic implications. Anticancer Research 2000; 20:1567-1576.

[88] Gombau L, Garcia F, Lahoz A, Fabre M, Roda-Navarro P, Majano P, Alonso-Lebrero JL, Pivel JP, Castell JV, Gomez-Lechon MJ, and Gonzalez S. Polypodium leucotomos extract: Antioxidant activity and disposition. Toxicology *in Vitro* 2006; 20:464-471.

[89] Philips N. An anti TGF-β increased the expression of transforming growth factor-β matrix metallproteinase-1 and elastin and its effects were antagonized by ultraviolet radiation in epidermal keratinocytes. J. Dermatol. Sci 2003; 33:177-179.

CHAPTER 5

Development of Novel Anti-Cancer Strategies Based on Angiogenesis Inhibition

Rajiv P. Gude*, Prachi Patil, Mohammad Zahid Kamran$ and Peeyush N. Goel$

Gude Lab, Advanced Centre for Treatment Research and Education in Cancer (ACTREC), Tata Memorial Centre, Kharghar, Navi-Mumbai, India

Abstract: Angiogenesis is the formation of new blood vessels from the pre-existing ones. It forms the core of the metastatic cascade, since the tumor cells needs nutrition for their growth at distant sites. The work by Folkman *et al.*, has led to a significant understanding of the angiogenesis process and development of novel anti-cancer strategies based on angiogenesis inhibition. In the present chapter we have discussed various pro-angiogenic and anti-angiogenic molecules that are involved in angiogenesis. Further, we have tried to summarise the various molecules both natural and synthetic that inhibit angiogenesis and are being used as novel strategies for the treatment of cancer.

Keywords: Angiogenesis, angiogenic switch, cancer, endothelial cells, growth factors, hypoxia, integrins, metastasis, natural inhibitors, signaling pathway, STAT-3, synthetic inhibitors, targeting, VEGF.

INTRODUCTION

Angiogenesis, a multistep process of sprouting new blood vessels from pre-existing ones involves various steps such as endothelial cell proliferation, invasion of basement membrane and blood vessel formation [1]. It is essential for patho-physiological processes such as wound healing, chronic inflammation and embryogenesis [2]. However, compelling evidence suggests that angiogenesis has a critical role in the progression of cancer [3].

The dynamic process of angiogenesis is tightly regulated by a large number of pro-angiogenic and anti-angiogenic factors [4]. Absence of this equilibrium leads to several devastating diseases such as cancer, that forms the focus of the present

*Corresponding author Rajiv P. Gude:** Gude Lab, Advanced Centre for Treatment Research and Education in Cancer (ACTREC), Tata Memorial Centre, Kharghar, Navi-Mumbai, India; Tel/Fax: +919869243267; E-mails: rgude@actrec.gov.in; rajivgude@yahoo.in

$Equal Contribution.

Atta-ur-Rahman and Muhammad Iqbal Choudhary (Eds)
Copyright © 2014 Bentham Science Publishers Ltd. Published by Elsevier Inc. All rights reserved.
10.1016/B978-0-12-803963-2.50005-3

study [5]. In this regard, angiogenesis has been a central core area for current investigation.

Several lines of evidence indicate that a cell becomes cancerous due to various genetic changes and cellular transformations [6-8]. These transformed cells give rise to smaller colonies of tumor cells and eventually grow until the supply for nutrients and oxygen becomes limited. As the tumor grows it needs additional vasculature to sustain continued growth, in order to receive adequate supply of metabolites, oxygen and for removal of waste products [5]. Thus, tumor either remains in dormant state and ultimately undergoes destruction by immune system or gets triggered by angiogenic factors, a process called as 'angiogenic switch'. Angiogenic switch refers to the shift in equilibrium between pro-angiogenic and anti-angiogenic factors [5]. Under normal physiological conditions there is presence of finely tuned equilibria between pro-angiogenic and anti-angiogenic factors (Fig. **1a**).

However, this equilibrium is disturbed under malignant conditions (Fig. **1b**). Continuous expression or up-regulation of various pro-angiogenic factors such as VEGF [9], bFGF [10], angiopoietin [11] have been reported in tumor cells and their microenvironment (Fig. **1b**). As a result, blood vessels grow around tumor cells that help in maintaining their continued growth, proliferation and metastasis formation. In contrast to this, compelling evidence support the notion that targeting tumor angiogenesis by small molecule inhibitors or drugs not only inhibits tumor growth but also tumor metastasis [12].

The process of angiogenesis is also regulated at both the genetic and epigenetic levels. These include genetic mutations, single nucleotide polymorphism or SNPs as well as epigenetic modifications such as DNA methylation and the altered histones [13, 14]. Mutation in genes involved in TGFß signalling pathway, such as activin receptor-like kinase (ALK1) and the endoglin have been associated with vascular disorders [15, 16]. SNPs associated with HIF-1α, VEGR and its receptors such as VEGFR2 have also been shown to be correlated with the angiogenic processes [13]. The process of DNA methylation can either result in transcriptional repression (hypermethylation) or genomic instability (hypomethylation). Hypermethylation of anti-angiogenic factors such as placental growth factor, thrombospondin-1 and ADAMTS-8 (protease) have been reported in various cancers of lung, colon and neuroblastoma [17-19]. In addition there are several reports that indicate hypomethylation occurs in pro-angiogenic factors like VEGF and its associated receptors VEGFR1/VEGFR2 in preeclampsia, leukemias [20, 21]. Histone modifying enzymes such as HDAC7 and SIRT1 have also shown to play an important role in regulating angiogenesis [22-24]. In light of these facts, both genetic and epigenetic alterations drive the process of angiogenesis.

Hence, targeting tumor angiogenesis may be used as a good therapeutic strategy for preventing tumour growth and metastasis.

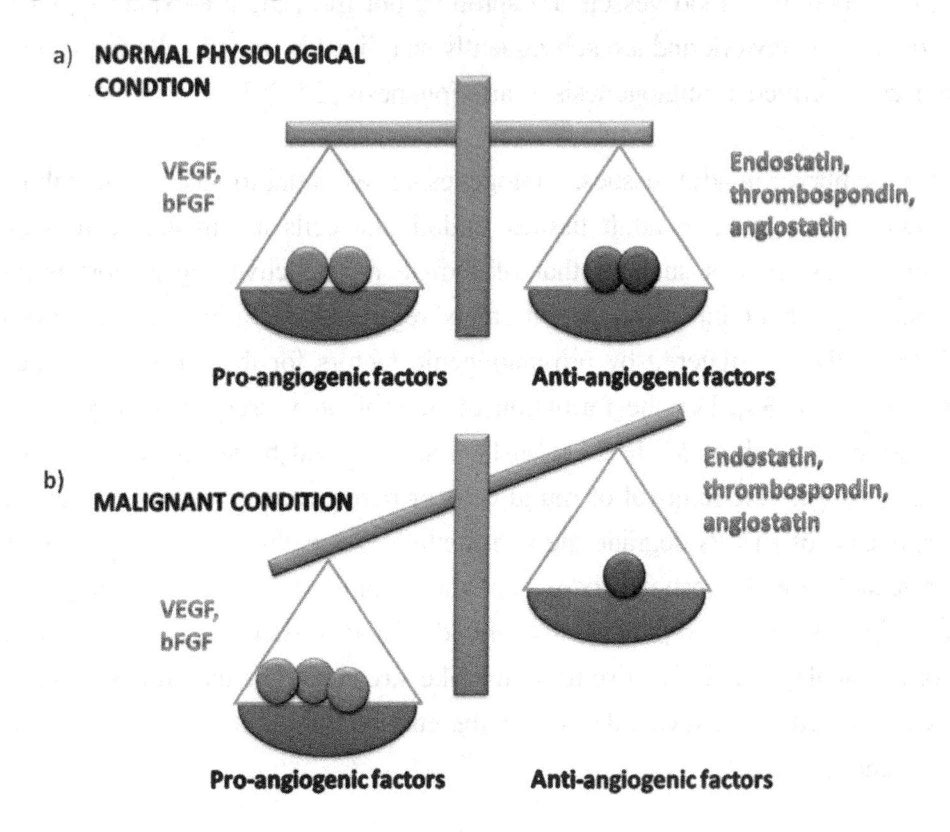

Figure 1: Balance between pro-angiogenic and anti-angiogenic factors: a) Under normal physiological conditions, there is an equilibrium maintained between pro-angiogenic (*e.g.* VEGF, bFGF) and anti-angiogenic factors (*e.g.* endostatin, thrombospondin, angiostatin). In contrast to this, b) under malignant conditions, there is up-regulation of pro-angiogenic factors which causes disruption of the balance leading to angiogenesis, and ultimately tumor growth.

PHYSIOLOGICAL ANGIOGENESIS

Angiogenesis is an essential process in human embryos and adult tissues for the formation and sprouting of new blood vessels [25, 26]. In embryos, vascular system is necessary for the development and formation of new organs [27]. Thus due to the need of oxygen and nutrient supply for the growing embryo, vascular system is developed [28]. In the embryo, vascular network is formed *de novo* from endothelial precursors (angioblasts) that assemble into primary capillary plexus whereby angioblasts proliferate and coalesce into primitive network of blood vessels, a process termed as vasculogenesis. After the formation of this primary capillary plexus, the additional blood vessels are sprouted out from the pre-existing ones to further develop a network and are subsequently stabilized by mural cells (pericytes). This process is termed as angiogenesis or arteriogenesis [27, 29].

Apart from embryo, in adult tissues, angiogenesis is essential to meet the metabolic requirements. Generally, in adult tissues, endothelial cells are in quiescent state. However, in few tissues such as that of female reproductive organs, or tissues undergoing growth or injury under extremely regulated conditions, the quiescent endothelial cells are triggered by pro-angiogenic factors for the formation of new blood vessels [29, 30]. For the formation of new blood vessels, primarily matrix metallo-proteases such as MMP-2 or MMP-9 are released by surrounding stromal cells which triggers the removal of mural cells or pericytes from branching vessels. Further, release of MMPs degrade the surrounding extracellular matrix (ECM) and synthesize and remodel fresh matrix, which when supplemented with soluble growth factors, helps migration and proliferation of endothelial cells. These endothelial cells then form monolayer and give rise to a tube-like structure. The maturation of blood vessels is attained by pericytes that cover the endothelium and thus blocks further development [29, 30].

PATHOLOGICAL ANGIOGENESIS

In contrast to the normal angiogenesis, excessive angiogenesis leads to various diseases such as arthritis, uterine bleeding, persistent hyperplastic vitreous syndrome and cancer [31]. However, the main focus of this review is its role in cancer. In tumors, angiogenesis is abruptly regulated by shift in the equilibrium of pro-angiogenic and anti-angiogenic factors. As a result, blood vessels are formed constantly without stopping and thus allow the excess growth of cells (3). Consequently, the blood vessel formation never undergo quiescent phase thus, differentiating it from the normal blood vessel system. Moreover, in contrast to normal blood vessels the structure of the blood vessels formed within tumors is irregular in shape, dilated and prone to hemorrhage [32]. The tumor cells secrete vascular endothelial growth factor (VEGF) in order to mainitain the formation of blood vessels continuously [33]. In response to this, the pericytes detach and pre-existing blood vessels get vasodilated and become more permeable to the recruitment of plasma proteins. These plasma proteins break down the extracellular matrix and allow migration of more endothelial cells towards chemotactic angiogenic stimuli [34].

FACTORS RESPONSIBLE FOR ANGIOGENESIS

Various factors together have role in angiogenesis that includes:

Significance of Endothelial Cells in Tumor Angiogenesis

Endothelial cells have major role in tumor angiogenesis. It is observed that blood vessels are primarily formed by endothelial cells [35]. Endothelial cells contain F-actin and non-muscle myosin filaments called as cytoskeleton cables that contract in response to growth factors [36]. In response to continuously secreted VEGF, endothelial cells present in the existing blood vessels get activated and secrete matrix metalloproteinase (MMP) enzymes to evade the extracellular matrix surrounding them [37, 38]. Further, growth factors released by tumor cell and tumor microenvironment induce formation and reorganization of these endothelial filaments. Reorganized endothelial cells then divide continuously to form a migration column, a rope like sequence to form hollow tubes, to allow blood flows. Endothelial cells adhere to each other to form a lumen while the basement membrane is formed. Finally, the blood vessel sprouts fuse with each other, to form a circulatory system (Fig. **2**) [30].

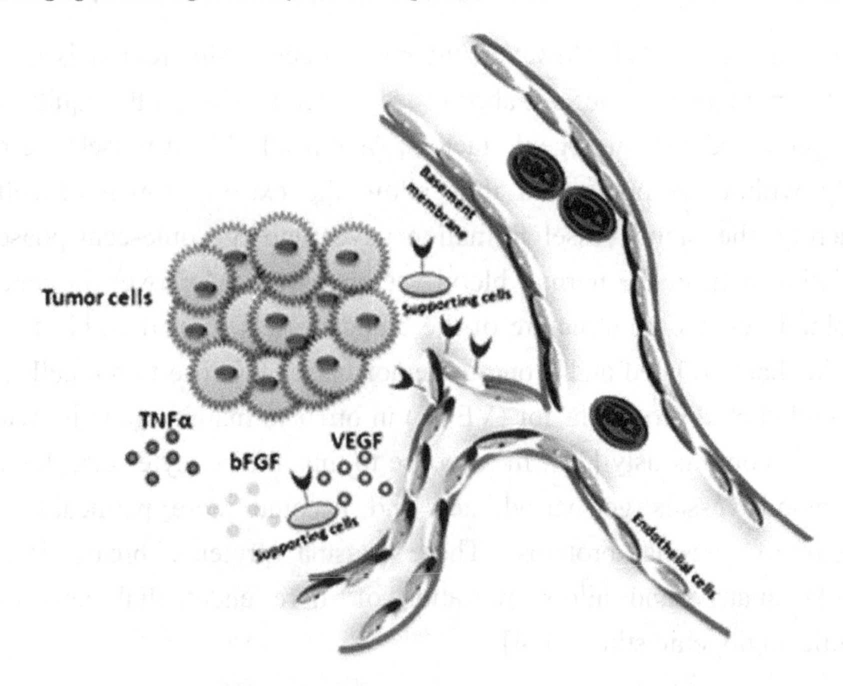

Figure 2: Tumor angiogenesis: In the tumor, due to hypoxic conditions, various cytokines such as TNFα, bFGF and VEGF are released. The specific receptors present on endothelial cells and the supporting cells such as pericytes, dendritic cells and smooth muscle cells are activated, thus causing the displacement of the activated pericytes from the basement membrane. Further, the endothelial cells undergo proliferation and reorganization to give rise to new blood vessels.

The Concept of Vascular Mimicry

Apart from endothelial cells, tumor cells themselves are also capable of forming a vasculature due to their plasticity. The tumor undergoes de-differentiation and adapts the properties of endothelial cells to build up their vascular system. This is known as 'vasculogenic mimicry' or vascular mimicry. The vascular mimicry is accomplished through various signal transduction pathways especially Notch and Wnt signaling pathway. The tumour cells capable of vascular mimicry exhibit higher level plasticity which helps the tumor cells to undertake multipotent phenotype resembling the embryonic stem cells. However, these tumor cells lack the regulatory markers needed for efficient regulation of de-differentiation followed by re-differentiation, thus differing from normal embryonic stem cells. As presence of vascular mimicry is associated with poor prognosis in cancer patients it could be a good target for novel anti-cancer therapies. Reports suggest that targeting endothelial cells using inhibitors such as endostatin, TNP-470, reduces the tumour vasculature built up by endothelial cells, however these inhibitors lack to effectively target the vascular network set up by vascular mimicry, as tumor cells lack the specific receptors.

Role of Matrix Metalloproteases in Angiogenesis

Matrix metalloproteases (MMPs) are a class of zinc dependent endopeptides that degrade extra cellular matrix enzymatically by breaking down the various components of the basement membrane such as, type IV collagen, laminin, fibronectin, collagen, elastin, and proteoglycans [39-41]. MMPs activation occurs *via* various pathways such as through the disruption of cysteine residue and zinc ion, *via* intracellular furin-like serine proteases, or pre-existing other types of MMPs activate the required enzyme [42, 43]. It is observed that on activation, MMPs regulate a number of processes, including cell migration, apoptosis and cell development [44, 45]. It is observed that MMPs are involved in various cell processes and hence, are tightly regulated by MMP inhibitors. One of the example of MMP inhibitor is α-2 macroglobulin which deactivated the activity of MMP by forming a complex with the enzyme [46]. Apart from α-2 macroglobulin, other commonly observed MMP inhibitors are thrombospondin-1 and -2, angiostatin and endostatin. Moreover, tissue inhibitors of metalloproteases (TIMPs) are the most important class of MMP inhibitors during normal conditions [45, 47, 48].

Unregulated continuous expression of MMPs leads to tumor development, angiogenesis and metastasis [44]. In angiogenesis, MMPs are involved in sprouting of blood vessel from the pr-existing ones or by eliminating the barriers by degrading extracellular matrix. It is observed that collagen IV, one of the key molecules of ECM, exposes a cryptic site which allows recruitment of endothelial cells [44]. Moreover, it is also observed that bioavailability of proangiogenic factor VEGF is also increased by MMPs. Surprisingly, MMP also downregulate the formation of new vessel formation by releasing angiogenesis inhibitors such as angiostatin and endostatin, through the cleavage of plasminogen and collagen type XVIII. Nonetheless, more study is required [49, 50].

However, recently, lot of work is being carried out to target MMP production as effective treatment against angiogenesis. The natural as well as synthetic inhibitors are used in order to block MMPs, which are summarized later in the text.

Role of Hypoxia in Angiogenesis

The growing tumor cells become devoid of oxygen and nutrients due to cell multiplication thus, undergo hypoxia. Hypoxic region release various chemoattractants such as chemo-attractants as Endothelin 2 and VEGF that enhances the rate of angiogenesis [51, 52]. Under hypoxic conditions, the mammalian cells induce transcription factors including hypoxia inducible factors (HIFs) to express the genes involved in angiogenesis [53].

The structure of HIF molecules has provided new insights of the role of hypoxia in tumor progression. It is noted that there are three isoforms of HIF transcription factor namely- HIF1, HIF2 and HIF3. HIF1 and HIF2 are predominantly expressed, while role of HIF3 is unknown being recently discovered transcription factor [54]. Structurally, HIF molecules have helix-loop-helix morphology and belong to PER-ARNT (aryl hydrocarbon receptor nuclear translocator)-SIM 2 transcription factors family [55]. These molecules have α and β subunits that dimerize to form a heterodimer in the absence of oxygen. The heterodimer further binds to Hypoxia responsive element (HRE) at its conserved sequence RCGTG [56, 57]. Evidence suggest that HIF1/2 α is oxygen labile while HIF1/2β is constitutively expressed inside the cell nucleus. Furthermore, the transcriptional activity of HIF1α resides in amino-terminal (NAD) region and that of HIF2α in C-terminal transactivation domain (CAD). Hence, a heterodimer formed by these two components forms an active molecule which can activate several aniogenic genes [58-60].

In general, under normoxic conditions, propyl-hydroxylase, an O_2 sensitive enzyme hydroxylates HIF1/2α subunit in the cytoplasm at proline residue of oxygen dependent degradation domain (ODD Domain) [61]. In response to this, a suppressor protein Von Hippel Lindau (VHL), then binds to the HIF1/2α-subunit, causing ubiquitination by ubiquitin proteosomal pathway and subjects HIF1/2α to proteasomal degradation. In contrast to this, under hypoxic conditions, in the absence of propyl-hydroxylase HIF1/2α does not undergo ubiquitination [62, 63]. Due to this, HIF1/2α gets accumulated within the cells and ultimately enters nucleus where it binds to constitutively expressed β subunit of transcription factors HIF1/2. This assembly fuses with P300 and interacts with HRE [54] to up-regulate the transcription of various genes such as VEGF-A, IL-1α, STAT-4 and STAT-6 [64-67]. Up-regulation of these genes depends upon the duration of exposure of cells to hypoxic conditions [53]. When the oxygen limitation is transient hypoxia is termed as acute while when oxygen limitation is permanent it is termed as chronic hypoxia [68] HIF-1 and 2 regulate both similar and unique genes which help in tumor progression.

MOLECULES INVOLVED IN ANGIOGENESIS

It is now known that there are different molecular players in the process of angiogenesis, which regulate vascular growth. The growth factors involved in angiogenesis are- eNOS [69, 70]. Vascular endothelial growth factor (VEGF), basic fibroblast growth factor (bFGF) [71], hepatocyte growth factor (HGF) [72], platelet derived growth factor (PDGF) [73], tumor necrosis factor-α (TNFα), tumor growth factor-β (TGFβ) [74], angiogenin, epidermal growth factor (EGF) [75], and angiopoietins. Amongst these, VEGF, bFGF and PDGF are most potent positive regulatory molecules [76].

Vascular Endothelial Growth Factor (VEGF)

One of the chief molecules involved in angiogenesis is vascular endothelial growth factor (VEGF). VEGF initially was known as vascular permeability factor (VPF) due to its ability to induce vascular leakage [77, 78]. VEGF is secreted by tumor cells as well as tumor microenvironment thus attracting endothelial cells to the site of tumor growth, for angiogenesis. It has been noted that there are six members in the VEGF family, namely- VEGF-A (which resemble the native form of VEGF), VEGF-B, VEGF-C, VEGF-D, placenta growth factor (PIGF) and VEGF identified in parapox orf named VEGF-E. Amongst these, VEGF A is found predominantly in five different isoforms as VEGF-A_{121}, VEGF-A_{145}, VEGF-A_{165}, VEGF-A_{189}, and VEGF-A_{206}. VEGF-A_{165} which contains heparin-binding disulfide linked homodimeric molecule, is found in variety of tumors, thus making it major angiogenic factor [79, 80].

Many factors lead to the upregulation of VEGF transcriptional level. Primarily, it is observed that the mRNA level of VEGF increases in the presence of hypoxia [81]. Not only hypoxia, but several growth factors such as epidermal growth factor (EGF-1], TGF-α/ TGF-β, FGF, keratinocyte growth factor, insulin-like growth factor, and platelet derived growth factor also upregulate the mRNA level of VEGF [82]. Up-regulation of VEGF mRNA leads to over-production of this angiogenic protein. VEGF has been found to bind specifically to 3 different forms of tyrosine kinases receptors (angiogenesis biologia) called receptors VEGFR-1(Flt-1), VEGFR-2(Flk-1) or VEGFR-3(Flt-4). Amonst these, VEGFR-1 and VEGFR-2 are, present primarily on endothelial cells of pre-existing blood vessels, thus making them specific vascular endothelial mitogens. VEGF-A is a specific ligand of these two receptors [80, 83]. Apart from these, VEGF-3 is present on lymphatic endothelial cells. Moreover, VEGFR2 predominantly has a role in angiogenesis. Further, VEGF also acts as potent pro-survival factor for endothelial cells in newly formed vessels [84, 85]. Moreover, VEGF expression shoots up in the presence of hypoxia [86]. It has been shown using knockout studies in mice that in the absence of VEGF cytokine, defects in the vascular system were observed, thus providing evidence that VEGF is necessary for angiogenesis [87, 88].

Basic Fibroblast Growth Factor (bFGF)

Another important growth factor, basic fibroblast growth factor (bFGF), is also regulated by tumor cells during angiogenesis. Currently, FGF has 20 members in the family out of which FGF-1 and FGF-2 are prominent. FGF-1 and FGF-2 bind with high affinity to heparan sulfate proteoglycans (HSPGs), receptors present on the cell surface and inside the extracellular matrix [89]. Along with these, FGF also binds with high affinity to four other tyrosine kinase receptors namely FGFR-1,-2,-3 and -4. FGFR-1 is important during the gastrulation of embryo. It helps in maintaining the vasculature in the embryo, FGFR-2 absence causes decreased vascular tone and low blood pressure, FGFR-3 is necessary for skeletal development whereas, role of FGFR-4 is not yet identified [90].

Platelet Derived Growth Factor (PDGF)

Platelet derived growth factor (PDGF) has a major role in vasculature formation. Similar to VEGF, this growth factor also is produced by neoplastic endothelial cells themselves or the microenvironment of the tumor cells. It binds specifically to tyrosine kinase receptor PDGF receptor [91, 92]. Structurally, PDGF exists in four different isoforms such as PDGF-A, PDGF-B, PDGF-C and PDGF-D. PDGF-A and PDGF-B can exist either as a homodimers (PDGF-AA, PDGF-BB) or heterodimers (PDGF-AB) with PDGF-BB having a predominant role in angiogenesis [92]. PDGF binds specifically to two tyrosine kinase receptors PDGFR- α and PDGFRβ in which PDGFRβ is specific to PDGF-BB with higher affinity and PDGF-AB with lower. On the other hand, PDGFRα binds to AA, BB and AB isoforms [92]. On binding to the receptors, PDGF activate the downstream signal transduction pathway that results in increased blood vessel formation. Data suggests that B-chain of PDGF when transfected, into human melanoma and then transplanting these cells into recipient animal shows increased vascularisation of connective tissue tumors [92]. Furthermore, studies undertaken in bovine aortic endothelial cells suggest the paracrine mechanism of PDGF in enhancing angiogenesis directly on PDGFRβ expressed by endothelial cells *in vitro* [92]. These reports suggest that binding of PDGF to its receptor promotes vascularization.

Number of signal transduction pathways including mitogen activating protein kinase (MAPK), extracellular signal regulated kinase (ERK), phosphoinositol 3-kinase (PI3K)/Akt require the presence of pro-angiogenic factor such as VEGF, bFGF and PDGF for their activation. Activation of these pathways promotes proliferation, migration and differentiation and survival of endothelial cells. Furthermore, other signal transduction pathways such as Ras-MAPK, b-FAR, Ras-GAP and STAT3 are activated and regulated by pro-angiogenic factor VEGF.

SIGNAL TRANSDUCTION PATHWAYS INVOLVED IN ANGIOGENESIS

Mitogen-Activated Protein-Kinase (MAPK) Pathway and Angiogenesis

MAPKs can be described as a family of ubiquitous proline-directed, protein-serine/threonine kinases. There are six distinct groups of MAPKs that have been characterized in mammals, *viz.* extracellular signal regulated kinase (ERK)1/2, ERK3/4, ERK5, ERK 7/8, c-JUN N-terminal kinase/stress-activated protein kinase (JNK/SAPK)1/2/3 and p38 isoforms α/β/γ(ERK6)/δ.

ERK signalling regulates diverse processes including proliferation, differentiation, survival, migration as well as angiogenesis [93]. The ERK pathway has been reported to be deregulated in approximately one-third of all human cancers [94]. ERK1/2 is activated by mitogenic stimuli such as growth factors, cytokines, phorbol esters. Specifically, ERK1/2 is activated upon dual phosphorylation of conserved tyrosine and threonine residues by MAPK kinases (MEK), which in turn are activated by phosphorylation of two serine residues by upstream Raf [95]. This is followed by translocation to the nucleus, where it is responsible for the activation of transcription factors such as activator protein (AP)-1, nuclear factor kappa-light-chain-enhancer of activated B cells (NF-κB), Myc, Bcl-2, cPL2 and the cytoskeletal scaffold paxillin [96].

The JNK and p38 pathways are stress-activated MAPK pathways. They are therefore activated by cellular stress such as UV light, inflammatory cytokines and osmotic and oxidative stress [97]. JNK activation requires the dual phosphorylation on tyrosine and threonine residues catalysed by MEK4 and MEK7 followed by translocation to the nucleus. A major substrate for activated JNK is c-JUN. Phosphorylation of c-JUN results in activation of AP-1. Other transcription factors include activation transcription factor (ATF)2, heat shock factor (HSF)2 and STAT3[98]. c-JUN N-terminal kinase/stress-activated protein kinase (JNK/SAPK) is critical for regulating basic fibroblastic growth factor (bFGF)-mediated angiogenesis in human umbilical vein endothelial cells (HUVECs) [99]. The p38 isoforms are characterized by the presence of conserved Thr-Gly-Tyr (TGY) motif [100]. The phosphorylation of this motif by MEK3 and MEK6 results in translocation of p38 to the nucleus where it is responsible for the activation of several transcription factors controlling apoptosis, cell motility and inflammation [94]. Reports suggest that p38 MAPK also regulates angiogenesis by regulating endothelial cells permeability [101] as well as later phase of angiogenesis *i.e.* tube formation phase [102].

From the point of view of angiogenesis, the ERK pathway plays the major role. Transcription of VEGF, a key regulator of angiogenesis, is induced by the direct phosphorylation of HIF1α by ERK1/2 [103-105]. In addition, sustained ERK activity is responsible for bFGF-induced angiogenesis [106]. Endothelial cell survival and blood vessel sprouting are promoted by ERK activity. A possible mechanism is *via* the suppression of Rho-kinase signalling [107].

STAT3 Signaling Pathway

STAT- Signal Transducer and Activator of Transcription- is a transcription factor present in the cytoplasm of normal cells. Initially, within a normal cell, STATs are in inactive state [108]. It undergoes activation in response to the binding of various cytokines to their specific receptors such as binding of VEGF, bFGF to their specicific receptors including VEGFR [109] and FGFR. On activation, STAT undergoes tyrosine phosphorylation, dimerizes by reciprocal SH2 phosphotyrosine interaction and translocates to the nucleus [110, 111] (Fig. **3**).

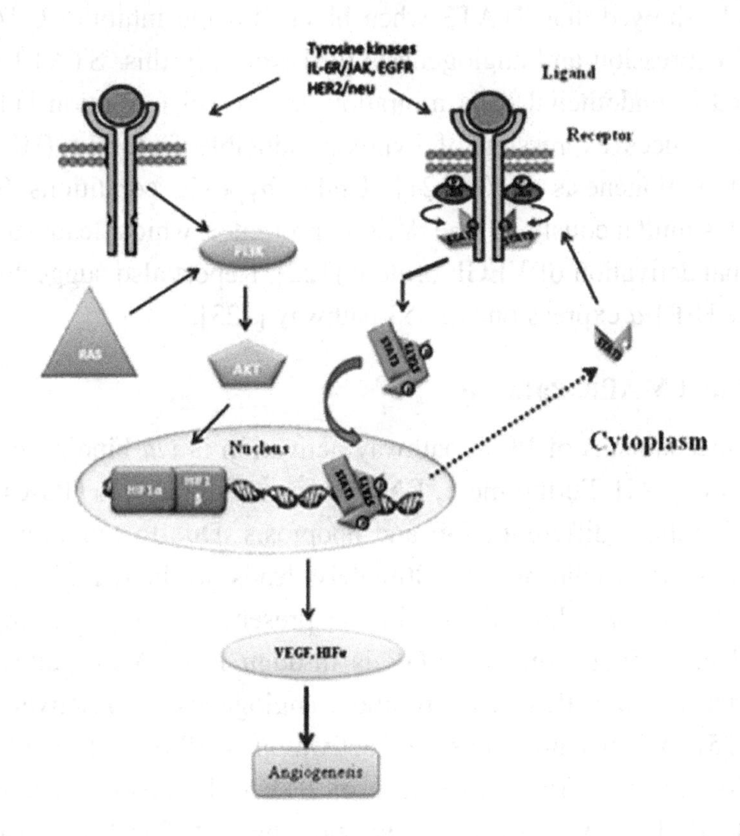

Figure 3: Role of STAT3/PI3K signaling pathway in angiogenesis: After STAT3 activation by various receptor and non-receptor tyrosine kinases such as JAKs or Src, STAT3 translocate to the nucleus and activates the transcription of various angiogenic molecules like VEGF, HIF1α. Parallel to STAT3, PI3K/AKT pathway also has a function in angiogenesis by regulating HIF1α expression.

In nucleus, it acts as a transcription factor and regulates vast number of genes including anti-apoptotic, cell proliferation, invasion survival and angiogenesis [72, 112-114]. Under normal physiological conditions, STAT3 activation is very transient and is tightly regulated by various protein phosphatases. However, persistent STAT3 activation has been detected in majority of human cancers such as breast, head and neck, lung, prostate cancer and melanoma as well as various blood malignancies. This persistent STAT3 activation plays an important role in tumorigenesis by enhancing cellular proliferation, migration invasion, angiogenesis and metastasis [115, 116]. STAT3 is constitutively active and acts as a direct transcriptional activator of VEGF gene [117]. In tumors, constitutively active STAT 3 up-regulates VEGF expression and thus, tumor angiogenesis [114, 118]. A study showed that STAT3 when blocked using inhibitor CPA-7 inhibits both VEGF expression and angiogenesis [66] Not only this, STAT3 signaling is also involved in endothelial cells migration and vessel formation [119]. STAT3 further also induces expression of hypoxia-inducible factor-1α (HIF1α), a key mediator of angiogenesis [120, 121]. Under hypoxic conditions STAT3 and HIF1α bind simultaneously to the VEGF promoter which leads to maximum transcriptional activation of VEGF protein [122]. Report also suggest that STAT3 also regulate HIF1 α expression *via* p53 pathway [123].

PI3K/AKT and MAPK Pathway

One of the mechanisms of PI3K pathway activation is *via* binding of EGF to its receptor EGFR [124]. Furthermore, Ras protein, a family of GTPases, has a vital role in proliferation, differentiation and apoptosis. Due to mutation in Ras, the gene becomes oncogenic which ultimately leads to increased production of VEGF. Mazure *et al.,* showed that in the presence of hypoxia, in the H-Ras transformed cells, induction of VEGF is through PI3K/AKT pathway (Fig. **4**) indicating that mutated Ras genes promote angiogenesis *via* activation of PI3K pathway [125]. Another mechanism of activation of PI3K is through binding of ANG1 to its receptor TIE2. Also, loss of PTEN, leads to activation of PI3K pathway [126]. However, interestingly, presence of PTEN is also promotes angiogenesis. Activation of PI3K results in phosphorylation of AKT at threonine 308 residue in the presence of PDK1 kinase enzyme. PDK is a essential component for the migration of endothelial cells and upregulation of VEGF in

PI3K dependent mechanism [127]. PI3K induces angiogenesis through the production of nitric oxide NO. This NO production is regulated by enzyme nitric oxide synthase (NOS), which has three isoforms- constitutively expressed nNOS and eNOS while expression of iNOS is inducible. Studies suggest that VEGF can induce nitric oxide production *via* phosphorylation of eNOS at serine 1177 residue by AKT. Furthermore, donors of nitric oxide increase the expression of HIF1α transcription factor. Thus, PI3K activation promotes angiogenesis [128].

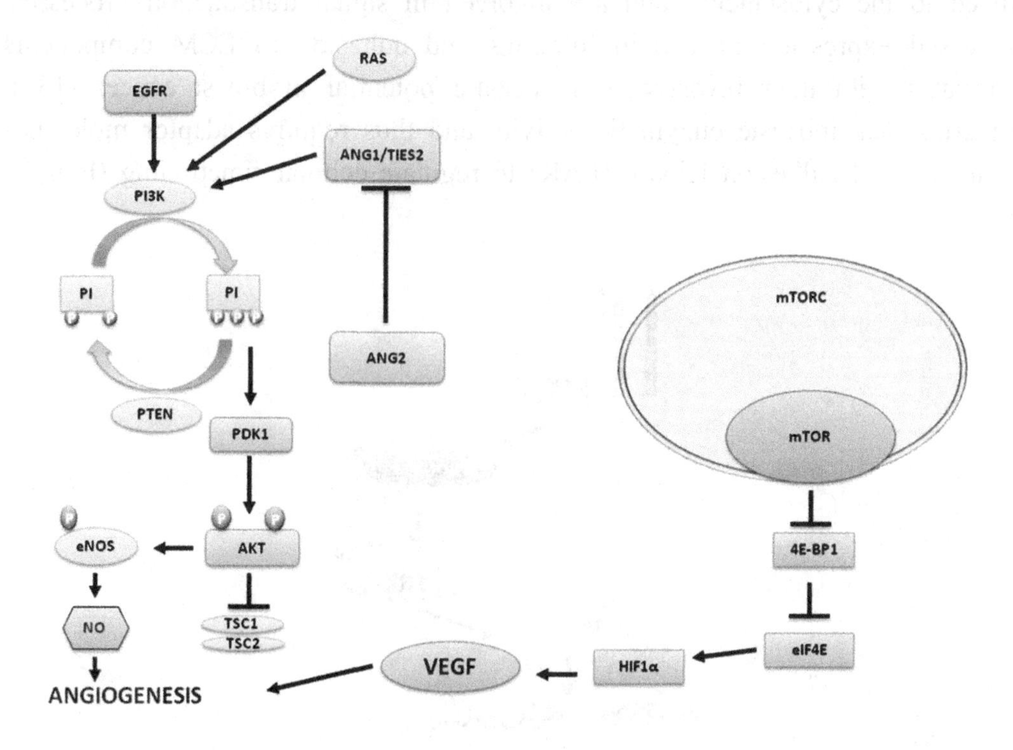

Figure 4: Involvement of PI3K/AKT/mTOR pathway in angiogenesis: These pathways are activated by various signaling molecules. Activation of PI3K pathway, activates various downstream signaling molecules like PDK1, pAKT which in turn activate eNOS that are ultimately responsible for angiogenesis. mTOR signaling pathway also regulates VEGF expression *via* modulating the activity of HIF1α.

Integrin Signalling

Adhesion molecules have traditionally been thought of simply as receptors that permit anchorage to other cells or to the underlying extracellular matrix (ECM) [129]. A typical integrin is a heterodimer made up of two chains *viz.* an α subunit non-covalently bound to a β subunit. Around 18 α subunits and 8 β subunits are known till date that makes up 24 combinations [130] The integrin signalling orchestrates diverse cellular processes such as the cellular proliferation, migration, differentiation, survival and angiogenesis. Increased expression of certain integrins have a strong correlation with the metastatic potential of cancerous cells [131, 132]. These act as receptors for extracellular matrix components such as Fibronectin, Vitronectin, Collagen, Laminin *etc.* Intracellularly, integrins are linked to the cytoskeleton and are involved in signal transduction processes. Increased expression of certain integrins and adhesion to ECM components correlates with their invasive or metastatic potential in breast cancer [133]. Integrins lack intrinsic enzymatic activity and thus requires adaptor molecules such as Focal Adhesion Kinase (FAK) to regulate cellular functioning (Fig. **5**).

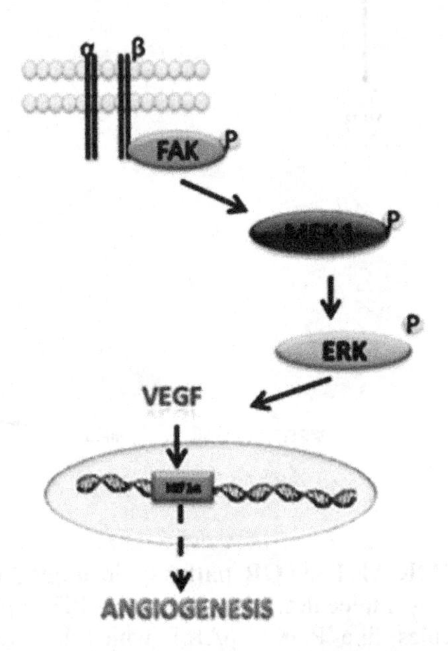

Figure 5: Integrin signaling: Integrin activated upon ECM engagement leads to activation of FAK by phosphorylating it at Tyrosine 397 residue. This leads to activation of MAPK signaling that causes increased transcription of VEGF gene, culminating in angiogenesis.

FAK is an intracellular tyrosine kinase that functions in signal transduction *via* the integrins [134]. It has been found to be involved in diverse cellular processes such as cell adhesion, migration, cell survival, proliferation and angiogenesis. [134]. It regulates the processes of cellular proliferation *via* the Mitogen Activated Protein Kinase (MAPK) as well as Akt pathways. The MAPK in turn regulates the process of angiogenesis *via* affecting the VEGF levels. In light of these available facts, targeting integrins and its downstream mediators such as FAK, forms a viable option for cancer therapy [135].

As angiogenesis is an essential step in tumor proliferation, growth, and metastasis, targeting angiogenesis using various inhibitors singly or in combination with chemotherapeutic agents can certainly result in effective treatment. Recently, various novel approaches are being investigated to block angiogenesis at various stages of the process. Further, works are in progress to understand the mechanism how various signaling pathways are involved in angiogenesis, so that effective novel anti-angiogenic strategies can be developed.

Transforming Growth Factor β (TGFβ)

Transforming growth factor β (TGFβ) another growth factor is observed to have a dual role in tumor development. TGFβ suppresses tumor growth through inhibition of cell proliferation and induction of apoptosis. However, TGFβ is also seen to promote tumor cell proliferation and thus enhanced tumor growth [136].

TGFβ are growth factors that comprises three different isoforms in mammals as TGFβ1,-2 and -3, TGFβ1 being the most commonly observed and studies isoform. All three isoforms have high degree similarity in their homology. TGFβ is secreted by various cells, platelets being the major source [137, 138]. TGFβ is secreted in an inactive state as a latent homodimeric polypeptide that is bound to other extracellular proteins. Bio-activation results due to cleavage of proteolytic latent complex and activated TGFβ then binds to the extracellular serine/threonine kinase receptors type I and type II (TβRI and TβRII) [139, 140].

Although TGFβ suppresses tumor development through a variety of modes such as cell proliferation, inducing apoptosis, genomic stability, tumor cell differentiation and various indirect effects on stroma [141-143]. TGFβ has major role in tumor angiogenesis. It stimulates the secretion of MMP2 and MMP9 and suppress the secretion of protease inhibitor TIMPs [143-145]. Apart from this, TGFβ promotes angiogenesis by upregulating the local growth factors such as

VEGF and connective tissue growth factor [143, 146, 147]. Experiments on mice model suggested that presence of TGFβ in tumor or their microenvironment causes amplification of angiogenesis, thus proving its strong role in tumor angiogenesis [143, 148].

STRATEGIES FOR TARGETING ANGIOGENESIS

Neutralizing VEGF has recently given success. Neutralizing VEGF using monoclonal antibody such as bevacizumab, prior to the attachment of VEGF to its receptor on endothelial cells, is one of the approaches. However, the effectiveness failed in the phase trials undertaken in breast cancer patients. Use of *VEGF-trap* has also gained attention to neutralize VEGF [149].

Next approach has been to block the binding of VEGF to its receptor by blocking the receptor using a monoclonal antibody against VEGFR-2, however, this approach has remained limited as extensive work is not yet done. For the same, Stopeck *et al.,* used an inhibitor SU5416 which showed both anti-tumor and anti-angiogenic properties on binding with VEGF receptor[112]. However, the drug has certain limitations. Its short plasma half life makes the frequent administration of the drug a necessity, which cannot be attained intravenously. Furthermore, its low bioavailability poses problems for frequent administration of the drug. Next, certain side effects such as headache and vomiting were observed during phase trials [150]. Apart from SU5416, other VEGF receptor inhibitors studied are SU6668 which binds to VEGF receptor, which was rejected due to its short plasma half life, PTK787/ZK22854 binds specifically to VEGFR-1 and-2, ZD6474 acting on VEGF as well as bFGF, and CP-547,632 specific for VEGFR-2 are under trials [150, 151].

Blocking endothelial cell proliferation is another novel strategy used to inhibit angiogenesis. Naturally occurring inhibitors including thrombospondin-1, endostatin, angiostatin, TNP-470 are extensively studied. However, the drugs are undergoing phase trials [152-154]. Vascular targeting agents such as combrestatin pro-drug (CA4P), ZD6126, DMXAA effectively disrupt endothelial shape, thus breaking the endothelium and causes loss of blood flow. Another molecule,

AVE8062-A is an analogue of CA4P having similar mechanism of action. All of these agents are effective treatments for angiogenesis [150].

Due to progress in research, a variety of inhibitors are identified that can be used to block angiogenesis. To simplify the understanding, they can be broadly divided into various groups. Primarily, microenvironment around the tumor cells and the host baring the tumor secrete various cytokines and growth factors, which themselves act as anti-angiogenic agents. These inhibitors are termed as endogenous inhibitors.

Angiostatin

Due to the cleavage of plasminogens, *via* proteases, a molecule known as angiostatin is produced [155]. It is a 38-45 kD polypeptide, containing triple disulphide bridged kringle domains from domain kringle-1 to -4 or kringle-1 to -3 [156]. This peptide molecule acts as anti-angiogenic factor. A study undertaken in murine Lewis lung carcinoma suggested that in the presence of primary tumor, inhibited the metastatic growth [157]. Angiostatin was purified from serum and urine of these tumor bearing animals. It is suggested that angiostatin inhibits angiogenesis in various ways. First of all, angiostatin binds to APT synthase present on the membrane of endothelial cells. The data by there is a possibility that due to this activity, the intracellular pH of endothelial cells might drop, leading to triggered response to apoptosis [158]. Furthermore, angiostatin binds to $\alpha v\beta 3$ integrin, thus inhibiting plasmin-induced cell migration [159, 160]. Secondly, angiomotin, an active molecule present in endothelial cells, binds and internalizes angiostatin, thus leading to RGD-independent induction of FAK activity. Thirdly, angiostatin also promotes apoptosis, however, in the absence of growth factors, and it inhibits endothelial migration, thus blocking the tube formation [161]. All these data suggest that presence of angiostatin is one of the many strategies in blocking angiogenesis.

Endostatin

Endostatin is a 20 kDa fragment produced by tumor cells themselves and is derived from the C-terminal region of collagen XVIII [162], It is known to show efficient anti-angiogenic properties. According to the Phase I studies undertaken with recombinant human endostatin show that daily intravenous injections do not cause drug-related toxicity. The mechanism of action of endostatin is that it inhibits tumor growth by interfering with FGF-2 induced signal transduction as well as VEGFR2- mediated signal transduction [163]. Further it also blocks endothelial cell motility by inducing G1 arrest in endothelail cells leading to endothelial cells apoptosis [164]. Apart from this, it is known that endostatin also block TNFα induced activation of JNK pathway, MAPK, and FAK pathway. Endostatin further regulates the upregulation of anti-angiogenic genes while down regulation of pro-angiogenic genes. Many researchers have shown interest in knowing the role of endostatin as anti-angiogenic agent and its uses. A recombinant immobilized human endostatin was studied by Rehn *et al.,* to show its role in binding of integrin α5 and αv [165]. Furthermore, work done by Sudhakar *et al.,* showed that endostatin binds to integrin α5β1 thus blocking the signal transduction pathways like ERK1, p38, Ras and Raf [166]. On the other hand, it is also known that endostatin also inhibits activation of MMPs especially MMP2 and MMP9 [167-169].

Thrombospondin

Thrombospondin is a molecule produced in many types of cells including the neoplastic cells [170, 171]. These are glycoprotein molecules that have direct effect on angiogenesis in tumor. it is known that release of thrombospondin-1 binds to the FGF-2 receptor thus inhibiting the attachment of FGF molecule to its receptor and further downstream signaling pathway [172]. Furthermore, thrombospondin-1 not only activates the apoptotic pathways in endothelial cells but also downregulate the survival pathways [173]. Moreover, thrombospondin-1 block the mobility of various pro-angiogenic factors like VEGF, FGF and PDGF, which ultimately results in reduced angiogenesis. Apart from the above mentioned role of thrombospondin-1 in blocking angiogenesis, thrombospondin-1 also binds the MMP thus reducing angiogenesis. A study carried out in transgenic mice with implanted tumours in mammary gland, showed that over expression of thrombospondin-1 in the tumors suppressed [174] the levels of MMP9 a factor involved in angiogenesis [175]. These results prove that use of thrombospondin-1 either alone or in combination with other chemotherapeutics could be an effective treatment for neoplastic disorders. However, there are various limitation for the use of this inhibitor. First of all, thrombospondin-1 even though inhibits angiogensis data also suggest that thrombospondin-1 supports angiogenesis through activation of various signaling pathways such as MAPK. The anti-angiogenic property of the inhibitor is restricted to only the N-terminal region of the protein. Role of thrombospondin-1 is also observed in tumor development which raises a question regarding the use of this inhibitor. Moreover, because of the large size of the inhibitor, its pharmacological use becomes difficult [176, 177].

Arresten

It is an endogenous inhibitor after the cleavage of basement membrane, generated from non-collagenous domain of a1 chain of type IV collagen [178, 179]. Arresten is known to selectively inhibit both the endothelial tube formation and formation of new blood vessels in mouse aortic endothelial cells. The inhibition of blood vessel formation is by suppressing the migration and proliferation of endothelial cells. However, the metastatic growth of tumor cells is not inhibited by this molecule. The mechanism of action of this inhibitor is that it block the binding of α1β1 integrin to type I collagen. α1β1 integrin activates Ras-Shc-mitogen-activated protein kinase MAPK pathway, thus promoting the tumor cell proliferation. However, arresten, efficiently blocks the activation of MAPK pathway by binding to α1β1 integrin fragment, thus reducing angiogenesis [180]. In addition to this, Arresten also functions by binding to heparin sulfate proteoglycan, which binds to another non-collagenase domain of type IV collagen called canstatin [181].

Apart from the above mentioned strategies, another strategy is to inhibit the signal transduction pathways involved angiogenesis. Our own laboratory data is suggestive of the results that STAT-3 has essential role in angiogenesis and metastasis. Thus, inhibition of these signal transduction pathways attenuates proliferation of tumor cells with little or no effect on normal cells. Various inhibitors include:

NATURALLY OCCURRING INHIBITORS

It is now known that various naturally available components have anti-tumor activities. These compounds act as inhibitors and thus suppress angiogenesis. In contrast to synthetic drugs, these products do not exhibit any cytotoxic effect on the host, rather these natural inhibitors are seen to be effective anti-angiogenic and anti-metastatic agents (Table **1**).

Table 1: Naturally occuring inhibitors of angiogenesis, their origin and role in tumor angiogenesis

Sr. No.	Naturally occurring compound	Origin	Activity
1.	Fumagillin	Aspergillus fumigatus	Anti-angiogenic activity
2.	Artemisinin	Artemisia annua (Chinese Wormwood)	Reduces VEGF expression and NFkB production
3.	Curcumin	Curcuma longa	Inhibition of VEGF , bFGF and nitric oxide
4.	Baicalin and baicalein	Scutellaria baicalensis	anti-angiogenic properties. Reduce VEGF, bFGF, 12-lipoxygenase activity, and MMP9 in prostate cancer
5.	Silibinin and silymarin	Silybum marianum	Downregulation of VEGF and bFGF in human ovarian cancer
6.	Rabdosia	Rabdosia rubescens	anti-angiogenic activity against prostate cancer
7.	Iscador	Viscum album	Downregulates VEGF
8.	Resveratrol	Grape seed extract	Disruption of src kinase activation and VE-cadherin tyrosine phophsrylation
9.	Epigallocatechin 3 gallate (EGCG)	Green tea	Inhibition of tumor cell proliferation
10.	Egb 761	Ginkgo biloba	
11.	Quercetin	Apples, onions, red grapes	Interaction with COX2, EGFR, HER-2 gene, NFkB
12.	Poria cocos	Mushrooms	Downregulation of NFkB. Inhibition of platelet aggregation
13.	Saponins	Panax ginseng	Induces apoptosis
14.	Squalamine	Squalus acanthios	Inhibition of VEGF induced MAPK activation
15.	Honokiol	Chinese Magnolia Tree	Inhibition of new blood vessel formation.
16.	Silibinin and silymarin	Silybum marianum	Down regulation of VEGF and EGFR

Fumigillin

It is a fungal metabolite obtained from *Aspergillus fumigates*. It is observed to block the endotheial cells in G1 phase of growth cycle, thus inhibits the neovascularisation. This inhibitor exhibits a unique structural feature of possessing etope group. The inhibitor functions *via* targeting the methionine aminopeptidase (MetAP2) specifically, as it targets MetAP2 and thus inhibits its activity ultimately leading to reduction in angiogenesis activity.

Artemisinin

It is obtained from plant artemisia annua or Chinese woodplant. Earlier used as anti-malaria drug, its role in tumor apoptosis has given wide scope for its use as anti-angiogenic agent [182, 183]. A study demonstrated that when the transplanted ovarian cancer cells (HO891) were injected in nude mice, in the presence of artemisinin, it reduced tumor growth and the micro vessel density by further reducing the VEGF expression in both tumors as well as endothelial cells. Further, it also inhibited NfkB production thus suppressing tumor progression. Artemisinin showed no negative effect on the host cells, and thus can be used as an effective anti-angiogenic drug [184].

Curcumin

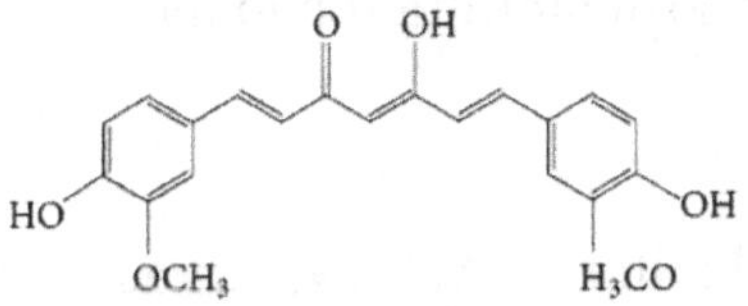

Curcumin, is a naturally occurring inhibitor which is an active component of turmeric, having anti-tumor, anti-inflammatory and anti-oxidant properties [185]. Its cytotoxic activity is mediated by down regulation of MMP2 and TIMP1, the enzymes involved in degradation of extracellular matrix. Curcumin further is also seen to reduce VEGF and bFGF expression, and reduce elevated level of endothelial nitric oxide which is known to promote tumor growth [185-187] Apart from that, curcumin also inhibits STAT3 activation in various cancer cell lines such as multiple myeloma, Hodgkin's lymphoma [188], pancreatic cancer cell lines [189, 190], primary effusion lymphoma [191], head and neck squamous cell carcinoma [192], human chronic myelogenous leukaemia [193], and ovarian cancer. Not only this, but curcumin is also effective in reducing the effects of vascular mimicry.

Resveratrol

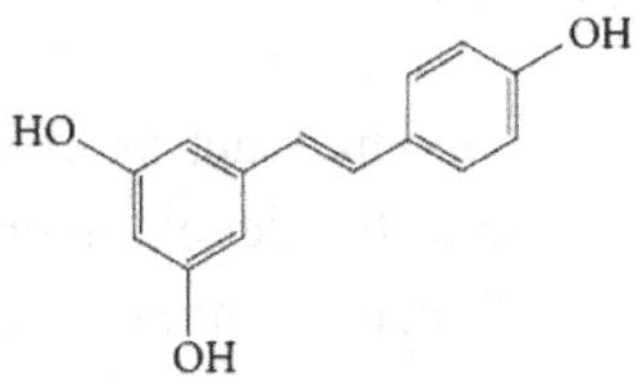

Resveratrol [3,5,4'-trihydroxystilbene), is a polyphenolic phytoalexin that is found in grapes and wine having anti-inflammatory and anti-oxidant properties [194]. Experimental studies on HUVEC cells demonstrated that Resveratrol, not only suppresses the lytic activity of MMP9 but also inhibits cell division [195]. (95 natural inhibitors) Resveratrol further disrupts the src kinase activation followed by VE-cadherin tyrosine phosphorylation thus inhibiting VEGF induced angiogenesis. Alternatively, in another study by Jang *et al.,* treatment with resveratrol inhibited the preneoplastic lesion in carcinogen-treated mouse mammary glands [194]. Further mechanistic studies revealed that resveratrol inhibits tumor growth and induces apoptosis by suppressing STAT3 signaling [196, 197]. Our own data suggests that resveratrol inhibits the activity of type II phosphatidylinositol 4-kinase, a component of phosphoinositide signalling pathway involved in lymphosite infiltration to the site of inflammation [198].

Neovastat

Neovastat is a natural inhibitor of MMP9, extracted from shark cartilage [199]. Neovastat were identified as anti-angiogenic and anti-metastatic molecules as it inhibits the enzymatic activity of MMP2 and MMP-1, 7 and 9 [200]. Furthermore, it also inhibits vascular endothelial growth factor, and its binding to endothelial cells [201]. Neostat being a natural substance has minimal dose-dependent side effects, which makes it an effective anti- angiogenic molecule. Neovastat is also known to stimulate angiostatin production, and promotes endothelial cell apoptosis. Neovastat causes chromatin condensation and fragmentation of DNA [202], further upregulates the caspase proteins in endothelial cell, ultimately leading to apoptosis [201, 203]. However, the exact mechanism of this up regulation of caspases is still unknown. Furthermore, oral administration and long term treatment of this inhibitor has made it ideal for traement either alone or in combination with other conventional drugs [203].

MIMICS FOR ANGIOGENESIS

As these naturally occurring compounds act as effective anti-angiogenic agents, more research on these have lead to the development of drugs that are developed from these natural compounds. Recently, number of bioactive naturally occuring compounds are being tested for their antiangiogenic potential which indicates a strong capacity of natural products in the discovery of new and effective antiangiogenic agents [204, 205].

Several Vinca alkaloid compounds including vincristine, vinblastine and vindesine have been observed to posses anti-tumor activity [206]. Recently, structural modifications in these compounds has also resulted in effective treatment. For example, structural changes in the velbenamine gave a new drug called vinflunine which not only showed anti-angiogenic properties but also delayed the multi-drug resistance of tumors. These potential anti-angiogenic and vascular disrupting properties support further clinical development [207]. Another example being Colchicine which is developed from natural product possesses tubulin-interacting potential. However, these drugs have antiangiogenic activity only at maximum tolerated dose (MTD) and thus need to be used in combination therapy with other chemotherapeutic drugs, for better efficacy [208].

Apart from alkaloids, biphenyl compounds also exhibit anti-angiogenic property. An example being a *Magnolia obovata* bark extract called Honokiol, that induces apoptosis in number of tumor cell lines [209-212]. Li *et al.,* proved the anti-angiogenic property of Honokiol by demonstrating that honokial dependent apoptosis was a result of decreased Bcl-Xl [213]. Furthermore, anti-tumor activity of Honokial was shown by using SKOV3 tumor xenograft nude mouse model, that showed suppressed tumor growth. The anti-angiogenic potential was demonstrated by a significant reduction in anti-angiogenic factors such as VEGF, TGFβ and TNFα [213]. All these results justifies the anti-tumorigenic and anti-angiogenic activities of Honokial. R-(–)-b-O-methylsynephrine (OMe-Syn) is another natural compound which is isolated from a plant of the *Rutaceae* family, also is a potential anti-angiogenic drug. Both *in vivo* assay on chick embryo and *in vitro* invasion assay on matrigel by Kim *et al.,* suggested that OMe-Syn inhibits VEGF induced angiogenesis [214].

Next group is that of quinones, a common compound having role in photosynthesis, respiration and many other biological processes [214-216]. Quinones are used as anti-angiogenic agents as many experimental models support their anti-angiogenic property. One example being that of HU-331, an anthracycline antibiotic isolated from *Cannabis sativa* [217]. Kogan *et al.*, showed using bovine aortic endothelial cells that HU-331 inhibits cell proliferation and angiogenesis in the *ex vivo* rat aortic ring assay [218].

Taken together, these examples demonstrate the potential of naturally occurring compounds as anti-angiogenic drugs.

SYNTHETIC INHIBITORS

Tyrosine Kinase Inhibitors

Tyrosine kinases are important receptors having role in signal transduction of various process including angiogenesis. Thus, a novel approach of targeting these receptors is an effective treatment for angiogenesis. There are small molecular compounds that penetrate the cell membrane which is hydrophobic in nature, and act on the downstream signalling cascade of signal transduction pathway. These molecules are the inhibitors of tyrosine kinase receptor. These compounds competitively inhibit the binding of VEGF, bFGF, and PDFG, the most important pro-angiogenic molecules, to their specific tyrosine kinase receptors. The various tyrosine kinase inhibitors are mentioned in Table **2**. Tyrosine kinase inhibitors are capable of inhibiting angiogenesis in tumors *via* multiple mechanisms, for example, SU6668 causes apoptosis in both endothelial as well as tumor cells [219, 220]. SU6668 exhibits anti-tumour activity against numerous tumour types including colon cancer [221]. Furthermore SU11248 exposure provided antitumour activity including tumour regression, reduced or arrested tumour growth [222]. The molecules resembles c-kit, structurally, and thus binds to the tyrosine kinase receptors, thus inhibiting the binding of tyrosine kinase ligands to their receptors. and acts as an anti-angiogenic agent, thus, studies also suggest that its action is effective on large tumors [223].

Vascular Targeting Agents

Another type of inhibitors belongs to Vascular targeting agents (VTAs) or vascular disrupting agents. These are the small molecule compounds that assure effective treatment against cancer by selectively suppressing the vasculature of tumors. VTAs disrupt the pre-existing tumor vasculature leading to tumor cell death due to extensive hemorrhagic necrosis. Furthermore, VTAs can also kill the tumour cells away from the blood vessels or in the hypoxic area where the conventional anti-tumour drugs do not reach effectively that leads to drug resistance [224, 225]. These agents target endothelial cells and pericytes present in tumour vasculature. However VTAs selectivity is a matter of question [226].

Inhibitors Against Endothelial Cell Proliferation

Next type of commonly observed inhibitor is endothelial cell proliferation inhibitors. This class of inhibitors include the heparin binding domain, which blocks the endothelial cell adhesion and chemotaxis to bFGF, thus inhibiting endothelil proliferation [172].

Thalidomide

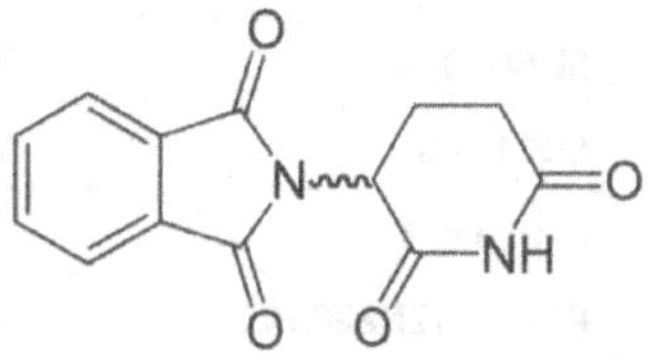

It was identified as an anti-angiogenic agent in multiple myeloma by Single and colleagues and was used as monotherapy drug. However, this drug is now given in combination with dexamethasone in order to get better treatment. Thalidomide is a glutamic acid derivative that exhibit linear and dose-dependent pharmacokinetics which do not change due to different age, sex or food. Thus it is used as a therapy against angiogenesis.

TNP-470

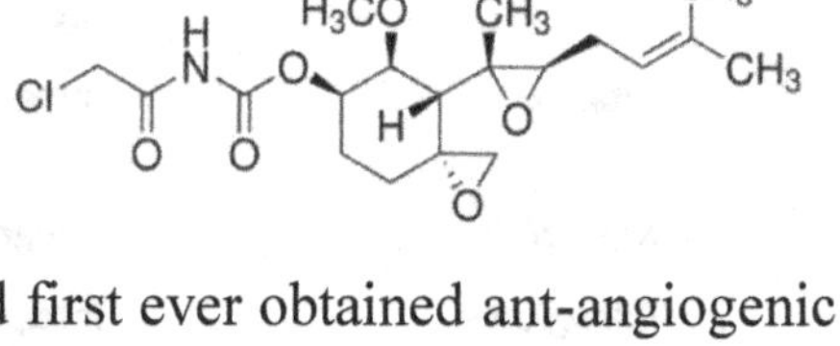

It is a synthetic drug_and first ever obtained ant-angiogenic drug. It is a derivative of fumagillin and shows effective results against various cancer types such as brain cancer, renal cell carcinoma, breast cancer and survical cancer. As this is the first drug against angiogenesis, it is used as a prototype for further development of other anti-angiogenic drugs.

The details of the other types and their inhibitors are described in Table **2**.

Table 2: Adopted from [124]. Clinically useful inhibitors: the following table describes about the type of the drug, its name and target molecule

Type	Drug	Target
Monoclonal antibodies targeting VEGF-A	Bevacizumab (Avastins)	VEGF-A
	VEGF-Trap	VEGF-A
Antibodies targeting VEGFR-2	IMC-1C11	VEGFR-2
Receptor tyrosine kinase inhibitors	SU5416	VEGFR-2
	SU6668	VEGFR-2, bFGFR, PDGFR
	SU11248	VEGFR-2, PDGFR,
	c-Kit, Flt-3	VEGFR-1, VEGFR-2
	PTK787/ZK22854	VEGFR-1, VEGFR-2
	ZD6474	VEGFR-2, EGFR
	CP-547,632	VEGFR-2, EGFR, PDGFR
Inhibitors of endothelial cell proliferation	ABT-510	Endothelial CD-36
	Angiostatin	Many
	Endostatin	Many
	TNP-470	Methionine aminopeptidase, cyclin dependent kinase 2
	Thalidomide	Reduction of TNF-a production
Inhibitors of integrin activity	Vitaxin	Integrin aVb3
	Medi-522	Integrin aVb3
	Cilengitide	Integrin aVb3
Vascular targeting agents	Combretastatin A4	Endothelial tubulin
	AVE8062A	Endothelial tubulin
	ZD6126	Endothelial tubulin
	DMXAA	Induction of TNF-a

Pentoxifylline

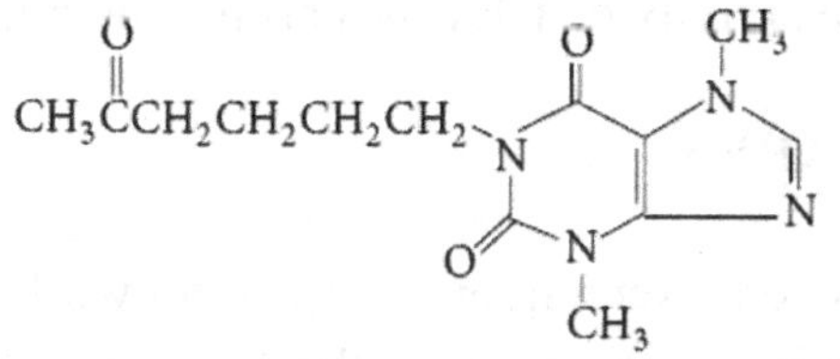

Pentoxifylline (PTX), a non-specific phosphodiesterase inhibitor belongs to methylxantine family. It is a FDA approved drug for the treatment of peripheral vascular disease [186, 187]. PTX augments erythrocyte flexibility and decreases blood viscosity resulting in increased microcirculatory blood flow [227]. It elevates c-AMP levels and has shown therapeutic effects in various other diseases and cellular processes including liver degeneration, sperm motility, chronic inflammatory diseases, muscular dystrophy, renal failure, HIV infection and cancer [228]. Earlier studies reported that PTX can inhibit cancer cell "stickness" and thus increase the circulating time of tumor cells [229]. This observation prompted further studies on the anti-tumorigenic effect of PTX. Studies from our laboratory have shown that PTX has potent anti-metastatic and anti-angiogenic activity against many cancers *in vitro* as well as *in vivo* [230]. PTX inhibits B16F10 experimental metastasis and growth of murine solid tumor [231]. Further mechanistic study revealed that PTX have these effects *via* its inhibitory action on cell adhesion, matrix metalloprotienase- 9 (MMP-9) secretions and tumor migration [232, 233]. We have also shown that PTX even at sub-toxic pharmacological doses acts as an effective anti-proliferative agent with significant anti-proteolytic and anti-adhesive effects in culture of B16F10 melanoma cells [234]. PTX also inhibits the proliferation, migration, invasion and adhesion of A375 human melanoma as well as MDA-MB-231 breast carcinoma [235, 236]. PTX inhibits human melanoma *in vivo* without showing any visible toxicity [195, 197]. Further, by using A375 human melanoma model, we have revealed that PTX inhibits tumor growth and angiogenesis by targeting STAT3 signaling pathway [237]. PTX inhibited STAT3 activation and STAT3 DNA binding in a dose dependent manner. Expression of various STAT3 regulated gene products such as Cyclin D1, cMyc, VEGF, HIFα and BclXL was down regulated following PTX treatment [197]. In MDA-MB-231 breast carcinoma, we have also shown that PTX delays tumor growth and angiogenesis [238]. PTX also shown to inhibit endothelial cell proliferation and tumor induced angiogenesis [239]. The present findings demonstrate the activity of PTX as a promising anti-angiogenic agent.

CONCLUDING REMARKS

Although there has been tremendous research work in order to inhibit angiogenesis, and thus tumour progression, there are many aspects still not clearly understood, thus hampering the use of new therapies. The current review first of all contrasts the physiological angiogenesis with angiogenesis in tumours. Furthermore, it puts light on various molecules that are involved in the upregulation of tumour angiogenesis. Moreover, the signal transduction pathways such as MAPK, PI3K, mTOR and STAT3 pathway have major role in the angiogenesis process and can be the targets for anti-angiogenic approaches. Finally the focus of this review is to explain the novel drugs that are used currently in angiogenesis. These include the naturally occurring inhibitors such as angiostatin, thrombospondin, endostatin and arrestin which are produced by tumour cells itself in the initial phase of the tumour development, and also some of the inhibitors such as Artemisinin, Curcumin, Resveratrol, Neovastat which are recently identified in plant and fungal sources. Apart from naturally occurring anti-angiogenic agents, synthetic angiogenesis inhibitors are also clinically approved and are used as anti-cancer therapies. However, there is still need to understand various mechanisms that regulate angiogenesis in order to have effective therapy against tumour angiogenesis. We believe that this approach of using rational drug design against angiogenesis in combination with high throughput screening should definitely lead to successful therapeutic strategy.

ACKNOWLEDGEMENTS

The authors are thankful to both CSIR-SRF and ICMR-SRF for providing financial support to Peeyush N Goel and Zahid Kamran respectively. The authors would also like to thank Shinjini Chaudhary for her help during the preparation of the text.

CONFLICT OF INTEREST

The authors confirm that this chapter contents have no conflict of interest.

REFERENCES

[1] Conway EM, Collen D, Carmeliet P. Molecular mechanisms of blood vessel growth. Cardiovascular research. 2001;49(3):507-21.

[2] Folkman J. Angiogenesis and angiogenesis inhibition: an overview. Regulation of angiogenesis: Springer; 1996. p. 1-8.

[3] Folkman J. Role of angiogenesis in tumor growth and metastasis. Seminars in oncology; 2002.

[4] O'Reilly MS, Boehm T, Shing Y, Fukai N, Vasios G, Lane WS, *et al.,* Endostatin: an endogenous inhibitor of angiogenesis and tumor growth. Cell. 1997;88(2):277-85.

[5] Carmeliet P, Jain RK. Angiogenesis in cancer and other diseases. Nature. 2000;407(6801):249-57.

[6] Knudson AG. Mutation and cancer: statistical study of retinoblastoma. Proceedings of the National Academy of Sciences. 1971;68(4):820-3.

[7] Hanahan D, Weinberg RA. The hallmarks of cancer. Cell. 2000;100(1):57-70.

[8] Fearon ER, Vogelstein B. A genetic model for colorectal tumorigenesis. Cell. 1990;61(5):759-67.

[9] Plate K, Breier G, Millauer B, Ullrich A, Risau W. Up-regulation of vascular endothelial growth factor and its cognate receptors in a rat glioma model of tumor angiogenesis. Cancer Research. 1993;53(23):5822-7.

[10] Smith K, Fox S, Whitehouse R, Taylor M, Greenall M, Clarke J, *et al.,* Upregulation of basic fibroblast growth factor in breast carcinoma and its relationship to vascular density, oestrogen receptor, epidermal growth factor receptor and survival. Annals of oncology. 1999;10(6):707-13.

[11] Giuliani N, Colla S, Lazzaretti M, Sala R, Roti G, Mancini C, *et al.,* Proangiogenic properties of human myeloma cells: production of angiopoietin-1 and its potential relationship to myeloma-induced angiogenesis. Blood. 2003;102(2):638-45.

[12] Kerbel R, Folkman J. Clinical translation of angiogenesis inhibitors. Nature Reviews Cancer. 2002;2(10):727-39.

[13] Buysschaert I, Schmidt T, Roncal C, Carmeliet P, Lambrechts D. Genetics, epigenetics and pharmaco-(epi)genomics in angiogenesis. J Cell Mol Med. 2008 Dec;12(6B):2533-51.

[14] Cooper MP, Keaney JF, Jr. Epigenetic control of angiogenesis *via* DNA methylation. Circulation. Jun 28;123(25):2916-8.

[15] Bertolino P, Deckers M, Lebrin F, ten Dijke P. Transforming growth factor-beta signal transduction in angiogenesis and vascular disorders. Chest. 2005 Dec;128(6 Suppl):585S-90S.

[16] McAllister KA, Grogg KM, Johnson DW, Gallione CJ, Baldwin MA, Jackson CE, *et al.,* Endoglin, a TGF-beta binding protein of endothelial cells, is the gene for hereditary haemorrhagic telangiectasia type 1. Nat Genet. 1994 Dec;8(4):345-51.

[17] Dunn JR, Reed JE, du Plessis DG, Shaw EJ, Reeves P, Gee AL, *et al.,* Expression of ADAMTS-8, a secreted protease with antiangiogenic properties, is downregulated in brain tumours. Br J Cancer. 2006 Apr 24;94(8):1186-93.

[18] Xu L, Jain RK. Down-regulation of placenta growth factor by promoter hypermethylation in human lung and colon carcinoma. Mol Cancer Res. 2007 Sep;5(9):873-80.

[19] Yang QW, Liu S, Tian Y, Salwen HR, Chlenski A, Weinstein J, *et al.,* Methylation-associated silencing of the thrombospondin-1 gene in human neuroblastoma. Cancer Res. 2003 Oct 1;63(19):6299-310.

[20] Quentmeier H, Eberth S, Romani J, Weich HA, Zaborski M, Drexler HG. DNA methylation regulates expression of VEGF-R2 (KDR) and VEGF-R3 (FLT4). BMC Cancer.12:19.

[21] Sundrani DP, Reddy US, Joshi AA, Mehendale SS, Chavan-Gautam PM, Hardikar AA, *et al.,* Differential placental methylation and expression of VEGF, FLT-1 and KDR genes in human term and preterm preeclampsia. Clin Epigenetics.5(1):6.

[22] Chang S, Young BD, Li S, Qi X, Richardson JA, Olson EN. Histone deacetylase 7 maintains vascular integrity by repressing matrix metalloproteinase 10. Cell. 2006 Jul 28;126(2):321-34.

[23] Mottet D, Bellahcene A, Pirotte S, Waltregny D, Deroanne C, Lamour V, *et al.,* Histone deacetylase 7 silencing alters endothelial cell migration, a key step in angiogenesis. Circ Res. 2007 Dec 7;101(12):1237-46.

[24] Potente M, Ghaeni L, Baldessari D, Mostoslavsky R, Rossig L, Dequiedt F, *et al.,* SIRT1 controls endothelial angiogenic functions during vascular growth. Genes Dev. 2007 Oct 15;21(20):2644-58.

[25] Folkman J. Blood vessel formation: what is its molecular basis. cell. 1996;87:1153-5.

[26] Risau W. Mechanisms of angiogenesis. Nature. 1997;386(6626):671-4.

[27] Carmeliet P. Mechanisms of angiogenesis and arteriogenesis. Nature medicine. 2000;6(4):389.

[28] Carmeliet P. Angiogenesis in life, disease and medicine. Nature. 2005;438(7070):932-6.

[29] Bergers G, Benjamin LE. Tumorigenesis and the angiogenic switch. Nature Reviews Cancer. 2003;3(6):401-10.

[30] Papetti M, Herman IM. Mechanisms of normal and tumor-derived angiogenesis. American Journal of Physiology-Cell Physiology. 2002;282(5):C947-C70.

[31] Carmeliet P. Angiogenesis in health and disease. Nature medicine. 2003;9(6):653-60.

[32] Benjamin LE, Golijanin D, Itin A, Pode D, Keshet E. Selective ablation of immature blood vessels in established human tumors follows vascular endothelial growth factor withdrawal. Journal of Clinical Investigation. 1999;103(2):159.

[33] Terris B, Scoazec J, Rubbia L, Bregeaud L, Pepper M, Ruszniewski P, *et al.,* Expression of vascular endothelial growth factor in digestive neuroendocrine tumours. Histopathology. 1998;32:133-8.

[34] Carmeliet P, Jain RK. Molecular mechanisms and clinical applications of angiogenesis. Nature. 2011;473(7347):298-307.

[35] Rafii S, Lyden D, Benezra R, Hattori K, Heissig B. Vascular and haematopoietic stem cells: novel targets for anti-angiogenesis therapy? Nature Reviews Cancer. 2002;2(11):826-35.

[36] Katoh K, Kano Y, Masuda M, Onishi H, Fujiwara K. Isolation and contraction of the stress fiber. Molecular Biology of the Cell. 1998;9(7):1919-38.

[37] Moses MA. The regulation of neovascularization by matrix metalloproteinases and their inhibitors. Stem cells. 1997;15(3):180-9.

[38] John A, Tuszynski G. The role of matrix metalloproteinases in tumor angiogenesis and tumor metastasis. Pathology oncology research. 2001;7(1):14-23.

[39] Nelson AR, Fingleton B, Rothenberg ML, Matrisian LM. Matrix metalloproteinases: biologic activity and clinical implications. Journal of Clinical Oncology. 2000;18(5):1135-.

[40] Śliwowska I, Kopczyński Z. Matrix metalloproteinases–biochemical characteristics and clinical value determination in breast cancer patients. Contemp Oncol. 2005;9:327-35.

[41] Vu TH, Werb Z. Matrix metalloproteinases: effectors of development and normal physiology. Science Signaling. 2000;14(17):2123.

[42] Van Wart HE, Birkedal-Hansen H. The cysteine switch: a principle of regulation of metalloproteinase activity with potential applicability to the entire matrix metalloproteinase gene family. Proceedings of the National Academy of Sciences. 1990;87(14):5578-82.

[43] Sternlicht MD, Werb Z. How matrix metalloproteinases regulate cell behavior. Annual review of cell and developmental biology. 2001;17:463.

[44] Egeblad M, Werb Z. New functions for the matrix metalloproteinases in cancer progression. Nature Reviews Cancer. 2002;2(3):161-74.

[45] Coussens LM, Fingleton B, Matrisian LM. Matrix metalloproteinase inhibitors and cancer—trials and tribulations. Science. 2002;295(5564):2387-92.

[46] Stamenkovic I. Extracellular matrix remodelling: the role of matrix metalloproteinases. The Journal of pathology. 2003;200(4):448-64.

[47] Gazzanelli G, Luchetti F, Burattini S, Mannello F, Falcieri E, Papa S. Matrix metalloproteinases expression in HL-60 promyelocytic leukemia cells during apoptosis. Apoptosis. 2000;5(2):165-72.

[48] Mannello F, Gazzanelli G. Tissue inhibitors of metalloproteinases and programmed cell death: conundrums, controversies and potential implications. Apoptosis. 2001;6(6):479-82.

[49] Ferreras M, Felbor U, Lenhard T, Olsen BR, Delaissé J-M. Generation and degradation of human endostatin proteins by various proteinases. FEBS letters. 2000;486(3):247-51.

[50] Cornelius LA, Nehring LC, Harding E, Bolanowski M, Welgus HG, Kobayashi DK, *et al.*, Matrix metalloproteinases generate angiostatin: effects on neovascularization. The Journal of Immunology. 1998;161(12):6845-52.

[51] Grimshaw MJ, Naylor S, Balkwill FR. Endothelin-2 Is a Hypoxia-induced Autocrine Survival Factor for Breast Tumor Cells 1 Supported, in part, by Oxford BioMedica United Kingdom Ltd.(to MJG). 1. Molecular cancer therapeutics. 2002;1(14):1273-81.

[52] Harris AL. Hypoxia—a key regulatory factor in tumour growth. Nature Reviews Cancer. 2002;2(1):38-47.

[53] Fang H-Y, Hughes R, Murdoch C, Coffelt SB, Biswas SK, Harris AL, *et al.*, Hypoxia-inducible factors 1 and 2 are important transcriptional effectors in primary macrophages experiencing hypoxia. Blood. 2009;114(4):844-59.

[54] Hu C-J, Wang L-Y, Chodosh LA, Keith B, Simon MC. Differential roles of hypoxia-inducible factor 1α (HIF-1α) and HIF-2α in hypoxic gene regulation. Molecular and cellular biology. 2003;23(24):9361-74.

[55] Wang GL, Jiang B-H, Rue EA, Semenza GL. Hypoxia-inducible factor 1 is a basic-helix-loop-helix-PAS heterodimer regulated by cellular O2 tension. Proceedings of the National Academy of Sciences. 1995;92(12):5510-4.

[56] Hu C-J, Iyer S, Sataur A, Covello KL, Chodosh LA, Simon MC. Differential regulation of the transcriptional activities of hypoxia-inducible factor 1 alpha (HIF-1α) and HIF-2α in stem cells. Molecular and cellular biology. 2006;26(9):3514-26.

[57] Higgins DF, Biju MP, Akai Y, Wutz A, Johnson RS, Haase VH. Hypoxic induction of Ctgf is directly mediated by Hif-1. American Journal of Physiology-Renal Physiology. 2004;287(6):F1223-F32.

[58] Gordan JD, Simon MC. Hypoxia-inducible factors: central regulators of the tumor phenotype. Current opinion in genetics & development. 2007;17(1):71-7.

[59] Kaelin Jr WG. ROS: really involved in oxygen sensing. Cell Metabolism. 2005;1(6):357-8.

[60] Huang LE, Bunn HF. Hypoxia-inducible factor and its biomedical relevance. Journal of Biological Chemistry. 2003;278(22):19575-8.

[61] Ahn J-E, Zhou X, Dowd SE, Chapkin RS, Zhu-Salzman K. Insight into Hypoxia Tolerance in Cowpea Bruchid: Metabolic Repression and Heat Shock Protein Regulation *via* Hypoxia-Inducible Factor 1. PLoS One. 2013;8(4):e57267.

[62] Lee J-W, Bae S-H, Jeong J-W, Kim S-H, Kim K-W. Hypoxia-inducible factor (HIF-1) α: its protein stability and biological functions. Experimental & molecular medicine. 2004;36(1):1-12.

[63] Cheng J, Kang X, Zhang S, Yeh ET. SUMO-specific protease 1 is essential for stabilization of HIF1α during hypoxia. cell. 2007;131(3):584-95.

[64] Kim HL, Cassone M, Otvos Jr L, Vogiatzi P. HIF-1α and STAT3 client proteins interacting with the cancer chaperone Hsp90: Therapeutic considerations. Cancer biology & therapy. 2008;7(1):10-4.

[65] Jung Y-J, Isaacs JS, Lee S, Trepel J, Neckers L. IL-1β-mediated up-regulation of HIF-1α *via* an NFκB/COX-2 pathway identifies HIF-1 as a critical link between inflammation and oncogenesis. The FASEB journal. 2003;17(14):2115-7.

[66] Xu Q, Briggs J, Park S, Niu G, Kortylewski M, Zhang S, *et al.*, Targeting Stat3 blocks both HIF-1 and VEGF expression induced by multiple oncogenic growth signaling pathways. Oncogene. 2005;24(36):5552-60.

[67] Saeij J, Coller S, Boyle J, Jerome M, White M, Boothroyd J. Toxoplasma co-opts host gene expression by injection of a polymorphic kinase homologue. Nature. 2006;445(7125):324-7.

[68] Vaupel P, Mayer A. Hypoxia in cancer: significance and impact on clinical outcome. Cancer and Metastasis Reviews. 2007;26(2):225-39.

[69] Brader S, Eccles SA. Phosphoinositide 3-kinase signalling pathways in tumor progression, invasion and angiogenesis. Tumori. 2004;90(1):2-8.

[70] Dimmeler S, Fleming I, Fisslthaler B, Hermann C, Busse R, Zeiher AM. Activation of nitric oxide synthase in endothelial cells by Akt-dependent phosphorylation. Nature. 1999;399(6736):601-5.

[71] Cross MJ, Claesson-Welsh L. FGF and VEGF function in angiogenesis: signalling pathways, biological responses and therapeutic inhibition. Trends in pharmacological sciences. 2001;22(4):201-7.

[72] Wagatsuma S, Konno R, Sato S, Yajima A. Tumor angiogenesis, hepatocyte growth factor, and c-Met expression in endometrial carcinoma. Cancer. 1998;82(3):520-30.

[73] Guo P, Hu B, Gu W, Xu L, Wang D, Huang H-JS, *et al.*, Platelet-derived growth factor-B enhances glioma angiogenesis by stimulating vascular endothelial growth factor expression in tumor endothelia and by promoting pericyte recruitment. The American journal of pathology. 2003;162(4):1083-93.

[74] Aguayo A, Kantarjian H, Manshouri T, Gidel C, Estey E, Thomas D, *et al.*, Angiogenesis in acute and chronic leukemias and myelodysplastic syndromes. Blood. 2000;96(6):2240-5.

[75] Fox SB, Leek RD, Smith K, Hollyer J, Greenall M, Harris AL. Tumor angiogenesis in node-negative breast carcinomas—relationship with epidermal growth factor receptor, estrogen receptor, and survival. Breast cancer research and treatment. 1994;29(1):109-16.

[76] Flier JS, Underhill LH, Folkman J. Clinical applications of research on angiogenesis. New England Journal of Medicine. 1995;333(26):1757-63.

[77] Dvorak H, Detmar M, Claffey K, Nagy J, Van de Water L, Senger D. Vascular permeability factor/vascular endothelial growth factor: an important mediator of angiogenesis in malignancy and inflammation. International archives of allergy and immunology. 2009;107(1-3):233-5.

[78] Senger DR, Van De Water L, Brown LF, Nagy JA, Yeo K-T, Yeo T-K, *et al.,* Vascular permeability factor (VPF, VEGF) in tumor biology. Cancer and Metastasis Reviews. 1993;12(3-4):303-24.

[79] DiSalvo J, Bayne ML, Conn G, Kwok PW, Trivedi PG, Soderman DD, *et al.,* Purification and characterization of a naturally occurring vascular endothelial growth factor· placenta growth factor heterodimer. Journal of Biological Chemistry. 1995;270(13):7717-23.

[80] Neufeld G, Cohen T, Gengrinovitch S, Poltorak Z. Vascular endothelial growth factor (VEGF) and its receptors. The FASEB Journal. 1999;13(1):9-22.

[81] Stavri GT, Zachary IC, Baskerville PA, Martin JF, Erusalimsky JD. Basic fibroblast growth factor upregulates the expression of vascular endothelial growth factor in vascular smooth muscle cells: synergistic interaction with hypoxia. Circulation. 1995;92(1):11-4.

[82] Robinson CJ, Stringer SE. The splice variants of vascular endothelial growth factor (VEGF) and their receptors. Journal of cell science. 2001;114(5):853-65.

[83] Ferrara N, Gerber H-P, LeCouter J. The biology of VEGF and its receptors. Nature medicine. 2003;9(6):669-76.

[84] Alon T, Hemo I, Itin A, Pe'er J, Stone J, Keshet E. Vascular endothelial growth factor acts as a survival factor for newly formed retinal vessels and has implications for retinopathy of prematurity. Nature medicine. 1995;1(10):1024-8.

[85] Benjamin LE, Keshet E. Conditional switching of vascular endothelial growth factor (VEGF) expression in tumors: induction of endothelial cell shedding and regression of hemangioblastoma-like vessels by VEGF withdrawal. Proceedings of the National Academy of Sciences. 1997;94(16):8761-6.

[86] Shweiki D, Itin A, Soffer D, Keshet E. Vascular endothelial growth factor induced by hypoxia may mediate hypoxia-initiated angiogenesis. Nature. 1992;359(6398):843-5.

[87] Takahashi H, Shibuya M. The vascular endothelial growth factor (VEGF)/VEGF receptor system and its role under physiological and pathological conditions. Clinical Science. 2005;109:227-41.

[88] Jain RK. Molecular regulation of vessel maturation. Nature medicine. 2003;9(6):685-93.

[89] Vlodavsky I, Bar-Shavit R, Ishar-Michael R, Bashkin P, Fuks Z. Extracellular sequestration and release of fibroblast growth factor: a regulatory mechanism? Trends in biochemical sciences. 1991;16:268-71.

[90] Yan G, Fukabori Y, McBride G, Nikolaropolous S, McKeehan W. Exon switching and activation of stromal and embryonic fibroblast growth factor (FGF)-FGF receptor genes in prostate epithelial cells accompany stromal independence and malignancy. Molecular and cellular biology. 1993;13(8):4513-22.

[91] Saik JE, Gould DJ, Watkins EM, Dickinson ME, West JL. Covalently immobilized platelet-derived growth factor-BB promotes angiogenesis in biomimetic poly (ethylene glycol) hydrogels. Acta biomaterialia. 2011;7(1):133-43.

[92] Battegay EJ, Rupp J, Iruela-Arispe L, Sage EH, Pech M. PDGF-BB modulates endothelial proliferation and angiogenesis *in vitro via* PDGF beta-receptors. The Journal of cell biology. 1994;125(4):917-28.

[93] Dunn KL, Espino PS, Drobic B, He S, Davie JR. The Ras-MAPK signal transduction pathway, cancer and chromatin remodeling. Biochemistry and cell biology. 2005;83(1):1-14.

[94] Dhillon A, Hagan S, Rath O, Kolch W. MAP kinase signalling pathways in cancer. Oncogene. 2007;26(22):3279-90.

[95] Qi M, Elion EA. MAP kinase pathways. Journal of cell science. 2005;118(16):3569-72.

[96] Pearson G, Robinson F, Gibson TB, Xu B-e, Karandikar M, Berman K, *et al.,* Mitogen-activated protein (MAP) kinase pathways: regulation and physiological functions. Endocrine reviews. 2001;22(2):153-83.

[97] Weston CR, Davis RJ. The JNK signal transduction pathway. Current opinion in genetics & development. 2002;12(1):14-21.

[98] Adler V, Franklin CC, Kraft AS. Phorbol esters stimulate the phosphorylation of c-Jun but not v-Jun: regulation by the N-terminal delta domain. Proceedings of the National Academy of Sciences. 1992;89(12):5341-5.

[99] Kaikai S, Yuchen S, Lili J, Zhengtao W. Critical role of c-Jun N-terminal kinase in regulating bFGF-induced angiogenesis *in vitro*. Journal of biochemistry. 2011;150(2):189-97.

[100] Kumar S, Boehm J, Lee JC. p38 MAP kinases: key signalling molecules as therapeutic targets for inflammatory diseases. Nature Reviews Drug Discovery. 2003;2(9):717-26.

[101] Issbrücker K, Marti HH, Hippenstiel S, Springmann G, Voswinckel R, Gaumann A, *et al.,* p38 MAP kinase—a molecular switch between VEGF-induced angiogenesis and vascular hyperpermeability. The FASEB journal. 2003;17(2):262-4.

[102] Sweeney SM, DiLullo G, Slater SJ, Martinez J, Iozzo RV, Lauer-Fields JL, *et al.,* Angiogenesis in collagen I requires α2β1 ligation of a GFP* GER sequence and possibly p38 MAPK activation and focal adhesion disassembly. Journal of Biological Chemistry. 2003;278(33):30516-24.

[103] Richard DE, Berra E, Gothié E, Roux D, Pouysségur J. p42/p44 mitogen-activated protein kinases phosphorylate hypoxia-inducible factor 1α (HIF-1α) and enhance the transcriptional activity of HIF-1. Journal of Biological Chemistry. 1999;274(46):32631-7.

[104] Lenormand P, Sardet C, Pages G, L'Allemain G, Brunet A, Pouysségur J. Growth factors induce nuclear translocation of MAP kinases (p42mapk and p44mapk) but not of their activator MAP kinase kinase (p45mapkk) in fibroblasts. The Journal of cell biology. 1993;122(5):1079-88.

[105] Berra E, Díaz-Meco MT, Lozano J, Frutos S, Municio M, Sanchez P, *et al.,* Evidence for a role of MEK and MAPK during signal transduction by protein kinase C zeta. The EMBO journal. 1995;14(24):6157.

[106] Eliceiri BP, Klemke R, Strömblad S, Cheresh DA. Integrin αvβ3 requirement for sustained mitogen-activated protein kinase activity during angiogenesis. The Journal of cell biology. 1998;140(5):1255-63.

[107] Mavria G, Vercoulen Y, Yeo M, Paterson H, Karasarides M, Marais R, *et al.,* ERK-MAPK signaling opposes Rho-kinase to promote endothelial cell survival and sprouting during angiogenesis. Cancer cell. 2006;9(1):33-44.

[108] Leaman DW, Leung S, Li X, Stark G. Regulation of STAT-dependent pathways by growth factors and cytokines. The FASEB Journal. 1996;10(14):1578-88.

[109] Bartoli M, PLATT D, Lemtalsi T, Gu X, Brooks SE, Marrero MB, *et al.,* VEGF differentially activates STAT3 in microvascular endothelial cells. The FASEB Journal. 2003;17(11):1562-4.

[110] Yuan Z-l, Guan Y-j, Chatterjee D, Chin YE. Stat3 dimerization regulated by reversible acetylation of a single lysine residue. Science Signaling. 2005;307(5707):269.

[111] Buettner R, Mora LB, Jove R. Activated STAT signaling in human tumors provides novel molecular targets for therapeutic intervention. Clinical Cancer Research. 2002;8(4):945-54.

[112] Stopeck A, Sheldon M, Vahedian M, Cropp G, Gosalia R, Hannah A. Results of a phase I dose-escalating study of the antiangiogenic agent, SU5416, in patients with advanced malignancies. Clinical Cancer Research. 2002;8(9):2798-805.

[113] Liang Q-C, Xiong H, Zhao Z-W, Jia D, Li W-X, Qin H-Z, *et al.,* Inhibition of transcription factor STAT5b suppresses proliferation, induces G1 cell cycle arrest and reduces tumor cell invasion in human glioblastoma multiforme cells. Cancer letters. 2009;273(1):164-71.

[114] Wei D, Le X, Zheng L, Wang L, Frey JA, Gao AC, *et al.,* Stat3 activation regulates the expression of vascular endothelial growth factor and human pancreatic cancer angiogenesis and metastasis. Oncogene. 2003;22(3):319-29.

[115] Yu C-L, Meyer DJ, Campbell GS, Larner AC, Carter-Su C, Schwartz J, *et al.,* Enhanced DNA-binding activity of a Stat3-related protein in cells transformed by the Src oncoprotein. Science. 1995;269(5220):81-3.

[116] Kamran MZ, Patil P, Gude RP. Role of STAT3 in Cancer Metastasis and Translational Advances. BioMed Research International. 2013;2013.

[117] Chen Z, Han ZC. STAT3: a critical transcription activator in angiogenesis. Medicinal research reviews. 2008;28(2):185-200.

[118] Niu G, Wright KL, Huang M, Song L, Haura E, Turkson J, *et al.,* Constitutive Stat3 activity up-regulates VEGF expression and tumor angiogenesis. Oncogene. 2002;21(13):2000.

[119] Yahata Y, Shirakata Y, Tokumaru S, Yamasaki K, Sayama K, Hanakawa Y, *et al.,* Nuclear translocation of phosphorylated STAT3 is essential for vascular endothelial growth factor-induced human dermal microvascular endothelial cell migration and tube formation. Journal of Biological Chemistry. 2003;278(41):40026-31.

[120] Semenza GL. Targeting HIF-1 for cancer therapy. Nature Reviews Cancer. 2003;3(10):721-32.

[121] Noman MZ, Buart S, Van Pelt J, Richon C, Hasmim M, Leleu N, *et al.,* The Cooperative Induction of Hypoxia-Inducible Factor-1α and STAT3 during Hypoxia Induced an Impairment of Tumor Susceptibility to CTL-Mediated Cell Lysis. The Journal of Immunology. 2009;182(6):3510-21.

[122] Jung JE, Lee HG, Cho IH, Chung DH, Yoon S-H, Yang YM, *et al.,* STAT3 is a potential modulator of HIF-1-mediated VEGF expression in human renal carcinoma cells. The FASEB Journal. 2005;19(10):1296-8.

[123] Yu H, Jove R. The STATs of cancer—new molecular targets come of age. Nature Reviews Cancer. 2004;4(2):97-105.

[124] Brunet A, Datta SR, Greenberg ME. Transcription-dependent and-independent control of neuronal survival by the PI3K-Akt signaling pathway. Current opinion in neurobiology. 2001;11(3):297-305.

[125] Mazure NM, Chen EY, Laderoute KR, Giaccia AJ. Induction of vascular endothelial growth factor by hypoxia is modulated by a phosphatidylinositol 3-kinase/Akt signaling pathway in Ha-ras-transformed cells through a hypoxia inducible factor-1 transcriptional element. Blood. 1997;90(9):3322-31.

[126] Karar J, Maity A. PI3K/AKT/mTOR pathway in angiogenesis. Frontiers in molecular neuroscience. 2011;4.

[127] Primo L, di Blasio L, Roca C, Droetto S, Piva R, Schaffhausen B, *et al.,* Essential role of PDK1 in regulating endothelial cell migration. The Journal of cell biology. 2007;176(7):1035-47.

[128] Tanimoto T, Jin Z-G, Berk BC. Transactivation of vascular endothelial growth factor (VEGF) receptor Flk-1/KDR is involved in sphingosine 1-phosphate-stimulated phosphorylation of Akt and endothelial nitric-oxide synthase (eNOS). Journal of Biological Chemistry. 2002;277(45):42997-3001.

[129] Aplin A, Howe A, Alahari S, Juliano R. Signal transduction and signal modulation by cell adhesion receptors: the role of integrins, cadherins, immunoglobulin-cell adhesion molecules, and selectins. Pharmacological reviews. 1998;50(2):197-264.

[130] Belkin AM, Stepp MA. Integrins as receptors for laminins. Microscopy research and technique. 2000;51(3):280-301.

[131] Humphries JD, Byron A, Humphries MJ. Integrin ligands at a glance. Journal of cell science. 2006;119(19):3901-3.

[132] Luo B-H, Springer TA. Integrin structures and conformational signaling. Current opinion in cell biology. 2006;18(5):579-86.

[133] Yao ES, Zhang H, Chen Y-Y, Lee B, Chew K, Moore D, *et al.,* Increased β1 integrin is associated with decreased survival in invasive breast cancer. Cancer Research. 2007;67(2):659-64.

[134] Clark EA, Brugge JS. Integrins and signal transduction pathways: the road taken. Science. 1995;268(5208):233-9.

[135] McLean GW, Carragher NO, Avizienyte E, Evans J, Brunton VG, Frame MC. The role of focal-adhesion kinase in cancer—a new therapeutic opportunity. Nature Reviews Cancer. 2005;5(7):505-15.

[136] Katz LH, Li Y, Chen JS, Munoz NM, Majumdar A, Chen J, *et al.,* Targeting TGF-beta signaling in cancer. Expert Opin Ther Targets. Jul;17(7):743-60.

[137] Childs CB, Proper JA, Tucker RF, Moses HL. Serum contains a platelet-derived transforming growth factor. Proc Natl Acad Sci U S A. 1982 Sep;79(17):5312-6.

[138] Mishra L, Derynck R, Mishra B. Transforming growth factor-beta signaling in stem cells and cancer. Science. 2005 Oct 7;310(5745):68-71.

[139] Derynck R, Zhang YE. Smad-dependent and Smad-independent pathways in TGF-beta family signalling. Nature. 2003 Oct 9;425(6958):577-84.

[140] Zu X, Zhang Q, Cao R, Liu J, Zhong J, Wen G, *et al.,* Transforming growth factor-beta signaling in tumor initiation, progression and therapy in breast cancer: an update. Cell Tissue Res. Jan;347(1):73-84.

[141] Massague J. TGFbeta in Cancer. Cell. 2008 Jul 25;134(2):215-30.

[142] Meulmeester E, Ten Dijke P. The dynamic roles of TGF-beta in cancer. J Pathol. Jan;223(2):205-18.

[143] Padua D, Massague J. Roles of TGFbeta in metastasis. Cell Res. 2009 Jan;19(1):89-102.

[144] Derynck R, Akhurst RJ, Balmain A. TGF-beta signaling in tumor suppression and cancer progression. Nat Genet. 2001 Oct;29(2):117-29.

[145] Hagedorn HG, Bachmeier BE, Nerlich AG. Synthesis and degradation of basement membranes and extracellular matrix and their regulation by TGF-beta in invasive carcinomas (Review). Int J Oncol. 2001 Apr;18(4):669-81.

[146] Kang Y, Siegel PM, Shu W, Drobnjak M, Kakonen SM, Cordon-Cardo C, *et al.,* A multigenic program mediating breast cancer metastasis to bone. Cancer Cell. 2003 Jun;3(6):537-49.

[147] Sanchez-Elsner T, Botella LM, Velasco B, Corbi A, Attisano L, Bernabeu C. Synergistic cooperation between hypoxia and transforming growth factor-beta pathways on human vascular endothelial growth factor gene expression. J Biol Chem. 2001 Oct 19;276(42):38527-35.

[148] Stearns ME, Garcia FU, Fudge K, Rhim J, Wang M. Role of interleukin 10 and transforming growth factor beta1 in the angiogenesis and metastasis of human prostate primary tumor lines from orthotopic implants in severe combined immunodeficiency mice. Clin Cancer Res. 1999 Mar;5(3):711-20.

[149] Dupont J, Camastra D, Gordon M, Mendelson D, Murren J, Hsu A, *et al.*, editors. Phase I study of VEGF Trap in patients with solid tumors and lymphoma. Proc Am Soc Clin Oncol; 2003.

[150] Eskens F. Angiogenesis inhibitors in clinical development; where are we now and where are we going&quest. British journal of cancer. 2004;90(1):1-7.

[151] Tamura T, Minami H, Yamada Y, Yamamoto N, Shimoyama T, Murakami H, *et al.,* A phase I dose-escalation study of ZD6474 in Japanese patients with solid, malignant tumors. Journal of Thoracic Oncology. 2006;1(9):1002-9.

[152] Bhargava P, Marshall JL, Rizvi N, Dahut W, Yoe J, Figuera M, *et al.,* A Phase I and pharmacokinetic study of TNP-470 administered weekly to patients with advanced cancer. Clinical Cancer Research. 1999;5(8):1989-95.

[153] Logothetis CJ, Wu KK, Finn LD, Daliani D, Figg W, Ghaddar H, *et al.,* Phase I trial of the angiogenesis inhibitor TNP-470 for progressive androgen-independent prostate cancer. Clinical Cancer Research. 2001;7(5):1198-203.

[154] Stadler WM, Kuzel T, Shapiro C, Sosman J, Clark J, Vogelzang NJ. Multi-institutional study of the angiogenesis inhibitor TNP-470 in metastatic renal carcinoma. Journal of Clinical Oncology. 1999;17(8):2541-.

[155] Gately S, Twardowski P, Stack MS, Patrick M, Boggio L, Cundiff DL, *et al.,* Human prostate carcinoma cells express enzymatic activity that converts human plasminogen to the angiogenesis inhibitor, angiostatin. Cancer Research. 1996;56(21):4887-90.

[156] Cao Y, Ji RW, Davidson D, Schaller J, Marti D, Söhndel S, *et al.,* Kringle domains of human angiostatin characterization of the anti-proliferative activity on endothelial cells. Journal of Biological Chemistry. 1996;271(46):29461-7.

[157] O'Reilly MS, Holmgren L, Shing Y, Chen C, Rosenthal RA, Moses M, *et al.,* Angiostatin: a novel angiogenesis inhibitor that mediates the suppression of metastases by a Lewis lung carcinoma. cell. 1994;79(2):315-28.

[158] Moser TL, Stack MS, Asplin I, Enghild JJ, Højrup P, Everitt L, *et al.,* Angiostatin binds ATP synthase on the surface of human endothelial cells. Proceedings of the National Academy of Sciences. 1999;96(6):2811-6.

[159] Tarui T, Majumdar M, Miles LA, Ruf W, Takada Y. Plasmin-induced Migration of Endothelial Cells A potential target for the anti-angiogenic action of angiostatin. Journal of Biological Chemistry. 2002;277(37):33564-70.

[160] Tarui T, Miles LA, Takada Y. Specific interaction of angiostatin with integrin αvβ3 in endothelial cells. Journal of Biological Chemistry. 2001;276(43):39562-8.

[161] Troyanovsky B, Levchenko T, Månsson G, Matvijenko O, Holmgren L. Angiomotin An Angiostatin Binding Protein That Regulates Endothelial Cell Migration and Tube Formation. The Journal of cell biology. 2001;152(6):1247-54.

[162] Sasaki T, Fukai N, Mann K, Göhring W, Olsen BR, Timpl R. Structure, function and tissue forms of the C-terminal globular domain of collagen XVIII containing the angiogenesis inhibitor endostatin. The EMBO Journal. 1998;17(15):4249-56.

[163] Kim Y-M, Hwang S, Kim Y-M, Pyun B-J, Kim T-Y, Lee S-T, *et al.,* Endostatin blocks vascular endothelial growth factor-mediated signaling *via* direct interaction with KDR/Flk-1. Journal of Biological Chemistry. 2002;277(31):27872-9.

[164] Hanai J-i, Dhanabal M, Karumanchi SA, Albanese C, Waterman M, Chan B, *et al.,* Endostatin causes G1 arrest of endothelial cells through inhibition of cyclin D1. Journal of Biological Chemistry. 2002;277(19):16464-9.

[165] Rehn M, Hintikka E, Pihlajaniemi T. Primary structure of the alpha 1 chain of mouse type XVIII collagen, partial structure of the corresponding gene, and comparison of the alpha 1 (XVIII) chain with its homologue, the alpha 1 (XV) collagen chain. Journal of Biological Chemistry. 1994;269(19):13929-35.

[166] Sudhakar A, Sugimoto H, Yang C, Lively J, Zeisberg M, Kalluri R. Human tumstatin and human endostatin exhibit distinct antiangiogenic activities mediated by $\alpha v\beta 3$ and $\alpha 5\beta 1$ integrins. Proceedings of the National Academy of Sciences. 2003;100(8):4766-71.

[167] Lee S-J, Jang J-W, Kim Y-M, Lee HI, Jeon JY, Kwon Y-G, *et al.,* Endostatin binds to the catalytic domain of matrix metalloproteinase-2. FEBS letters. 2002;519(1):147-52.

[168] Nyberg P, Heikkilä P, Sorsa T, Luostarinen J, Heljasvaara R, Stenman U-H, *et al.,* Endostatin inhibits human tongue carcinoma cell invasion and intravasation and blocks the activation of matrix metalloprotease-2,-9, and-13. Journal of Biological Chemistry. 2003;278(25):22404-11.

[169] Kim Y-M, Jang J-W, Lee O-H, Yeon J, Choi E-Y, Kim K-W, *et al.,* Endostatin inhibits endothelial and tumor cellular invasion by blocking the activation and catalytic activity of matrix metalloproteinase 2. Cancer Research. 2000;60(19):5410-3.

[170] Sargiannidou I, Zhou J, Tuszynski GP. The role of thrombospondin-1 in tumor progression. Experimental Biology and Medicine. 2001;226(8):726-33.

[171] Bornstein P, Armstrong LC, Hankenson KD, Kyriakides TR, Yang Z. Thrombospondin 2, a matricellular protein with diverse functions. Matrix Biology. 2000;19(7):557-68.

[172] Vogel T, Guo NH, Krutzsch HC, Blake DA, Hartman J, Mendelovitz S, *et al.,* Modulation of endothelial cell proliferation, adhesion, and motility by recombinant heparin-binding domain and synthetic peptides from the type I repeats of thrombospondin. Journal of cellular biochemistry. 1993;53(1):74-84.

[173] Volpert O. Modulation of endothelial cell survival by an inhibitor of angiogenesis thrombospondin-1: a dynamic balance. Cancer and Metastasis Reviews. 2000;19(1-2):87-92.

[174] Lawler J, Detmar M. Tumor progression: the effects of thrombospondin-1 and-2. The international journal of biochemistry & cell biology. 2004;36(6):1038-45.

[175] Lawler J. Thrombospondin-1 as an endogenous inhibitor of angiogenesis and tumor growth. Journal of cellular and molecular medicine. 2002;6(1):1-12.

[176] Armstrong LC, Bornstein P. Thrombospondins 1 and 2 function as inhibitors of angiogenesis. Matrix Biology. 2003;22(1):63-71.

[177] Ren B, Yee KO, Lawler J, Khosravi-Far R. Regulation of tumor angiogenesis by thrombospondin-1. Biochimica et Biophysica Acta (BBA)-Reviews on Cancer. 2006;1765(2):178-88.

[178] Madri JA. Extracellular matrix modulation of vascular cell behaviour. Transplant immunology. 1997;5(3):179.

[179] Colorado PC, Torre A, Kamphaus G, Maeshima Y, Hopfer H, Takahashi K, *et al.,* Anti-angiogenic cues from vascular basement membrane collagen. Cancer Research. 2000;60(9):2520-6.

[180] Senger DR, Claffey KP, Benes JE, Perruzzi CA, Sergiou AP, Detmar M. Angiogenesis promoted by vascular endothelial growth factor: regulation through α1β1 and α2β1 integrins. Proceedings of the National Academy of Sciences. 1997;94(25):13612-7.

[181] Nyberg P, Xie L, Kalluri R. Endogenous inhibitors of angiogenesis. Cancer Research. 2005;65(10):3967-79.

[182] Mueller MS, Runyambo N, Wagner I, Borrmann S, Dietz K, Heide L. Randomized controlled trial of a traditional preparation of Artemisia annua L.(Annual Wormwood) in the treatment of malaria. Transactions of the Royal Society of Tropical Medicine and Hygiene. 2004;98(5):318-21.

[183] Singh NP, Lai HC. Artemisinin induces apoptosis in human cancer cells. Anticancer research. 2004;24(4):2277-80.

[184] Aldieri E, Atragene D, Bergandi L, Riganti C, Costamagna C, Bosia A, *et al.,* Artemisinin inhibits inducible nitric oxide synthase and nuclear factor NF-kB activation. FEBS letters. 2003;552(2):141-4.

[185] Sen S, Sharma H, Singh N. Curcumin enhances Vinorelbine mediated apoptosis in NSCLC cells by the mitochondrial pathway. Biochemical and biophysical research communications. 2005;331(4):1245-52.

[186] Narayan S. Curcumin, a multi-functional chemopreventive agent, blocks growth of colon cancer cells by targeting β-catenin-mediated transactivation and cell–cell adhesion pathways. Journal of molecular histology. 2004;35(3):301-7.

[187] Khafif A, Hurst R, Kyker K, Fliss DM, Gil Z, Medina JE. Curcumin: a new radio-sensitizer of squamous cell carcinoma cells. Otolaryngology-Head and Neck Surgery. 2005;132(2):317-21.

[188] Mackenzie GG, Queisser N, Wolfson ML, Fraga CG, Adamo AM, Oteiza PI. Curcumin induces cell-arrest and apoptosis in association with the inhibition of constitutively active NF-κB and STAT3 pathways in Hodgkin's lymphoma cells. International journal of cancer. 2008;123(1):56-65.

[189] Dhillon N, Aggarwal BB, Newman RA, Wolff RA, Kunnumakkara AB, Abbruzzese JL, *et al.,* Phase II trial of curcumin in patients with advanced pancreatic cancer. Clinical Cancer Research. 2008;14(14):4491-9.

[190] Glienke W, Maute L, Wicht J, Bergmann L. Curcumin inhibits constitutive STAT3 phosphorylation in human pancreatic cancer cell lines and downregulation of survivin/BIRC5 gene expression. Cancer investigation. 2009;28(2):166-71.

[191] Uddin S, Hussain AR, Manogaran PS, Al-Hussein K, Platanias LC, Gutierrez MI, *et al.,* Curcumin suppresses growth and induces apoptosis in primary effusion lymphoma. Oncogene. 2005;24(47):7022-30.

[192] Chakravarti N, Myers JN, Aggarwal BB. Targeting constitutive and interleukin-6-inducible signal transducers and activators of transcription 3 pathway in head and neck squamous cell

carcinoma cells by curcumin (diferuloylmethane). International journal of cancer. 2006;119(6):1268-75.
[193] Reuter S, Charlet J, Juncker T, Teiten MH, Dicato M, Diederich M. Effect of curcumin on nuclear factor κB signaling pathways in human chronic myelogenous K562 leukemia cells. Annals of the New York Academy of Sciences. 2009;1171(1):436-47.
[194] Jang M, Cai L, Udeani GO, Slowing KV, Thomas CF, Beecher CW, *et al.,* Cancer chemopreventive activity of resveratrol, a natural product derived from grapes. Science. 1997;275(5297):218-20.
[195] Oh WK, George DJ, Hackmann K, Manola J, Kantoff PW. Activity of the herbal combination, PC-SPES, in the treatment of patients with androgen-independent prostate cancer. Urology. 2001;57(1):122.
[196] Kotha A, Sekharam M, Cilenti L, Siddiquee K, Khaled A, Zervos AS, *et al.,* Resveratrol inhibits Src and Stat3 signaling and induces the apoptosis of malignant cells containing activated Stat3 protein. Molecular cancer therapeutics. 2006;5(3):621-9.
[197] Bhardwaj A, Sethi G, Vadhan-Raj S, Bueso-Ramos C, Takada Y, Gaur U, *et al.,* Resveratrol inhibits proliferation, induces apoptosis, and overcomes chemoresistance through down-regulation of STAT3 and nuclear factor-κB–regulated antiapoptotic and cell survival gene products in human multiple myeloma cells. Blood. 2007;109(6):2293-302.
[198] Srivastava R, Ratheesh A, Gude RK, Rao K, Panda D, Subrahmanyam G. Resveratrol inhibits type II phosphatidylinositol 4-kinase: A key component in pathways of phosphoinositide turn over. Biochemical pharmacology. 2005;70(7):1048-55.
[199] Barber R, Delahunt B, Grebe SK, Davis PF, Thornton A, Slim GC. Oral shark cartilage does not abolish carcinogenesis but delays tumor progression in a murine model. Anticancer research. 2001;21(2A):1065-9.
[200] Falardeau P, Champagne P, Poyet P, Hariton C, Dupont É, editors. Neovastat, a naturally occurring multifunctional antiangiogenic drug, in phase III clinical trials. Seminars in oncology; 2001: Elsevier.
[201] Béliveau R, Gingras D, Kruger EA, Lamy S, Sirois P, Simard B, *et al.,* The antiangiogenic agent neovastat (AE-941) inhibits vascular endothelial growth factor-mediated biological effects. Clinical Cancer Research. 2002;8(4):1242-50.
[202] Boivin D, Gendron S, Beaulieu É, Gingras D, Béliveau R. The Antiangiogenic Agent Neovastat (Æ-941) Induces Endothelial Cell Apoptosis 1 Supported by Æterna Laboratories, Québec City, Québec, Canada. 1. Molecular cancer therapeutics. 2002;1(10):795-802.
[203] Gingras D, Batist G, Béliveau R. Æ-941 (Neovastat®): a novel multifunctional antiangiogenic compound. Expert review of anticancer therapy. 2001;1(3):341-7.
[204] Sagar SM, Yance D, Wong RK. Natural health products that inhibit angiogenesis: a potential source for investigational new agents to treat cancer-Part 1. Curr Oncol. 2006 Feb;13(1):14-26.
[205] Sagar SM, Yance D, Wong RK. Natural health products that inhibit angiogenesis: a potential source for investigational new agents to treat cancer-Part 2. Curr Oncol. 2006 Jun;13(3):99-107.
[206] Jordan MA, Wilson L. Microtubules as a target for anticancer drugs. Nat Rev Cancer. 2004 Apr;4(4):253-65.
[207] Fahy J. Modifications in the "upper" velbenamine part of the Vinca alkaloids have major implications for tubulin interacting activities. Curr Pharm Des. 2001 Sep;7(13):1181-97.

[208] Kruczynski A, Hill BT. Vinflunine, the latest Vinca alkaloid in clinical development. A review of its preclinical anticancer properties. Crit Rev Oncol Hematol. 2001 Nov;40(2):159-73.

[209] Fong WF, Tse AK, Poon KH, Wang C. Magnolol and honokiol enhance HL-60 human leukemia cell differentiation induced by 1,25-dihydroxyvitamin D3 and retinoic acid. Int J Biochem Cell Biol. 2005 Feb;37(2):427-41.

[210] Hibasami H, Achiwa Y, Katsuzaki H, Imai K, Yoshioka K, Nakanishi K, *et al.,* Honokiol induces apoptosis in human lymphoid leukemia Molt 4B cells. Int J Mol Med. 1998 Dec;2(6):671-3.

[211] Wang XH, Wu SY, Zhen YS. Inhibitory effects of emodin on angiogenesis. Yao Xue Xue Bao. 2004 Apr;39(4):254-8.

[212] Yang SE, Hsieh MT, Tsai TH, Hsu SL. Down-modulation of Bcl-XL, release of cytochrome c and sequential activation of caspases during honokiol-induced apoptosis in human squamous lung cancer CH27 cells. Biochem Pharmacol. 2002 May 1;63(9):1641-51.

[213] Li Z, Liu Y, Zhao X, Pan X, Yin R, Huang C, *et al.,* Honokiol, a natural therapeutic candidate, induces apoptosis and inhibits angiogenesis of ovarian tumor cells. Eur J Obstet Gynecol Reprod Biol. 2008 Sep;140(1):95-102.

[214] Kim NH, Pham NB, Quinn RJ, Kwon HJ. R-(-)-beta-O-methylsynephrine, a natural product, inhibits VEGF-induced angiogenesis *in vitro* and *in vivo*. Biochem Biophys Res Commun. Aug 13;399(1):20-3.

[215] Asche C. Antitumour quinones. Mini Rev Med Chem. 2005 May;5(5):449-67.

[216] Thomson IV R. Natural Occurring Quinones IV: Recent Advances. Chapman and Hall: London; 1997.

[217] Kogan NM, Rabinowitz R, Levi P, Gibson D, Sandor P, Schlesinger M, *et al.,* Synthesis and Antitumor Activity of Quinonoid Derivatives of Cannabinoids. Journal of Medicinal Chemistry. 2004 2013/10/10;47(15):3800-6.

[218] Kogan NM, Blazquez C, Alvarez L, Gallily R, Schlesinger M, Guzman M, *et al.,* A cannabinoid quinone inhibits angiogenesis by targeting vascular endothelial cells. Mol Pharmacol. 2006 Jul;70(1):51-9.

[219] Hoekman K. SU6668, a multitargeted angiogenesis inhibitor. Cancer journal (Sudbury, Mass). 2001;7:S134.

[220] Laird AD, Vajkoczy P, Shawver LK, Thurnher A, Liang C, Mohammadi M, *et al.,* SU6668 is a potent antiangiogenic and antitumor agent that induces regression of established tumors. Cancer Research. 2000;60(15):4152-60.

[221] Shaheen RM, Davis DW, Liu W, Zebrowski BK, Wilson MR, Bucana CD, *et al.,* Antiangiogenic therapy targeting the tyrosine kinase receptor for vascular endothelial growth factor receptor inhibits the growth of colon cancer liver metastasis and induces tumor and endothelial cell apoptosis. Cancer Research. 1999;59(21):5412-6.

[222] Mendel DB, Laird AD, Xin X, Louie SG, Christensen JG, Li G, *et al., In Vivo* Antitumor Activity of SU11248, a Novel Tyrosine Kinase Inhibitor Targeting Vascular Endothelial Growth Factor and Platelet-derived Growth Factor Receptors Determination of a Pharmacokinetic/Pharmacodynamic Relationship. Clinical Cancer Research. 2003;9(1):327-37.

[223] Godl K, Gruss OJ, Eickhoff J, Wissing J, Blencke S, Weber M, *et al.,* Proteomic characterization of the angiogenesis inhibitor SU6668 reveals multiple impacts on cellular kinase signaling. Cancer Research. 2005;65(15):6919-26.

[224] Thorpe PE. Vascular targeting agents as cancer therapeutics. Clinical Cancer Research. 2004;10(2):415-27.

[225] Ferrara N, Kerbel RS. Angiogenesis as a therapeutic target. Nature. 2005;438(7070):967-74.

[226] Hinnen P, Eskens F. Vascular disrupting agents in clinical development. British journal of cancer. 2007;96(8):1159-65.

[227] Ward A, Clissold SP. Pentoxifylline. Drugs. 1987;34(1):50-97.

[228] Bessler H, Gilgal R, Djaldetti M, Zahavi I. Effect of pentoxifylline on the phagocytic activity, cAMP levels, and superoxide anion production by monocytes and polymorphonuclear cells. Journal of leukocyte biology. 1986;40(6):747-54.

[229] Gastpar H. The inhibition of cancer cell stickiness by the methylxanthine derivative pentoxifylline (BL 191). Thrombosis Research. 1974;5(3):277-89.

[230] Dua P, Ingle A, Gude RP. Suramin augments the antitumor and antimetastatic activity of pentoxifylline in B16F10 melanoma. International journal of cancer. 2007;121(7):1600-8.

[231] Gude R, Ingle A, Rao S. Inhibition of lung homing of B16F10 by pentoxifylline, a microfilament depolymerizing agent. Cancer letters. 1996;106(2):171-6.

[232] Gude R, Binda MM, Presas HL, pez o, Klein-Szanto A, P es, *et al.,* Studies on the mechanisms responsible for inhibition of experimental metastasis of B16-F10 murine melanoma by pentoxifylline. Journal of biomedical science. 1999;6(2):133-41.

[233] Dua P, Gude RP. Pentoxifylline impedes migration in B16F10 melanoma by modulating Rho GTPase activity and actin organisation. European Journal of Cancer. 2008;44(11):1587-95.

[234] Dua P, Gude RP. Antiproliferative and antiproteolytic activity of pentoxifylline in cultures of B16F10 melanoma cells. Cancer chemotherapy and pharmacology. 2006;58(2):195-202.

[235] Kamran MZ, Gude RP. Preclinical evaluation of the antimetastatic efficacy of Pentoxifylline on A375 human melanoma cell line. Biomedicine & Pharmacotherapy. 2012;66(8):617-26.

[236] Goel PN, Gude RP. Unravelling the antimetastatic potential of pentoxifylline, a methylxanthine derivative in human MDA-MB-231 breast cancer cells. Molecular and cellular biochemistry. 2011;358(1-2):141-51.

[237] Kamran MZ, Gude RP. Pentoxifylline inhibits melanoma tumor growth and angiogenesis by targeting STAT3 signaling pathway. Biomedicine & Pharmacotherapy. 2013.

[238] Goel PN, Gude RP. Curbing the focal adhesion kinase and its associated signaling events by pentoxifylline in MDA-MB-231 human breast cancer cells. Eur J Pharmacol. 2013;714(1-3):432-41.

[239] Gude RP, Binda MM, Boquete AL, Bonfil DR. Inhibition of endothelial cell proliferation and tumor-induced angiogenesis by pentoxifylline. Journal of cancer research and clinical oncology. 2001;127(10):625-30.

CHAPTER 6

Discovery and Development of Antiangiogenetic Drugs in Ovarian Cancer

Madon M. Maile[1], Evelyn Y. T. Wong[2], Daphne Suzin[1], Nicole E. Birrer[1] and Richard T. Penson[1,*]

[1]Massachusetts General Hospital, Boston, MA, USA and [2]Yong Loo Lin School of Medicine, Singapore

Abstract: It is more than 30 years since the seminal observations by Folkman of the development of new blood vessels (angiogenesis) in tumors. Ovarian cancer remains the most lethal gynecologic malignancy in the US, and angiogenesis is a particularly important target as VEGF levels are high, manifest as ascites and pleural effusions, and the response rates to single agent bevacizumab, a recombinant humanized monoclonal antibody directed against VEGF, are the highest (16-25%) of any reported in oncology. Antiangiogenics have generally been well tolerated, but are associated with gastrointestinal perforation in 1-2%. New angiogenesis targets are being identified (ANG-2, PDGFR, FGFR, inflammation and the microenvironment), and a plethora of new agents is in clinical development: tyrosine-kinase inhibitors (sunitinib, cediranib, pazopanib), multitargeted agents (XL-184), anti-angiopoietins (trebananib), novel anti-vascular approaches (VB-111 and ombrabulin). Antiangiogenic therapy appears to impact PFS, but does not impact cure. In subsets of patients, it may improve overall survival (OS), and its use remains costly and controversial. Although approved in Europe, the pathway to approval of bevacizumab for ovarian cancer in the US is currently still unclear. There is a clinical need to define the role of these drugs in ovarian cancer management and to identify robust predictive biomarkers.

Keywords: Aflibercept, angiogenesis, bevacizumab, blood, cediranib, disruption, dormancy, endothelium, factor, inhibitor, kinase, maintenance, normalization, novel, pazopanib, progression, remission, survival, trebananib, tyrosine, vascular, vessel.

INTRODUCTION

Angiogenesis is the formation of new blood vessels and lymphatics from existing

***Corresponding author Richard T. Penson:** Massachusetts General Hospital, 55 Fruit Street, Boston, MA, 02114, Mailstop: Yawkey 9064, USA; Tel: 617-726-0845; Fax: 617-724-6898; E-mail: rpenson@partners.org

Atta-ur-Rahman and Muhammad Iqbal Choudhary (Eds)
Copyright © 2014 Bentham Science Publishers Ltd. Published by Elsevier Inc. All rights reserved.
10.1016/B978-0-12-803963-2.50006-5

vasculature. It is an integral part in the evolution of tumor development, crucial for solid tumor growth beyond 100-200μm. It promotes metastases and contributes to tumor behavior such as ascites production. Because of the critical role angiogenesis plays in tumor progression, inhibition of its pathways is currently an extensive target for novel therapies, many of which are Food and Drug Administration (FDA) approved and advancing in clinical trials which present promising results. Theoretical advantages of antiangiogenic drugs include:1) novel mechanism of action, 2) less resistance as the target genome of the drug is stable, 3) applicable to many different tumor types, 4) normalizes or compromises tumor vasculature, and 5) hopes to maintain a stable state of the disease [1].

In particular, bevacizumab (Avastin™), an anti-VEGF humanized monoclonal antibody, is the most well studied antiangiogenic to date. Other antiangiogenic therapies include soluble decoy receptors, tyrosine kinase inhibitors (sunitinib, cediranib, pazopanib), and vascular disrupting agents. Newer agents are targeting other drivers of angiogenesis, such as ANG-2 or multiple targets, to try and anticipate physiologic escape.

MECHANISMS OF ACTION

Angiogenesis is the physiological process through which new blood vessels are formed from pre-existing vessels. Angiogenesis is a normal and crucial process in healing and growth. However, it is also noted to play an important role in cancer growth. On a cellular scale, angiogenesis is initiated and maintained by various pro-angiogenic proteins, including several growth factors. There are a large variety of growth factors of which vascular endothelial growth factor (VEGF) appears to be the primary angiogenic protein [1, 2]. Other growth factors include fibroblast growth factors (FGF), which promote the proliferation of fibroblasts, endothelial cells and Ang1, and Ang2 proteins that stabilize newly formed vessels. Platelet derived growth factor (PDGF) promotes the maturing of vasculature by the development of pericytes [1, 2].

The vascular endothelial growth factors (VEGF A- VEGF E) are considered to be the main proangiogenic signaling proteins[2]. VEGF-A is the main mediator of tumor angiogenesis and subsequently the main focus of targeted VEGF therapy. VEGF promotes cell proliferation, migration, stabilization, and activation of endothelial progenitor cells from the bone marrow to form new blood vessels [2]. Elevated serum levels of VEGF have been shown to be of prognostic value and correlate with the development of ascites. Additionally, increased VEGF expression has been reported in women with advanced stages and more poorly differentiated tumors [3], such that VEGF acts as an autocrine growth factor for tumor cells [4]. Inhibiting VEGF expression is believed to downgrade the production of tumor proliferative factors by endothelial and stromal cells, suggesting effects such as ceased tumor growth, shrinkage of tumor size, and decreased ascites production [5]. It remains unclear if the combination of VEGF inhibitors and cytotoxic drugs is more efficacious than a single anti VEGF agent. In the GOG-218 study, it was noted that patients on single agent bevacizumab suffered from more adverse gastrointestinal side effects like gastric perforation as compared to the patients on combination therapy [6]. The most efficacious effects have been seen when VEGF inhibitors are administered in combination with cytotoxic therapies, rather than as a single agent treatment. A proposed explanation for this success proposes that VEGF "normalizes" tumor vasculature creating a tumor microenvironment more accessible for chemotherapy drug and oxygen delivery.

VEGF signals through three main tyrosine kinase receptors (VEGFR-1, VEGFR-2, and VEGFR-3], which provides another antiangiogenics target, specifically the main receptor VEGFR-2. Tyrosine kinase Inhibitors (TKIs) are small molecules that can pass through the cell membrane, allowing intracellular inhibition of downstream signaling pathways including Raf/MeK/Erk and PI3K/Akt. TKIs such as sutinib and sorafenib compete with ATP binding sites either directly or allosterically respectively, both of which have demonstrated anti tumor activity in early clinical trials. Ang are known as angiopoietins, which are are protein growth factors that are known to promote angiogenesis. They bind to TIE receptors (tyrosine kinase) and activate angiogenesis through cell signaling. Other growth factors include platelet-derived growth factors (PDGF), which bind to its own unique set of tyrosine kinase receptors.

RESISTANCE

Even with the promising activity antiangiogenic agents have demonstrated in clinical trials, drug resistance poses a current challenge. Many patients either do not respond to antiangiogenic therapy or show disease progression. Major mechanisms of resistance include: redundant receptor signaling that results in the up-regulation of alternative pro-angiogenic pathways, epigenetic mechanisms of resistance, and tumor cell metastases to normal tissues that can benefit from normal tissue vasculature rather than requiring neovascularization [7]. Furthermore, hypoxia resistant clones (as seen with the loss of the p53 gene) have been observed in antiangiogenic therapies, thus further promoting tumor metastases [8]. For example, in breast cancer and malignant melanoma, antiangiogenic treatment inhibited tumor growth; however, it also promoted metastases [8]. Other evidence suggests that a prominent means of resistance is through increased pericyte production-inducing tumor vasculature stabilization and maturation [9]. Of note, in some cases after bevacizumab treatment was stopped, VEGF levels were observed to increase dramatically, eliminating the advances made during treatment [10]. Anti VEGF drug resistance mechanisms warrant further investigation to identify the circumstances and the patient profiles in which this resistance are likely to take place. Because of the synergistic effect of VEGF signaling, the use of combination antiangiogenic therapies may bring about the greatest benefit.

GENERAL TOXICITIES

Bevacizumab and other antibody therapies are generally well tolerated. Bevacizumab shares distinct toxicities with other antiangiogenic agents. These include hypertension, proteinuria, impaired wound healing and other serious complications such as arterial (ATE) and venous thromboembolism (VTE), as well as intestinal perforations. Among all the non-gynecological solid tumors, the most common side effects of bevacizumab are hypertension and proteinuria[53]. ATEs estimated occurrence rate is 3-5% and are of increasing risk with increasing age [>65 years) and prior occurrence of ATEs. In comparison, there have been no significant differences in VTEs reported[53]. Gastrointestinal (GI) perforations are of major concern in the use of bevacizumab, and although their overall occurrence rate is low, their severity and high prevalence in ovarian cancer stresses concern. In comparison to antibody therapies, TKIs often exhibit a wider range of toxicities, such as diarrhea, fatigue, nausea, and skin toxicities [11].

Despite bevacizumab's novel mechanism, studies have also been done to show its limitations in efficacy. In the study by de Gramont A *et al.*, it was seen that bevacizumab did not increase the period of disease-free survival when it was added to adjuvant chemotherapy in patients with resected stage 3 colorectal cancer [12]. More importantly, it was noted in this phase III study that patients who had bevacizumab on top of their typical chemotherapy scored poorly in overall survival data [12].

AGENTS BEVACIZUMAB

Trials

Currently bevacizumab is FDA approved for cancer of the lung, kidney, brain and colon and was recently approved in Europe for the treatment of ovarian cancer (OC). Recently, two major phase III trials have been done to assess the potential of bevacizumab as a first line treatment in OC. GOG 218 was a three arm, double-blind, placebo controlled study of patients presenting advanced stage III or IV OC, peritoneal cancer (PPC), or fallopian tube (FT) cancer who had undergone surgical debulking [6]. A sizeable percentage (40%) of stage III patients were suboptimally debulked. Patients were randomly assigned to [3] standard chemotherapy treatment (carboplatin and paclitaxel), [13] standard treatment plus bevacizumab (15mg/kg), and [4] standard treatment bevacizumab followed by maintenance. A statistically significant 4-month increase in progression free survival (PFS) of the maintenance therapy arm in comparison to the control group resulted. However, no significant difference in PFS was reported between those who received chemotherapy alone *vs.* chemotherapy and bevacizumab. Most importantly, no advantage for median OS was found. Prevalent toxicities included grade 2 and grade 3 hypertension (23% and 10% respectively) [14]. Grade 3 or greater GI toxicities were reported in 2.3% of patients [14]. It should be noted that patients with a history of small bowel obstruction were excluded, possibly contributing to the low rate of GI events.

Bevacizumab was also evaluated in the International Collaborative Neoplasm 7 (ICON7) study. This two armed, non-placebo controlled study randomized individuals to receive only standard chemotherapy (carboplatin and paclitaxel) or chemotherapy with bevacizumab (7.5mg/kg), followed by 12 cycles of maintenance. Extended PFS was found in the bevacizumab treated arm (19.8 months *vs.* 17.4 months).

Given the high recurrence rate of epithelial ovarian cancer (EOC), bevacizumab has additionally been examined as a viable treatment in platinum sensitive and resistant patients with recurrent disease. OCEANS, a double blind, placebo controlled trial studied bevacizumab plus carboplatin and gemcitabine against carboplatin and gemcitabine alone in platinum sensitive patients with recurrent EOC, PPC, or FTC. The individuals treated with bevacizumab exhibited longer PFS (12.4 months *vs.* 8.4 months). A need exists for treatment in platinum resistant EOC patients, imparting great value to the ongoing AURELIA trial, which will measure the effects of bevacizumab either alone or in combination with different platinum compounds (paclitaxel, topotecan, or liposomal doxorubicin) in platinum resistant patients.

Toxicities

Hypertension

The pathogenesis of the class effect hypertension is currently unknown. It has been proposed that vascular rarefaction or a decrease in the production of nitric oxide production and metabolism is the cause. Among breast, lung, colorectal, and kidney cancer patients receiving a low dose of bevacizumab (3, 5, or 7 mg/kg/dose), the relative risk of hypertension was 3.0 *versus* 7.5 in the high dose (10 or 15 mg/kg/dose) group [15] suggesting a dose dependent relationship. Higher rates of hypertension have also been noted with the addition of angiogenesis inhibiting cytotoxic agents such as taxanes and topotecan to bevacizumab or aflibercept [16]. In a phase III study of combination paclitaxel and bevacizumab *versus* paclitaxel alone, 15% of women with metastasized breast cancer receiving combination therapy developed grade 3 or 4 hypertension compared to zero in the monotherapy paclitaxel group [36].

Proteinuria

Proteinuria is another class-effect of antiangiogenic agents thought to occur as a result of damaged or dysfunctional glomerular endothelium. Similar to hypertension, the relative risk of proteinuria differed between the same low dose and high dose with a risk of 1.4 and 2.2 respectively [15]. In one study, cediranib showed a rapid onset on proteinuria [17]. Of all of the proteinuria incidences in this study (30%), 50% developed by week 2, 92% by week 4, and 100% by week 6. Because of an association with rapid onset and the asymptomatic behavior of proteinuria, monitoring by periodic dipstick analysis for patients on cediranib is recommended.

Thrombotic Events

Rare but serious, both arterial (ATE) and venous thromboembolic events (VTE) have been observed with the use of bevacizumab. Under normal physiological conditions, platelet secreted VEGF is thought to promote vascular repair. Consequently, VEGF blockage may compromise this process, threatening cell lining and promoting platelet aggregation. Pooled data from five trials of 1,745 colorectal (CRC), lung, and breast cancer patients randomized to receive either standard chemotherapy or chemotherapy in combination with bevacizumab reported ATE rates of 1.7% and 3.8% respectively but no statistically significant differences in VTE rates (9.97% *vs.* 9.85%) [18]. Other large meta analyses have reported small increases in VTE rates (6.3% *vs.* 4.2%), however, the results were not statistically significant [19].

Gastrointestinal Perforations

Bevacizumab-related GI perforations were first observed in metastasized CRC and have been further observed in other diseases such as ovarian cancer. GI perforation risk varies in relation to the primary tumor site. In particular, ovarian cancer has demonstrated a relatively high GI perforation risk of 11% [20]. OC characteristics such as diffuse peritoneal disease and bowel obstructions may play a role in increased perforation risk [20]. Additionally, it has also been shown that patients who undergo≥3 previous chemotherapy regimens are more likely to have bowel perforations.

Han has reported an overall GI perforation risk of 5.4% from collective data analysis [21]. In 2005, the National Cancer Institute (NCI) released a warning letter when 5/44 patients in a phase II trial exhibited GI perforations with 1 of the 5 patients dying [21]. One retrospective study of recurrent OC patients receiving chemotherapy alone or chemotherapy with bevacizumab reported perforation risks of 6.5% and 7.2% respectively and a relative risk of 1.09 [22] suggesting that bevacizumab, when added to salvage chemotherapy, may pose a low risk to benefit ratio. However, combination therapy of two antiangiogenic drugs may pose an escalated risk. In a phase II study of bevacizumab and erlotinib, 4/26 patients experience GI perforations of which 2 were fatal [23].

At the Cleveland Clinic, a retrospective study was done that excluded patients with 1) clinical bowel obstruction symptom, 2) evidence of recto sigmoid on pelvic exam and 3) evidence of bowel involvement on CT [44]. Out of 25 platinum resistant patients with a median of 5 previous chemotherapy regimens, zero patients developed GI perforations. This emphasizes the importance of patient selection and its potential impact to avoid adverse effects. A summary of potentially important antiangiogenic agents in clinical trial is listed in Table **1**.

AFLIBERCEPT (VEGF TRAP)

VEGF trap is a decoy receptor created from the fusion of the Fc segment of IgG1 with the extracellular domains of VEGFR-1 and VEGFR-2. It has strong binding affinity and specificity for VEGF-A as well as placenta growth factor (PGF). Successes from preclinical data include reductions in, vasculature normalization and regression, as well as inhibition of new tumor growth and ascites production. These all support VEGF trap's efficacious role in treating cancer. VEGF trap monotherapy was examined in a randomized phase II trial of 215 advanced stage EOC patients who have had more than 3 previous treatment regimens. Patients were allocated IV administration of either 2 mg/kg or 4 mg/kg periodically every two weeks. Clinical investigators response rates of the 4 mg/kg *vs.* 2 mg/kg were 7.3% and 3.8% respectively, while the IRS showed a statistically insignificant response rate of 4.6% *vs.* 0.9% [24].

A valuable phase II study of VEGF trap as combination therapy established that 6 mg/kg of aflibercept could be given safely in addition to docetaxel [75 mg/m^2] with an objective response rate of 54% [25]. Two out of 9 patients had to stop treatment due to high-grade hypertension and ulceration toxicities. The results of this trial warrant further evaluation.

In the phase III VELOUR trial comparing combination therapy of folinic acid, fluoruracil, and irinotecan (FOLFIRI) with either bevacizumab or aflibercept in metastasized colorectal cancer (mCRC) patients, significant differences were observed among the rates of hypertension and proteinuria [26]. In the aflibercept arm, proteinuria of all grades occurred in 62% (8% grade 3/4] of patients *versus* only 3.3% (1.9% grade 3/4) of patients in the bevacizumab arm. All grade hypertension rates were 41% (19% grade 3/4) *versus* 28 % (5%) respectively.

IMC-1121B

IMC-1121B is a humanized monoclonal antibody targeting VEGFR-2 expression either through autocrine or paracrine mediated pathways [24]. A phase I trial of patients with advanced cancer showed well tolerated doses of 6 mg/kg/wk and ≥ grade 2 toxicities of anorexia, vomiting, anemia, depression, fatigue, and insomnia [27]. However, it remains questionable whether the toxicities resulted directly from the study drug. Studies are continuing into phase II trials which are currently recruiting patients for recurrent OC, metastatic renal, and metastatic melanoma and non-small cell lung cancer [24].

Tyrosine Kinase Inhibitors (TKIs)

Multiple TKI drugs are under current exploration because of the major role kinases play in growth factor signaling and kinase mutations' notable contribution to tumor genesis [28].

Cediranib

Cediranib (AZD2171] is an oral TKI that targets all three VEGFRs as well as c-kit. A phase II open label trial of 46 total patients with either recurrent OC, FT or PPC showed an overall RR of 17% and a median PFS of 5.2 months [29]. Twenty-three percent of the patients were removed for high-grade toxicities before cycle 2, most commonly hypertension (46%), fatigue (24%), and diarrhea (13%). ICON6, a 3 stage double blind, placebo-controlled trial also investigated the potential of cediranib in platinum sensitive OC patients with recurrent disease.

Data from previous trials reported significant toxicities at a 30 mg dose. As a result, a dose reduction to 20mg was established. Stage I of ICON6 analyzed toxicity levels in 60 patients given 20mg of cediranib plus combination chemotherapy [30]. Overall, there was a 55% response of grade 3/4 toxicities. Adverse Events (AEs) were manageable by dose reductions.

The development of cediranib was halted by AstraZeneca, but one of their phase III trials ICON-6, recently reported promising results. ICON-6 is a three-arm randomized placebo controlled maintenance trial of cediranib for 18 months or longer if the patient is clinically benefitting. It enrolled 456 participants with ovarian cancer once in remission and preliminary results were presented at the 2013 European Cancer Congress (ECC) on October 29, 2013, unexpectedly showing, a statistically significant increase in overall survival [31]. Patients received chemotherapy alone, concurrent cediranib 20 mg per oral once daily (PO QD) followed by placebo, or concurrent followed by maintenance cediranib (maintenance arm). Chemotherapy was at physician discretion (up to #6 carboplatin/paclitaxel, carboplatin/gemcitabine, or single-agent carboplatin). Median PFS was 8.7 months in the chemotherapy alone arm and 11.1 months in the cediranib maintenance arm (HR: 0.57; $P = 0.00001$). Overall survival improved from 20.3 months to 26.3 months with cediranib maintenance (HR: 0.70; $P = 0.042$).

The cediranib maintenance dose was reduced from 30 mg PO QD to 20 mg PO QD during the duration of the study. Whether there is a dose response is unclear. Cediranib has been given at 30-45 mg PO QD, but the higher dose is intolerable in this population [31]. The only side effect that was significantly worse with cediranib maintenance was diarrhea ($p < 0.001$).

What this means for approval is not clear, but it is encouraging that antiangiogenic benefit in PFS may translate into overall survival gains in the right subsets of patients in the right context.

Sorafenib

Sorafenib targets the RAF pathways and multiple kinase receptors. Promising activity in advanced renal cell carcinoma and hepatocellular carcinoma led to FDA approval [32]. A Gynecologic Oncology Group (GOG) study evaluated sorafenib oral monotherapy (400 mg BID) in patients with recurrent OC or PPC [33]. PFS of 6 months was reported in 24% of patients, but only 3.4% of patients had a partial response. Significant grade 3/4 toxicities noted included rash, hand-foot syndrome, metabolic, GI, cardiovascular, and pulmonary toxicities. Another phase II trial assessing sorafenib in combination with gemcitabine in recurrent OC also showed similar results [32]. At 6 months, 23% of patients were PFS [6], and 4.7% had a partial response.

Sunitinib

Sunitinib is another multikinase inhibitor blocking receptors and rearranged during transfection (RET). Sunitinib treatment as a single agent was examined in a phase II trial of 30 eligible patients receiving 50 mg daily [34]. Stable disease presented in 53% of patients, 3.3% showed a partial response, and the median PFS was 4.1 months. Reported toxicities were those common to antiangiogenics: fatigue, GI, hand-foot syndrome, and hypertension. Of note, toxic fluid accumulation in a subset of patients resulted in a patient division receiving a dose reduction of 37.5 mg daily. Higher toxicities were associated with a higher dosage; however, activity was only shown in patients receiving the 50 mg dose.

Pazopanib

Pazopanib, currently FDA approved for the treatment of advanced renal cell carcinoma, inhibits VEGFR-1, VEGFR-2, VEGFR-3, PDGFF-α and PDGFR-β. Patients with recurrent OC, FT, or PPC were evaluated in a phase II trial of pazopanib monotherapy [35]. Out of 36 patients, eleven demonstrated a cancer antigen (CA)- 125 response (31%) and an 18% overall response rate was reported. The most common AE were elevated alanine aminotransferase (ALT) and aspartate aminotransferase (AST) levels (8% for each). Further studies continue, in particular, a phase III trial in women who have achieved a partial or complete response to primary platinum-based adjuvant chemotherapy in ovarian cancer [36].

The POIZE study randomized 940 patients with OC, PPC, or FTC without progression after surgery and ≥5 cycles of platinum-taxane chemotherapy to pazopanib 800 mg PO QD or placebo for up to 24 months. Most of the patients had advanced (stage III/IV 91%) optimally cytoreduced disease (58%). Pazopanib prolonged PFS from a median 12.3 to 17.9 months (HR = 0.766; 95% CI: 0.64-0.91; p = 0.0021) [37]. The overall survival data is not yet mature. Pazopanib was associated with more serious adverse events (26% *vs.* 11% with placebo) including hypertension, diarrhea, nausea, headache, fatigue, and neutropenia, but not GI perforation.

Pazopanib may be approved for ovarian cancer in Europe based on this data.

BIBF1120

BIBF1120 works by simultaneously inhibiting VEGFRs, PDGFs, and FGFRs. Current clinical trials are underway for various carcinomas, such as Non-small cell lung cancer (NSCLC), prostate, ovarian and colorectal cancer. A phase II, randomized double blind, controlled trial compared maintenance therapy of BIBF 1120 *versus* placebo in patients who had just completed chemotherapy for relapsed OC [38]. PFS rates of BIBF 1120 and the placebo arm were 16.3% and 5.0% respectively. Grade 3 and 4 toxicity rates were reported as similar between the two groups (34.9% for BIBF 1120, 27.5% for placebo). Diarrhea, nausea, and/or vomiting were all more common among patients given BIBF 1120. Grade 3 and 4 hepatotoxicity was much higher in BIBF 1120 (51.2%) compared to the placebo (7.5%). However, this was of no clinical significance.

Cabozatinib (XL 184]

Cabozatinib (XL184) is a dual inhibitor of the MNG Hos transforming gene (MET) and VEGF pathways. Overexpression of MET has been observed in advanced ovarian cancers [39] and because of the achievements with anti VEGF therapies, a combination target yields exciting interest. A phase II randomized discontinuation trial showed an overall RR of 24% [40]. In particular, significant RRs occurred in patients with platinum refractory (18%), platinum resistant (22%) and platinum sensitive (28%) disease [41]. Adverse events [3] (grade 3) were hand and foot syndrome (10%), diarrhea (8%), and fatigue (4%). Two deaths were reported, one due to an enterocutaneous fistula and the other due to intestinal perforation. Cabozatinib exhibits clinical activity in ovarian cancer patients regardless of platinum status, posing it as an effective treatment possibility. However, the potential severity of its side effects in a small population must be considered. More studies are needed concerning safety and dosage. A recent phase II trial of single agent cabozatinib administered in metastatic, hormone refractory prostate cancer showed motivating response rates [42]. In patients with bone metastases, 76% showed complete or partial bone resolution. There was also measurable tumor regression in 74% of patients, and a disease control rate of 68%.

Brivanib

Brivanib is a small molecule dual TKI of FGF and VEGF administered orally. Preclinical and phase I studies have showed promising antitumor activity. In one early study of four patients with advanced tumors given single dose of brivanib, one patient showed stable disease [43]. This study reported Brivanib as well tolerated, with fatigue as the most frequent AE occurring in all four patients. GI events such as nausea, diarrhea, and constipation also occurred in all patients. Promising results from an open label phase II trial of brivanib in hepatocellular cancers (HCC) has advanced investigation into more studies. [44]

ADDITIONAL ANTIANGIOGENICS

Trebananib (AMG 386]

Trebananib is a peptide-Fc fusion protein that blocks binding of both angiopoietin-1 and angiopoietin-2 with their receptor, Tie2. Preclinical trials showed inhibition of both angiopoietin 1/2 exhibited vessel regression, while single inhibition of angiopoietin did not [44]. A significant phase II trial randomized patients to one of three treatment arms all treated with weekly paclitaxel plus: (A) 10 mg/kg trebananib, (B), 3 mg/kg, or (C) placebo [44]. Reported PFS was 7.2 months for arm A as compared to 4.6 months in arm C. Reported response rates by Response Evaluation Criteria in Solid Tumors (RECIST) were 37%, 19% and 27% in arms A, B and C respectively [44]. Additional analyses demonstrated a higher response rate in trebananib treated patients with platinum refractory disease. Adverse events included hypokalemia, venous and arterial thromboembolic events, hypertension (all ≥ grade 3), and peripheral edema. Disease progression was the most common cause for patient discontinuation. Notably different from bevacizumab toxicities is the lack of GI perforations. These results have led to three phase III trials, the first of which, TRINOVA-1, evaluated the combination of weekly paclitaxel +/- trebananib which reported a PFS advantage at the ECC meeting in Amsterdam on October 1, 2013 [45]. PFS was improved from 5.4 to 7.2 months (HR: 0.66, p<0.001), but OS was not statistically superior 19.0 for the combination *vs.* 17.3 (HR: 0.86, p=0.19).

In TRINOVA-1, trebananib was well tolerated and barely increased the incidence of Grade ≥3 adverse events (56%) compared to paclitaxel alone (54%). With the different mechanism of action, the toxicity profile is very different. Trebananib did not cause the typical VEGF/VEGFR targeting side effects with no increase in hypertension (6% *vs.* 4%), proteinuria (3% in both arms), arterial thrombotic events (<1% in both arms), and impaired wound healing (<1% in both arms), but still had a 1% incidence of GI perforation or fistula. Curiously, it causes edema, effusions, and ascites through an unclear mechanism [46].

MEDI3617

The high expression of ANG-2 in tumors, its possible role in metastases and inflammation, and the results of AMG-386 (angiopoietin antagonist peptide Fc fusion protein) have brought an interest in targeting angiopoietin pathways and the development of MEDI3617. MEDI3617 is a human anti-Ang2 monoclonal antibody that prevents Ang2 binding to the Tie2 receptor. Currently, a phase I trial is being conducted to evaluate the safety, tolerability, efficacy, and pharmokinetics of MEDI3617 in patients with advanced tumors where no successful treatment exists [47]. This trial will examine the effects of MEDI3617 as a single agent and in combination with bevacizumab, paclitaxel, paclitaxel plus carboplatin, or gemcitabine combination chemotherapy. Xenograft tumor models have shown MEDI3617 monotherapy to reduce tumor angiogenesis and increase tumor hypoxia [48]. Furthermore, when combined with bevacizumab, tumor growth was delayed. Preliminary data from the Phase I trial has reported toxicities common to most antiangiogenics drugs. Most frequent adverse events reported from 21 patients included: fatigue (24%), diarrhea (19%), nausea (19%), dyspepsia (14%), and headache (14%) [49].

Vascular Disrupting Agents (V*DA*s)

VDAs differ from other antiangiogenic drugs in their mechanism of action. Rather than inhibiting the formation of new vasculature, VDAs destroy existing tumor vasculature, leaving the tumor without the blood and oxygen necessary to survive [50]. Oxi4503, a potent VDA was evaluated in a phase I trial of patients with solid tumors [51]. In this study, one patient with heavily pretreated ovarian cancer showed a partial response by RECIST criteria. Frequent adverse events included hypertension, tumor pain, anemia, lymphopenia, nausea, vomiting, and fatigue. Phase II trials are to continue. In addition, zybrestat, a drug being evaluated in combination with carboplatin and paclitaxel among advanced platinum resistant ovarian cancer is moving onto stage II trials after 3 out of 18 patients showed a partial response in stage I [52].

NOVEL DELIVERY OF ANTIANGIOGENICS

Vascular Biogenics Laboratory's VTS™ (Vascular Targeting System) enables control of gene expression to areas in which angiogenesis is taking place, and allows a drug to be both antiangiogenic and a vascular disrupting agent (VDA). This is promoted as a "biological knife" to destroy tumor vasculature. Preclinical pharmacological and toxicology studies on VB-111 show tissue specificity for tumor through the adenoviral vector for the murine PPE-1 promoter, no significant damage to normal non-cancerous tissues or to the normal blood vessels in the body, and dramatic (> 90%) reduction in the Lewis Lung Cancer Model by activation of the TNFα-1 and Fas from the plasmid. To date, in phase 1 clinical trials, more than 100 patients have been treated with an excellent safety profile (fever and chills), and promising clinical efficacy.

Angiogenesis and Immunosuppression

Much research exists and suggests that the limitations of antiangiogenics therapies are due to the failure of drug developers to target overlapping mechanisms of immunosuppression and angiogenesis. Common cell types have been identified in the functioning of both immunosuppression and angiogenesis in tumors, suggesting an integrated promotion of one another. Similar to wound healing, tumor development requires neovascularization and suppression of excessive inflammation, indicating that dual regulation of tumor angiogenesis and immunosuppression could result in an efficacious treatment [15].

LIMITATIONS AND FUTURE CLINICAL USE

Optimal Use

The optimal timing and use of antiangiogenic agents is presently unknown. Fig. **1** illustrates the present biases in treatment. There are multiple questions to be addressed, such as: should these agents be used as a first or second line treatment, as single agent therapy in consolidating a response or in combination with chemotherapy?

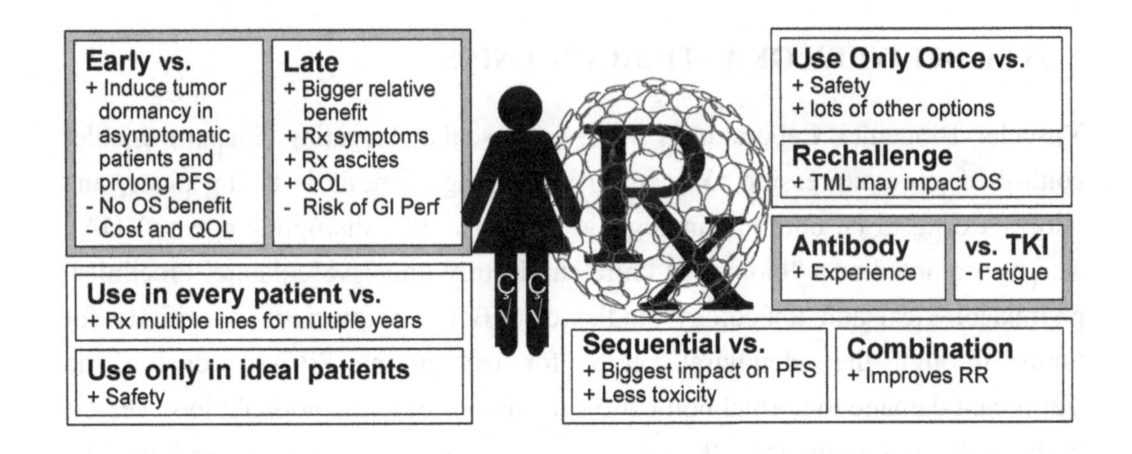

Figure 1: Decisions in Antiangiogenic Therapy.

Breast Cancer Story

In February 2008, bevacizumab was approved under the FDA accelerated approval program for treatment of serious or life threatening metastatic breast cancer, after results from a key study (E2100) were published. In this phase III study, although a statistically significant improvement in PFS was reported when patients were treated with bevacizumab (11.8 months *vs.* 5.9 months) [53], no differences in OS were found (26.7 months *versus* 25.2 months), and the bevacizumab treated group showed no improvement in quality of life (QOL) accompanied by higher levels of toxicities. The fast track approval of bevacizumab required that additional studies be completed to validate the results from E2100.

Two subsequent phase III studies were performed (AVADO and RIBBON-1). No OS or QOL improvements were reported in either study [54, 55]. Moreover, as the gains in PFS were modest and tied to an increase in potentially serious toxicities, the FDA advisory panel voted to revoke bevacizumab use for breast cancers.

Regardless of the FDA's revoking of bevacizumab in the treatment of breast cancer, the potential of bevacizumab is still recognized, and there is encouragement to conquer the challenge in developing identification markers of the subset population of individuals who can benefit and minimize the safety risks.

REGISTRATION STRATEGIES

Optimal results and successes in extending the OS of ovarian cancer patients by the use of antiangiogenic drugs is likely to be accomplished through different strategies and by targeting multiple angiogenesis pathways. One approach of interest is to target the PDGF signaling pathway as it plays an important role in pericyte regulation [36]. In a preclinical study of ovarian cancer mouse models, a combination of bevacizumab plus a PDGF-B inhibitor (AX102) demonstrated higher efficacy than single agent bevacizumab [36].

Another targeting area of concentration is the notch-signaling pathway. Vascular endothelial cells primarily express the Delta-like ligand-4 (Dll4] that binds to notch cell surface receptors, notch 1 and notch 4 [54]. Experiments have shown this receptor-ligand pair as vital to the vascular development in mice embryos [13]. Compelling results from a Dll4 blockade in animal xenograft models showed an increase in tumor vasculature but a decrease in tumor growth [54].

Focal adhesion kinase (FAK) is overexpressed in a variety of tumor types, providing another promising combination antiangiogenic therapy. FAK has an identified role in angiogenesis, contributing to adhesion, mobility, cell survival and growth by linking signals from integrins or growth factors to intracellular cytoskeletal and signaling proteins [55]. Recent research has studied the effects of two FAK inhibitors on primary human umbilical vein endothelial cells (HUVEC) [56]. Both inhibitors reduced HUVEC viability, migration, and tube formation, while one FAK inhibitor also induced endothelial cell apoptosis.

COMBINATION ANTIVASCULAR THERAPIES

Due to drug resistance and limited improvements in OS, combination antivascular therapies targeting various angiogenic mechanisms are of interest. A phase I trial of combination sorafenib and bevacizumab showed a 43% partial response rate in ovarian cancer patients [57]. However, toxicity was much higher than with single agent anti VEGF therapy. The majority (74%) of patients required a dose reduction of sorafenib. Proteinuria and thrombocytopenia presented as dose limiting toxicities, while other adverse effects included hypertension, hand-foot syndrome, diarrhea, transaminitis, and fatigue.

EGFR expression has been shown to upregulate VEGF secretion [58]. Currently, various studies are investigating the dual inhibition of these two pathways. A phase II clinical trial of heavily pretreated patients with recurrent ovarian, PP, and FTC evaluated the safety of bevacizumab plus erlotinib in thirteen patients [23]. No advantages were noted compared to single agent bevacizumab. Two fatal GI perforations were reported along with higher rates of grade 3 and 4 toxicities.

The addition of VDAs to bevacizumab is also of interest. A synergistic effect of their inhibition has been observed in preclinical xenograft models of two VDAs (CA4P or OXi4503) [59]. When bevacizumab was co-administered with either VDA, significantly greater response rates and tumor growth delays were achieved when compared to any single agent therapy. CA4P in combination with bevacizumab recently revealed exciting results in a phase I trial of 15 patients with advanced cancers and is continuing onto phase II [60, 61]. Common toxicities were hypertension, headache, lymphopenia, pruritus, and pyrexia [62].

CONCLUSION

The anticipated success of antiangiogenic drugs as a promising new strategy in treating ovarian cancer, has fallen short of approval in the US to date because of a lack of a demonstrated overall survival advantage. A challenging learning curve has emerged as researchers continue to seek a better mechanistic understanding of the complexity of angiogenic pathways, try and identify predictive markers, and attempt to reach broad agreement on what represent clinically meaningful endpoint for ovarian cancer trials in a fiscally tighter era. There remain challenges in selecting optimal dosages, drugs, combinations, investigating new targets and overcoming physiologic escape or drug resistance. With a superior understanding of the toxicity profiles, we are much better at avoiding GI perforations. The two key issues are: identifying predictive markers to target optimal patient groups, and establishing paths to registration in the US for drugs with clear clinical benefit.

Table 1: Clinical development of Antiangiogenic Agents

Generic Name	**Proprietary Name**	**Target(s)**	**Categories of the Therapeutic Agent**	**Route**	**Clinical Development**
Bevacizumab	Avastin	VEGF	Antibody	IV	Approved EU- III GOG-218 ICON-7 GOG-213 OCEANS GOG-252 GOG262 GOG-241 AURELIA
VEGF Trap Aflibercept Ziv-aflibercept	ZIV- Zaltrap	VEGF PlGF	Fusion protein Aptamer	IV	III
GW786034	Pazopanib	VEGFR PDGFR cKit	TKI	PO	III OVAR-16
BIBF1120	Nintedanib	VEGFR PDGFR FGFR	TKI	PO	III
SU11248	Sunitinib	VEGFR PDGFR, cKit	TKI	PO	II
Sunitinib, SU11248	Sutent	VEGFR PDGR	TKI	PO	III
Zactima ZD6474	Vandetanib	VEGFR1-3 EGFR	TKI	PO	III
AG013736	Axitinib	VEGFR1-3 PDGFR	TKI	PO	III

Table 1: contd....

AZD2171	Cediranib	VEGFR1-3 PDGFR	TKI	PO	III ICON-6
AMG706	Motesanib	VEGFR1-3 PDGFR cKit	TKI	PO	III
Axitinib	Inlyta	VEGFR1-3 PDGFR cKit	TKI	PO	III
PTK787/ZK 222584	Vatalanib	VEGFR2[-3] PDGFR cKit	TKI	PO	II
AMG-386	Trebananib	Angiopoietin	TKI	IV	III
MEDI-3617		Angiopoietin	TKI	IV	I/II
REGN-421		DLL4/Notch	TKI	IV	I
Vandetanib	Vandetanib	EGFR targeting	TKI	PO	II-III
MM-121 AR256212		EGFR targeting	Antibody	IV	I-II
BMS-354825	Dasatinib	EphrinA2 Src/Fak	TKI	PO	III
TKI258	Dovitinib	FGFR PDGFR	TKI	PO	III
E7080	Lenvatinib	FGFR PDGFR VEGFR	TKI	PO	II
BIBF 1120	Vargatef	FGFR PDGFR VEGFR	TKI	PO	III
BMS-582664	Brivanib	FGFR VEGFR	TKI	PO	I
MK-0646		IGF1-R	MK-0646	IV	I-II
CNTO 328	Siltuximab	IL-6	Antibody	IV	II
XL-184	Cabozantinib	MET VEGFR-2	TKI	PO	III
3G3		PDGFR	Antibody	IV	II
BKM-120		PI3K/mTOR	TKI	PO	I
PKC412	Midostaurin	Protein kinase C VEGFR2	Multitargeted protein kinase inhibitor	PO	I
Sorafenib	Nexavar	RAF VEGFR2-3 PDGFR ckit	TKI	PO	II
MLN518	Tandutinib	Type III RTK	TKI	PO	I
AVE8062	Ombrabulin	Vascular disrupting agent	VDA	IV	II
VB-111		Vascular disrupting agent and antiangiogenic	Gene therepy viral vector	IV	I
MK-2206		AKT	TKI	PO	II

ACKNOWLEDGEMENTS

We appreciate the patients at the MGH and our team at the DF/HCC Gynecologic Cancer and their tireless commitment to advancing the field.

CONFLICT OF INTEREST

The authors confirm that this chapter contents have no conflict of interest.

COMPETING INTERESTS

Dr. Penson has served on scientific advisory boards for: Genentech, Inc., Eisai Inc., and Vascular Biogenics Ltd, was the Data Safety Monitoring Board Chair for the OCEANS study, and has received research funding from: Genentech, Inc., ImClone Systems, Inc., Endocyte, Inc., AstraZeneca., Eisai Inc., Amgen Inc., Vascular Biogenics Ltd, and PDL BioPharma, Inc.

REFERENCES

[1] Scappaticci FA. Mechanisms and future directions for angiogenesis-based cancer therapies. J Clin Oncol. Sep 15 2002; 20(18): 3906-3927.

[2] Spannuth WA SA, Coleman RL. Angiogenesis as a strategic target for ovarian cancer therapy. Nat Clin Pract Oncol. 2008; 5(4): 194-204.

[3] Li L, Wang L, Zhang W, *et al.* Correlation of serum VEGF levels with clinical stage, therapy efficacy, tumor metastasis and patient survival in ovarian cancer. Anticancer research. May-Jun 2004; 24(3b): 1973-1979.

[4] Kerbel RS. Tumor angiogenesis. N Engl J Med 2008; 358(19): 2039-2049.

[5] Lai GG, Penson RT. Bevacizumab and ovarian cancer. Drugs Today (Barc). Sep 2011; 47(9): 669-681.

[6] Kroep JR, Nortier JW. The role of bevacizumab in advanced epithelial ovarian cancer. Current pharmaceutical design. 2012; 18(25): 3775-3783.

[7] Amini A, Masoumi Moghaddam S, Morris DL, Pourgholami MH. Utility of vascular endothelial growth factor inhibitors in the treatment of ovarian cancer: from concept to application. J Oncol. 2012; 2012: 540791.

[8] Abdullah SE, Perez-Soler R. Mechanisms of resistance to vascular endothelial growth factor blockade. Cancer. Jul 15 2012; 118(14): 3455-3467.

[9] Bergers G, Hanahan D. Modes of resistance to anti-angiogenic therapy. Nat Rev Cancer. Aug 2008; 8(8): 592-603.

[10] Han ES, Monk BJ. Bevacizumab in the treatment of ovarian cancer. Expert Rev Anticancer Ther. Oct 2007; 7(10): 1339-1345.

[11] Mario E Lacouture DJL, Susan E Quaggin. Antiangiogenic Therapy: Tolerability and Management of Side Effects. Targeting Tumor Angiogenesis. 2009.

[12] de Gramont A, Van Cutsem E, Schmoll HJ, *et al.* Bevacizumab plus oxaliplatin-based chemotherapy as adjuvant treatment for colon cancer (AVANT): a phase 3 randomised controlled trial. The lancet oncology. Dec 2012; 13(12): 1225-1233.

[13] Sato S, Itamochi H. Bevacizumab and ovarian cancer. Curr Opin Obstet Gynecol. Feb 2012; 24(1): 8-13.

[14] Burger RA. Experience with bevacizumab in the management of epithelial ovarian cancer. J Clin Oncol. Jul 10 2007; 25(20): 2902-2908.

[15] Zhu X, Wu S, Dahut WL, Parikh CR. Risks of proteinuria and hypertension with bevacizumab, an antibody against vascular endothelial growth factor: systematic review and meta-analysis. Am J Kidney Dis. Feb 2007; 49(2): 186-193.

[16] Stone RL, Sood AK, Coleman RL. Collateral damage: toxic effects of targeted antiangiogenic therapies in ovarian cancer. Lancet Oncol. May 2010; 11(5): 465-475.

[17] Robinson ES, Matulonis UA, Ivy P, *et al.* Rapid development of hypertension and proteinuria with cediranib, an oral vascular endothelial growth factor receptor inhibitor. Clin J Am Soc Nephrol. Mar 2010; 5(3): 477-483.

[18] Scappaticci FA, Skillings JR, Holden SN, *et al.* Arterial thromboembolic events in patients with metastatic carcinoma treated with chemotherapy and bevacizumab. J Natl Cancer Inst. Aug 15 2007; 99(16): 1232-1239.

[19] Helen X Chen JNC. Adverse effects of anticancer agents that target the VEGF pathway. Nature Reviews Clinical Oncology. 2009; 6: 465-477.

[20] Abu-Hejleh T, Mezhir JJ, Goodheart MJ, Halfdanarson TR. Incidence and management of gastrointestinal perforation from bevacizumab in advanced cancers. Curr Oncol Rep. Aug 2012; 14(4): 277-284.

[21] Han ES MB. What is the risk of bowel perforation associated with bevacizumab therapy in ovarian cancer? Gynecologic oncology. 2007; 105(1): 3-6.

[22] Sfakianos GP NT, Halverson CB, Panjeti D, Kendrick JE 4th, Straughn JM Jr. The risk of gastrointestinal perforation and/or fistula in patients with recurrent ovarian cancer receiving bevacizumab compared to standard chemotherapy: a retrospective cohort study. Gynecol Oncol. 2009; 114(3): 424-426.

[23] Nimeiri HS, Oza AM, Morgan RJ, *et al.* Efficacy and safety of bevacizumab plus erlotinib for patients with recurrent ovarian, primary peritoneal, and fallopian tube cancer: a trial of the Chicago, PMH, and California Phase II Consortia. Gynecologic oncology. Jul 2008; 110(1): 49-55.

[24] Macromolecular Anticancer Therapies. (Chapter 14): 480.

[25] Coleman RL, Duska LR, Ramirez PT, *et al.* Phase 1-2 study of docetaxel plus aflibercept in patients with recurrent ovarian, primary peritoneal, or fallopian tube cancer. Lancet Oncol. Nov 2011; 12(12): 1109-1117.

[26] Wang TF, Lockhart AC. Aflibercept in the treatment of metastatic colorectal cancer. Clin Med Insights Oncol. 2012; 6: 19-30.

[27] D. R. Camidge SGE, S. Diab, L. Gore, L. Chow, C. O'Bryant, E. Temmer, A. Ervin-Haynes, T. Katz, F. Fox and R. B. Cohen. A phase I dose-escalation study of weekly IMC-1121B, a fully human anti-vascular endothelial growth factor receptor 2 (VEGFR2) IgG1 monoclonal antibody (Mab), in patients (pts) with advanced cancer. Journal of Clinical Oncology, 2006 ASCO Annual Meeting Proceedings (Post-Meeting Edition). 2006; 24(No 18S (June 20 Supplement)): 2006: 3032.

[28] Karlan BY, Oza AM, Richardson GE, *et al.* Randomized, double-blind, placebo-controlled phase II study of AMG 386 combined with weekly paclitaxel in patients with recurrent ovarian cancer. J Clin Oncol. Feb 1 2012; 30(4): 362-371.

[29] Monk BJ PA, Vergote I, *et al.* Abstract LBA41. European Cancer Congress 2013.

[30] Monk BJ, Minion L, Lambrechts S, Vergote IB, Devoogdt N, Karlan BY. Incidence and management of edema associated with trebananib (AMG 386). Gynecologic oncology. Sep 2013; 130(3): 636-641.

[31] Zhang J, Yang PL, Gray NS. Targeting cancer with small molecule kinase inhibitors. Nature reviews. Cancer. Jan 2009; 9(1): 28-39.

[32] Matulonis UA, Berlin S, Ivy P, *et al.* Cediranib, an oral inhibitor of vascular endothelial growth factor receptor kinases, is an active drug in recurrent epithelial ovarian, fallopian tube, and peritoneal cancer. J Clin Oncol. Nov 20 2009; 27(33): 5601-5606.

[33] Raja FA, Griffin CL, Qian W, *et al.* Initial toxicity assessment of ICON6: a randomised trial of cediranib plus chemotherapy in platinum-sensitive relapsed ovarian cancer. British journal of cancer. Sep 27 2011; 105(7): 884-889.

[34] Randomised double-blind phase III trial of cediranib (AZD 2171) in relapsed platinum sensitive ovarian cancer: Results of the ICON6 trial. 2013 European Cancer Congress (ECC2013), Presidential session III. Monday 30 September 2013; Abstract no: LBA10.

[35] Welch SA, Hirte HW, Elit L, *et al.* Sorafenib in combination with gemcitabine in recurrent epithelial ovarian cancer: a study of the Princess Margaret Hospital Phase II Consortium. Int J Gynecol Cancer. Jul 2010; 20(5): 787-793.

[36] Matei D, Sill MW, Lankes HA, *et al.* Activity of sorafenib in recurrent ovarian cancer and primary peritoneal carcinomatosis: a gynecologic oncology group trial. J Clin Oncol. Jan 1 2011; 29(1): 69-75.

[37] Biagi JJ, Oza AM, Chalchal HI, *et al.* A phase II study of sunitinib in patients with recurrent epithelial ovarian and primary peritoneal carcinoma: an NCIC Clinical Trials Group Study. Ann Oncol. Feb 2011; 22(2): 335-340.

[38] Friedlander M, Hancock KC, Rischin D, *et al.* A Phase II, open-label study evaluating pazopanib in patients with recurrent ovarian cancer. Gynecologic oncology. Oct 2010; 119(1): 32-37.

[39] Zand B, Coleman RL, Sood AK. Targeting angiogenesis in gynecologic cancers. Hematol Oncol Clin North Am. Jun 2012; 26(3): 543-563, viii.

[40] Du Bois A FA, Kim JW, *et al.* Randomized, double-blind, phase III trial of pazopanib versus placebo in women who have not progressed after first-line chemotherapy for advanced epithelial ovarian, fallopian tube, or primary peritoneal cancer (AEOC): Results of an international Intergroup trial (AGO-OVAR16). J Clin Oncol 31, 2013 (suppl; abstr LBA5503).

[41] Ledermann JA, Hackshaw A, Kaye S, *et al.* Randomized phase II placebo-controlled trial of maintenance therapy using the oral triple angiokinase inhibitor BIBF 1120 after chemotherapy for relapsed ovarian cancer. J Clin Oncol. Oct 1 2011; 29(28): 3798-3804.

[42] Ayhan A, Ertunc D, Tok EC, Ayhan A. Expression of the c-Met in advanced epithelial ovarian cancer and its prognostic significance. Int J Gynecol Cancer. Jul-Aug 2005; 15(4): 618-623.

[43] R. J. Buckanovich RB, A. Sella, B. I. Sikic, X. Shen, D. A. Ramies, D. C. Smith, I. B. Vergote; University of Michigan, Ann Arbor, MI; Sheba Medical Center, Tel HaShomer, Israel; Assaf Harofeh Medical Center, Zerifin, Israel; Stanford University School of

Medicine, Stanford, CA; Exelixis, South San Francisco, CA; University Hospital Leuven, Leuven, Belgium. Activity of cabozantinib (XL184) in advanced ovarian cancer patients: Results from a phase II randomized discontinuation trial. 2011 ASCO Annual Meeting. 2011; Abstract No. 5008.

[44] 2011 ASCO: Exelixis Reports Expanded Cabozantinib (XL184) Phase II Data For Advanced Ovarian Cancer; Six Deaths Reported.

[45] M. Hussain MRS, C. Sweeney, P. G. Corn, A. Elfiky, M. S. Gordon, N. B. Haas, A. L. Harzstark, R. Kurzrock, P. Lara, C. Lin, A. Sella, E. J. Small, A. I. Spira, U. N. Vaishampayan, N. J. Vogelzang, C. Scheffold, M. D. Ballinger, F. Schimmoller, D. C. Smith; University of Michigan Comprehensive Cancer Center, Ann Arbor, MI; Massachusetts General Hospital Cancer Center, Boston, MA; Dana-Farber Cancer Institute, Boston, MA; University of Texas M. D. Anderson Cancer Center, Houston, TX; Pinnacle Oncology Hematology, Scottsdale, AZ; Abramson Cancer Center, University of Pennsylvania, Philadelphia, PA; University of California, San Francisco, San Francisco, CA; University of California, Davis, Sacramento, CA; National Taiwan University Hospital, Taipei, Taiwan; Assaf Harofeh Medical Center, Zerifin, Israel; US Oncology Research, LLC, The Woodlands, TX; Virginia Cancer Specialists, PC, Fairfax, VA; Karmanos Cancer Institute, Wayne State University, Detroit, MI; Comprehensive Cancer Centers of Nevada, Las Vegas, NV; Exelixis, South San Francisco, CA; University of Michigan Cancer Center, Ann Arbor, MI. Cabozantinib (XL184) in metastatic castration-resistant prostate cancer (mCRPC): Results from a phase II randomized discontinuation trial. J Clin Oncol 29: 2011 (suppl; abstr 4516). 2011.

[46] Motz GT, Coukos G. The parallel lives of angiogenesis and immunosuppression: cancer and other tales. Nat Rev Immunol. Oct 2011; 11(10): 702-711.

[47] Lippert JW, 3rd. Vascular disrupting agents. Bioorg Med Chem. Jan 15 2007; 15(2): 605-615.

[48] Patterson DM, Zweifel M, Middleton MR, *et al.* Phase I clinical and pharmacokinetic evaluation of the vascular-disrupting agent OXi4503 in patients with advanced solid tumors. Clin Cancer Res. Mar 1 2012; 18(5): 1415-1425.

[49] OXiGENE Reports Positive Results in Phase 2 Study of ZYBRESTAT(TM) in Platinum-Resistant Ovarian Cancer.

[50] Phase 1/1b, Open-Label, Dose-Escalation and Expansion Study to Evaluate the Safety and Antitumor Activity of MEDI3617 as a Single-Agent or in Combination Therapy in Adult Subjects with Advanced Solid Tumors. Oct 15, 2010.

[51] Leow CC, Coffman K, Inigo I, *et al.* MEDI3617, a human anti-angiopoietin 2 monoclonal antibody, inhibits angiogenesis and tumor growth in human tumor xenograft models. Int J Oncol. May 2012; 40(5): 1321-1330.

[52] Mekhail T, Masson E, Fischer BS, *et al.* Metabolism, excretion, and pharmacokinetics of oral brivanib in patients with advanced or metastatic solid tumors. Drug Metab Dispos. Nov 2010; 38(11): 1962-1966.

[53] Miller K, Wang M, Gralow J, *et al.* Paclitaxel plus bevacizumab versus paclitaxel alone for metastatic breast cancer. N Engl J Med. Dec 27 2007; 357(26): 2666-2676.

[54] Miles DW, Chan A, Dirix LY, *et al.* Phase III study of bevacizumab plus docetaxel compared with placebo plus docetaxel for the first-line treatment of human epidermal growth factor receptor 2-negative metastatic breast cancer. J Clin Oncol. Jul 10 2010; 28(20): 3239-3247.

[55] Robert NJ, Dieras V, Glaspy J, *et al*. RIBBON-1: randomized, double-blind, placebo-controlled, phase III trial of chemotherapy with or without bevacizumab for first-line treatment of human epidermal growth factor receptor 2 negative, locally recurrent or metastatic breast cancer. J Clin Oncol. Apr 1 2011; 29(10): 1252-1260.

[56] Kumaran GC, Jayson GC, Clamp AR. Antiangiogenic drugs in ovarian cancer. British journal of cancer. Jan 13 2009; 100(1): 1-7.

[57] Azad NS, Posadas EM, Kwitkowski VE, *et al*. Combination targeted therapy with sorafenib and bevacizumab results in enhanced toxicity and antitumor activity. J Clin Oncol. Aug 1 2008; 26(22): 3709-3714.

[58] Cabrita MA, Jones LM, Quizi JL, Sabourin LA, McKay BC, Addison CL. Focal adhesion kinase inhibitors are potent anti-angiogenic agents. Mol Oncol. Dec 2011; 5(6): 517-526.

[59] MT. R. Focal adhesion kinase and angiogenesis. Where do we go from here? Cardiovasc Res. 2004; 64(3): 377-378.

[60] Tabernero J. The role of VEGF and EGFR inhibition: implications for combining anti-VEGF and anti-EGFR agents. Mol Cancer Res. Mar 2007; 5(3): 203-220.

[61] Siemann DW, Shi W. Dual targeting of tumor vasculature: combining Avastin and vascular disrupting agents (CA4P or OXi4503). Anticancer Res. Jul-Aug 2008; 28(4B): 2027-2031.

[62] Nathan P, Zweifel M, Padhani AR, *et al*. Phase I trial of combretastatin A4 phosphate (CA4P) in combination with bevacizumab in patients with advanced cancer. Clin Cancer Res. Jun 15 2012; 18(12): 3428-3439.

CHAPTER 7

STAT3 Signaling in Cancer: Small Molecule Intervention as Therapy?

John S. McMurray[1] and Jim Klostergaard[2,*]

[1]Departments of Experimental Therapeutics and [2]Molecular and Cellular Oncology, The University of Texas, MD Anderson Cancer Center, Houston, TX, USA

Abstract: Signal Transducer and Activator of Transcription 3 (STAT3) transmits signals from receptors of the IL-6 family of cytokines and from receptors for several growth factors, including vascular endothelial growth factor (VEGF). Transmission occurs directly to the nucleus where STAT3 participates in the expression of numerous genes involved in tumor cell survival, cell cycling, invasion, and angiogenesis. Mechanistic studies have established that upon cytokine or growth factor binding to their respective receptors, STAT3 is recruited to pTyr residues on those receptors *via* its Src homology 2 (SH2) domain where it becomes phosphorylated on Tyr705. Reciprocal SH2-pTyr705 interactions lead to STAT3 dimerization, followed by its nuclear translocation and a resultant cascade of gene transcription. STAT3 is activated (phosphorylated on Tyr705) in numerous tumor types. Many small molecule inhibitors of STAT3 phosphorylation have been reported. These include natural products such as cryptotanshinone, resveratrol and analogues, quercetin, curcumin and analogues, and 2-methoxystypandrone, as well as synthetic compounds: *e.g.*, Stattic, STA-21, sorafenib, STX-0119, CPA-7, LLL12, PM-73G, sorafenib, and others. For most of these agents, associated with their capacity to inhibit STAT3 phosphorylation is expression of cytotoxicity for tumor cells, raising the issue of whether they also inhibit pathways other than those responsible for STAT3 phosphorylation. Such off-target toxicities could lead to dose-limiting side effects and become a developmental impediment when evaluated in the clinic. However, recent reports on JAK inhibitors, such as Pyridone P6, AZD1480 and ruxolotinib, and SH2-targeted inhibitors suggest that inhibiting STAT3 phosphorylation is not in and of itself cytotoxic to tumor cells. *In vivo*, administration of AZD1480 or the STAT3 phosphorylation inhibitor, PM-73G, result in reduced tumor volume and microvessel density in human tumor xenografts in mice, suggesting blockade of VEGF signaling. Collectively, these data support the hypothesis that specific, small molecule-mediated inhibition of STAT3 activation may be a viable anti-tumor treatment strategy by inhibiting interactions between the tumor cells and the stromal compartment. This review will focus on key developments, spanning more than the last decade, that create the foundation for this pursuit.

***Corresponding author Jim Klostergaard:** Department of Molecular and Cellular Oncology, The University of Texas, MD Anderson Cancer Center, Houston, TX, USA; Tel: 713-792-8962; Fax: 713-794-3270; E-mail: jkloster@mdanderson.org

Atta-ur-Rahman and Muhammad Iqbal Choudhary (Eds)
Copyright © 2014 Bentham Science Publishers Ltd. Published by Elsevier Inc. All rights reserved.
10.1016/B978-0-12-803963-2.50007-7

Keywords: Angiogenesis, STAT3, VEGF, STAT3 phosphorylation, Janus Kinase, VEGF receptor, VEGFR.

INTRODUCTION: ANGIOGENESIS, VEGF AND STAT3

Angiogenesis is a sequential, multistep process of endothelial cell (EC) proliferation, degradation of basement membrane (BM) and extracellular matrix (ECM) components, and EC migration, culminating in the formation of tubular structures--precursors of actual blood vessels. Angiogenesis is the net result of a dynamic competition between angiogenic and angiostatic mechanisms, among the former being the vascular endothelial growth factor (VEGF) pathway, identified as a key pro-angiogenic driver. There is increasing evidence that Signal Transducer and Activator of Transcription 3 (STAT3) participates in regulating VEGF-promoted angiogenesis: hence, the focus on STAT3 inhibition in this Review.

STAT3 transmits signals from receptors for the IL-6 family of cytokines, growth factors such as EGF, PDGF, ALK, and VEGF (as mentioned above), directly from the receptor to the nucleus where it is involved in the transcription of genes. It is involved in cell cycling, metastasis, angiogenesis and immune cell evasion in most human cancers and its activity has been reviewed extensively [1-3]. Upon activation of the cytokine or growth factor receptor, STAT3, *via* is SH2 domain, is recruited to pTyr residues on the cytoplasmic domains of the receptor. Following receptor docking, Tyr705 of STAT3 becomes phosphorylated by associated Janus kinases (JAK), Src kinase, or the kinase activity of the receptor. Tyr705 phosphorylation, termed activation, results in STAT3 dimerization by reciprocal pTyr-SH2 domain interactions. The dimer is translocated to the nucleus where STAT3 behaves as a transcription factor and participates in the expression of numerous genes including those in the acute phase response, angiogenesis, and cell cycling. STAT3 has been found to be activated in a large percentage of human tumors and tumor cell lines. Several academic and industry groups have developed small molecule and phosphopeptide-based compounds targeting the SH2 domain of STAT3 and these have been recently reviewed [4-7].

Constitutively activated STAT3 has been shown to upregulate VEGF expression and thereby induce tumor angiogenesis, one of multiple tumorigeneic mechanisms thus impacted; for a spectrum of diverse tumor histiotypes, VEGF expression has been directly correlated with constitutively elevated STAT3 activity. The causality underlying this correlation is supported by immunoprecipitation assays demonstrating that STAT3 protein binds directly to the VEGF promoter [8]. Mutation of this STAT3-binding site caused loss of STAT3-induced VEGF promoter activity--indicating that the VEGF gene is a direct target of STAT3. Transfection studies with constitutively activated mutant STAT3 demonstrated increased VEGF expression and enhanced tumor angiogenesis; in contrast, blockade of STAT3 signaling with dominant-negative STAT3 inhibited VEGF expression, with resultant downstream reductions in angiogenesis and tumor growth.

Activation of upstream Src and epidermal growth factor receptor (EGFR) tyrosine kinases is frequently observed in many cancers. STAT3 has, in fact, been shown necessary for VEGF upregulation by the v-Src kinase since downstream blocking of STAT3 signaling inhibited Src-driven VEGF expression. Further, melanoma differentiation-associated gene-7 protein was shown to inhibit c-Src kinase activity and abolished STAT3 binding to the VEGF promoter--suppressing VEGF expression.

Increased expression of EGFRs was shown to persistently activate STAT3 in both breast cancer and head and neck carcinoma cells [8]. In the latter, it was observed that epigallocatechin-3-gallate (EGCG)—a key neutraceutical in green tea--decreased VEGF production by inhibiting constitutive activation of EGFR and its downstream signaling [9]. EGFR-induced constitutive activation of STAT3 was inhibited by EGCG and significantly decreased VEGF promoter activity and VEGF production. Persistent activation of STAT3 signaling in cancer cells provides an abundance of cytokines and growth factors that induce VEGF expression and angiogenesis.

This Review will cover the status of small molecules that target the STAT3 SH2 domain, those that act upstream of STAT3 at JAK and affect downstream STAT3, and those with diverse molecular targets. Emphasis will of course be placed on those compounds reported to affect angiogenesis.

Small Molecules Targeting the STAT3 SH2 Domain

Targeting the SH2 domain of STAT3 is intended to prevent recruitment to phosphotyrosine residues on activated growth factor or cytokine receptors. Such compounds would have the effect of preventing phosphorylation of Tyr705 and subsequent dimerization, nuclear translocation, and expression of downstream genes. Further, such compounds have been shown to disrupt STAT3-DNA complex formation.

STA-21, LLL12, and Analogs

Several groups have employed high-throughput screens to identify small molecules with conventional drug-like properties to target the SH2 domains of STAT3. Using computational methods, Song *et al.* docked a virtual library to the SH2 domain of STAT3 and discovered lead compound STA-21 [10] (Fig. **1**). This compound inhibited STAT3-dependant luciferase activity in human MDA-MB-435 breast cancer cells as well as the expression of canonical STAT3 downstream genes. Preferential growth inhibition was also observed in breast cancer cell lines harboring constitutively active STAT3 phosphorylation. In spite of this cellular data, no biochemical evidence was reported that demonstrated binding of this material to the SH2 domain of STAT3. Further development of STA-21 resulted in 1-acetyl-5-hyroxyanthraquinone, 1-acetyl-8-hydroxyanthraquinone and LLL12 (Fig. **1**), analogs that very potently inhibited cancer cell proliferation, induced apoptosis, *etc.* concomitantly with reductions in pSTAT3 levels in cultured cells [11, 12]. The new materials were also docked to the SH2 domain of STAT3 using computational methods, but again no biochemical evidence was presented to show that they bind to the SH2 domain of STAT3. STA-21 and its analogs are benzoquinones and anthraquinones, which are known to exert cytotoxic effects through redox recycling and arylation of essential nucleophiles, such as glutathione and protein thiols [13]. Further, they are structural analogs of plumbagin, a well-known phytochemical which has been shown to have antibacterial properties, alter redox potential in cells, inhibit NF-κB, alter mitotic spindles, and chelate heavy metals, among other effects [14]. In light of the lack of cytotoxicity of JAK inhibitors discussed below, care must be exercised when interpreting the effects of these compounds as the quinone moieties themselves may be responsible for the anti-proliferative activity of STA-21 and its analogs.

LLL12 exhibited impressive anti-angiogenic activity. It inhibited proliferation, tube formation and migration of human umbilical vascular endothelial cells (HUVEC) *in vitro*. Further, this material ablation the growth of OS-1 osteosarcoma xenografts that was accompanied by extensive reduction in microvessel density (MVD) and STAT3 phosphorylation. Growth inhibition was the result of inhibition of VEGF signaling as well as the inhibition of the production of angiogenic factors such as VEGF, MMP-9, angiopoetin, TF and FGF1 which was attributed to the near complete ablation of pSTAT3 in the tumors [15].

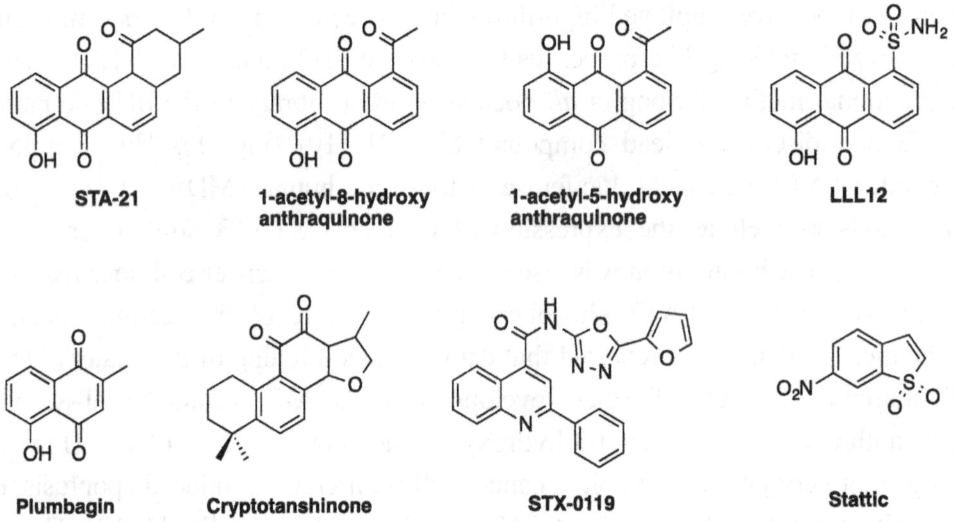

STA-21 | 1-acetyl-8-hydroxy anthraquinone | 1-acetyl-5-hydroxy anthraquinone | LLL12

Plumbagin | Cryptotanshinone | STX-0119 | Stattic

Figure 1: Small molecules hypothesized to bind to the SH2 domain of STAT3.

STX-0119

STX-0119, a N-[2-(1,3,4-oxadiazolyl)]-4-quinolinecarboxamide derivative, was first identified as a novel small-molecule inhibitor of STAT3 dimerization using a virtual screen [16]. STX-0119 selectively blocked the DNA binding activity of STAT3 suggesting binding to the SH2 or the DNA binding domains, and suppressed the expression of STAT3 downstream genes, *e.g.*, c-myc, cyclin D1, survivin and Bcl-xL, as well as the growth of several hematological cancer cell lines, lymphomas in particular [17]. Oral administration of STX-0119 effectively abrogated the growth of the human SCC-3 lymphoma model in a SCC-3 s.c. xenograft model, and suppressed tumor levels of c-myc, Ki67, and pSTAT3.

This same group extended these studies to glioblastoma multiforme stem-like cells (GBM-SC) derived from recurrent GBM patients [18]. STAT3 phosphorylation was present more in the GBM-SC lines than serum-derived GB cell lines. The growth inhibitory effect of STX-0119 on GBM-SCs was moderate, but stronger compared to that of WP1066, a tyrphostin analog reported to inhibit STAT3 phosphorylation in cancer cells [19, 20]; however, the effect of temozolomide was weak in all the cell lines. STX-0119 strongly inhibited expression of STAT3 target genes, including HIF-1α and VEGF; VEGFR2 mRNA was also inhibited by STX-0119. In a transplantable model of GBM-SC *in vivo*, STX-0119 was active, although anti-angiogenic activity was not investigated.

Cryptotanshinone

The natural product, Cryptotanshinone, was recently identified as a potent STAT3 inhibitor [21] (Fig. **1**). Cryptotanshinone inhibited STAT3 phosphorylation at Tyr705 in a human prostate cancer cell line and decreased the expression of its downstream target proteins including cyclin D1, survivin, and Bcl-xL. Upstream of STAT3, phosphorylation of JAK2 was only inhibited by Cryptotanshinone in a protracted manner, whereas inhibition of STAT3 phosphorylation occurred rapidly, suggesting a JAK2-independent mechanism--suppression of JAK2 phosphorylation being secondary. Other evidence consistent with rapid inhibition suggested that Cryptotanshinone directly binds to STAT3, as they colocalized in the cytoplasm and formation of STAT3 dimers was suppressed. Computational modeling revealed Cryptotanshinone could bind to the SH2 domain of STAT3, although no direct biochemical evidence was presented. These results were paralleled in human glioma cell lines [22]. The phosphorylation of STAT3 Tyr705--but not Ser727--was inhibited by Cryptotanshinone, and STAT3 nuclear translocation was attenuated. Overexpression of a constitutively active mutant of STAT3 overrode this inhibitory effect, and STAT3 knockdown had similar inhibitory effects as Cryptotanshinone treatment. It should be noted that Cryptotanshinone has several other activities such as inhibition of NF-kB and COX-2, and activation of the PI3K/Akt pathway [23, 24].

Evidence for anti-angiogenic effects of tanshinones was first derived in investigations of the regulation of hypoxia-inducible factor-1 (HIF-1) by monitoring hypoxia-induced reporter gene expression in AGS human gastric cancer cells and Hep3B human hepatocarcinoma cells [25]. Suppression of HIF-1α accumulation and of VEGF mRNA under hypoxia was observed.

Tanshinones, including Cryptotanshinone, downregulated Aurora A kinase expression in DU145 human prostate cancer cells, and demonstrated potent anti-angiogenic activity *in vitro* and *in vivo*, the latter associated with growth inhibition, apoptosis induction, decreased proliferation, and downregulation of Aurora A [26]. Cryptotanshinone inhibited lymphangiogenesis in an *in vitro* model of tube formation [27], attributed to inhibiting expression of VEGF receptor 3 (VEGFR-3) in murine lymphatic endothelial cells (LEC). Overexpression of VEGFR-3 conferred resistance to Cryptotanshinone-mediated inhibition of tube formation, and downregulation of VEGFR-3 simulated Crytotanshinone in blocking tube formation. Cryptotanshinone also inhibited phosphorylation of the extracellular signal-related kinase 1/2 (ERK1/2), and VEGFR-3 overexpression reversed Cryptotanshinone-mediated inhibition of ERK1/2 phosphorylation--whereas downregulation of VEGFR-3 inhibited ERK1/2 phosphorylation in LECs. Expression of constitutively active MAPK kinase 1 resulted in activation of ERK1/2 and tended to negate Cryptotanshinone effects on LEC tube formation. Thus, Cryptotanshinone inhibits LEC tube formation, in part, by inhibiting VEGFR-3-mediated ERK1/2 phosphorylation.

Using the human PC-3 prostate cancer model, Cryptotanshinone was shown to be cytotoxic against hypoxic cells and suppressed their hypoxia-induced accumulation of HIF-1α [28]. Cryptotanshinone reduced VEGF levels in these cells, impeded tube formation of HUVECs, and inhibited the binding of HIF-1α to the VEGF promoter. Cryptotanshinone administration suppressed PC-3 xenograft growth, and decreased expression of Ki-67, CD34 and VEGF. Similar effects were observed in lung cancer models [29]. Tanshinones inhibited proliferation of lung cancer cell lines *in vitro via* cell cycle arrest and apoptosis induction; Aurora A knockdown by siRNA virtually eliminated these activities, suggesting it as a target. Human NSCLC xenografts in mice showed reduced tumor weight associated with inhibiting proliferation, inducing apoptosis, inhibiting lung tumor angiogenesis, and reducing Aurora A expression in response to tanshinone administration.

Stattic

Stattic, a nonpeptidic small molecule, was first identified using screening of chemical libraries and was claimed to selectively inhibit the function of the STAT3 SH2 domain regardless of the STAT3 activation state *in vitro* [30] (Fig. **1**). Stattic selectively inhibited activation, dimerization, and nuclear translocation of STAT3 and increased the apoptotic rate of STAT3-dependent breast cancer cell lines. The claimed specificity was recently called into question following comparative *in silico* docking studies that determined SH2-binding specificity of Stattic and of fludarabine, a STAT1 inhibitor. By principally targeting the highly conserved phosphotyrosine (pY+0) SH2 binding pocket, Stattic is not a specific STAT3 inhibitor, but is equally effective towards STAT1 and STAT2 [31]. In human micro-vascular endothelial cells (HMECs), Stattic inhibited interferon-α-induced phosphorylation of all three STATs. Fludarabine inhibited both STAT1 and STAT3 phosphorylation, but not that of STAT2. In HMECs, fludarabine inhibited cytokine and lipopolysaccharide-induced phosphorylation of STAT1 and STAT3, but not STAT2. Further, multiple sequence alignment of STAT-SH2 domain sequences confirmed high conservation of Stattic binding sites between STAT1 and STAT3, but not STAT2.

NK4 is a cleavage fragment of and competitive antagonist for hepatocyte growth factor (HGF)/c-Met signaling, and was observed to suppress murine CT26 colon tumor growth *via* inhibiting angiogenesis rather than through HGF antagonism [32]; NK4 inhibited the angiogenic responses induced by VEGF and basic fibroblast growth factor (bFGF), as well as those of HGF [33]. CT26 cells were genetically-modified to produce NK4, and VEGF expression was found to be reduced in s.c. homografts of these cells, compared to those of the control CT26 cells [34]. HGF-induced VEGF expression was inhibited by anti-HGF antibody, NK4 and by several kinase inhibitors, including Stattic *in vitro.* Real-time RT-PCR demonstrated that HGF-induced HIF-1α mRNA expression was inhibited by Stattic.

Cpd188 and Analogs

An independent computational screen of small molecules identified Cpd3, Cpd30 and Cpd188 (Fig. **2**). These materials competed with a phosphopeptide derived from EGF pTyr1068 in a surface plasmon resonance (SPR) assay for binding to STAT3, thus providing evidence that they actually bind to the SH2 domain [35]. Further, they inhibited IL-6 stimulated STAT3 phosphorylation in HepG2 cells. Similarity screening of other compound databases revealed that Cpd3-7, Cpd3-12, and Cpd30-20 were capable of inhibiting phosphopeptide binding to STAT3 as well as IL-6 stimulated STAT3 phosphorylation. This group of compounds was assayed for the ability to induce apoptosis in a panel of breast cancer cell lines. Compounds Cpd3, Cpd30 and Cpd188 were selective for cell lines with constitutively phosphorylated STAT3: MDA-MB-468, MDA-MB-231 and MDA-MB-435. No effect was observed on the control lines, MDA-MB-453 and MCF7, not harboring constitutively phosphorylated STAT3. Of all the compounds, Cpd188 was the most potent, causing apoptosis in the phosphoSTAT3 lines with IC_{50} values of 0.73 - 7.01 μM. Cross reactivity with the control lines was observed, but with reduced potencies: IC_{50} values were 15 and 17μM. Cpd188 was found to decrease tumor initiating cell populations and mammosphere formation in breast tumor xenografts *in vivo* [4].

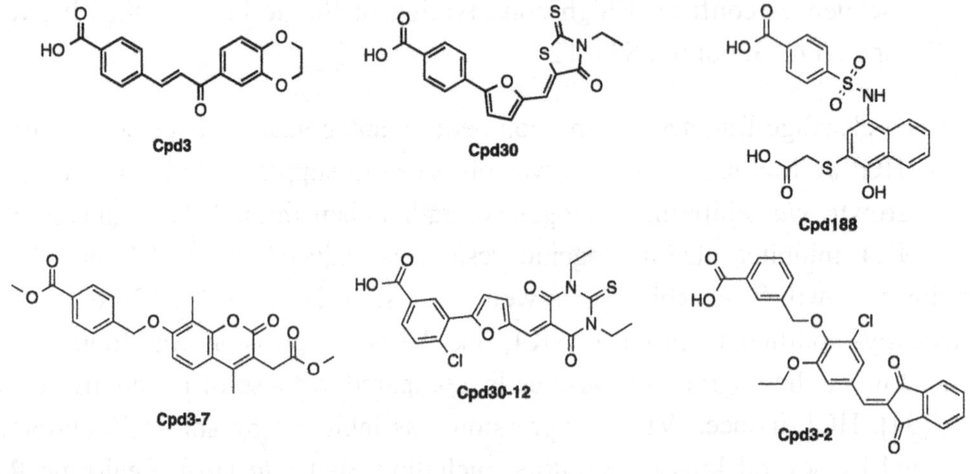

Cpd3

Cpd30

Cpd188

Cpd3-7

Cpd30-12

Cpd3-2

Figure 2: Small molecule antagonists of phosphopeptide-STAT3 SH2 domain binding.

Peptidomimetic Inhibitors Targeting the SH2 Domain of STAT3

ISS-610 and Analogues

The sequence surrounding phosphotyrosine 705, (Pro-pTyr-Leu-Lys-Thr-Lys), which participates in the SH2 domain-mediated dimerization of STAT3 [36], served as the starting point for peptidomimetic inhibitors [37]. Computational analysis of the interactions of tripeptide mimic ISS-610 [38] and the SH2 domain of STAT3, combined with structural information derived from the X-ray crystal structure of STAT3β [36], led to the development of S3I-M2001 (Fig. **3**) [39]. This compound inhibited STAT3 phosphorylation in NIH 3T3/v-Src and MDA-435-MB cells at high uM concentrations and strongly blocked tumor growth in human MDA-MB-231 TNBC xenografts.

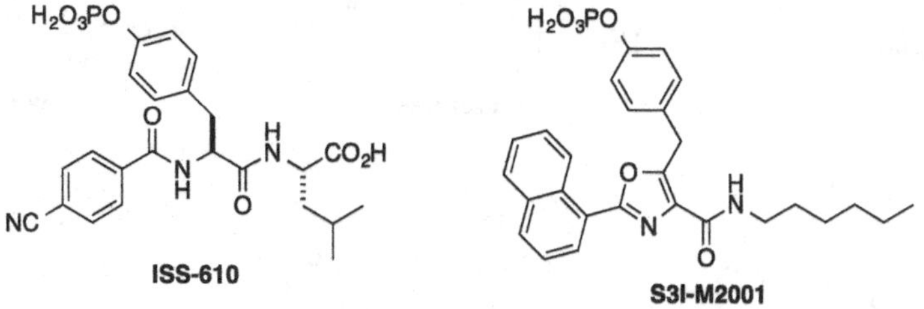

Figure 3: Peptide mimetic inhibitors of STAT3 derived from the sequence surrounding Tyr705.

Computer-aided screening of compound libraries from the National Cancer Institute (NCI) identified S3I-201 (Fig. **4**), which inhibited STAT3 activation and decreased survival of Src-transformed mouse fibroblasts and human breast cancer cells lines MDA-MB-231, MDA-MB-468 and MDA-MB-435 harboring constitutive STAT3 activation. Inhibition of tumor growth was also observed in MDA-MB-231 xenografts [40]. Modifications to improve binding affinity led to S3I-201.1066 (Fig. **4**) that inhibited STAT3 in NIH-3T3/v-Src, Panc-1 and MDA-MB-231 cells as well as causing reduction of tumor growth in human MDA-MB-231 breast cancer xenografts. Further modification led to an orally bioavailable analog, BP-1-102 (Fig. **4**) which exhibited increased potency and suppression of survival of tumor cells with constitutive active STAT3 phosphorylation [41]. BP-1-102 inhibited growth of several human tumor xenografts bearing constitutively activated STAT3.

Structure activity studies of S3I-201, carried out independently by Urlam *et al.* [42], led to S3I-1757 [43] (Fig. **4**). This material (but not S3I-1756, a closely related, but non-inhibitory analog), inhibited selectively the phosphorylation of STAT3 over AKT1 and ERK1/2 (MAPK3/1), the nuclear accumulation of P-Y705-STAT3, STAT3–DNA binding, and transcriptional activation and expression of STAT3 target genes, such as Bcl-xL (BCL2L1), survivin (BIRC5), cyclin D1 (CCND1), and matrixmetalloproteinase (MMP)-9. S3I-1757, but not S3I-1756, inhibited anchorage-dependent and -independent growth, migration, and invasion of human cancer cells which depend on STAT3.

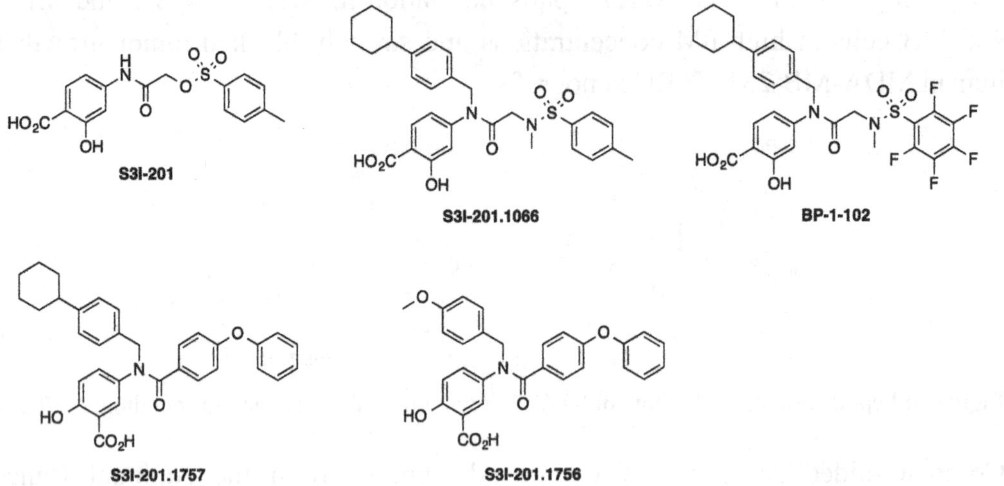

Figure 4: S3I-201 and subsequent analogs.

Phosphopeptide Mimic Prodrugs

Following the screening of a panel of phosphopeptides derived from known STAT3 docking sites on cytokine receptors, the lead phosphopeptide Ac-pTyr-Leu-Pro-Gln-Thr-Val-NH_2 was selected [44] for a separate program of phosphopeptide mimetic inhibitor development. The lead, derived from Tyr904 of the IL-6 co-receptor gp130, possessed the recognition determinant for STAT3, Tyr-Xaa-Yaa-Gln [45, 46]. Similar sequences were identified independently using a combinatorial peptide library [47]. Extensive studies carried out to probe the phosphopeptide binding site led to a series of peptidomimetic inhibitors, *e.g.* PM-66F (Fig. **5**) [48-54]. To adapt these phosphopeptide mimetics for studies in intact cells, the phosphate group was substituted with the phosphatase-stable phosphonodifluoromethyl group [50, 55]. The oxygens were capped with carboxyesterase-labile pivaloyloxymethyl groups in a prodrug approach, thereby blocking the negative charges of the phosphonate and allowing cell penetration, [50, 55]. Cellular potency was observed to be influenced by structural features that had little effect on affinity for the STAT3 protein [50, 53].

A series of prodrugs (Fig. **5**) was tested for the ability to inhibit the phosphorylation of STAT3 in several cancer cell lines [50]. The prodrugs were very potent and inhibited constitutive STAT3 phosphorylation significantly at 100 nM with complete inhibition observed at 500 nM. At 5 μM the inhibitors were selective for STAT3 over EGF-induced phosphorylation of STAT5, PI3K, and FAK, and were 10-fold more selective for STAT3 over IFN-γ-induced STAT1. In contrast to the dogma that STAT3 is required for cancer cell growth and survival, daily dosing with the inhibitor at 5 μM, a 10-fold higher concentration than that required for complete inhibition STAT3 phosphorylation, resulted in no significant cytotoxicity for breast cancer lines MDA-MB-468 and MCF7, as well as the lung cancer line HCC827 and the ovarian line SKOV3-ip. The IC_{50} for MDA-MB-468 cells was 12 μM for PM-72G-1 and between 25 μM and 50 μM for PM-73G and PM-274G-1. However, at 25 μM PM-73G, 2 hr treatment showed significant inhibition of pSTAT5, pAkt, and pFAKTyr861[50]. Thus, high concentrations of inhibitor resulted in off-target effects that correlated with cytotoxicity of cancer cells.

Intratumoral injection of PM-73G into female nude mice bearing MDA-MB-468 orthotopic xenografts caused reduction in tumor growth concomitant with decreased MVD and immunohistochemically-detectable VEGF protein [56, 57]. Of note and of greater therapeutic relevance, these effects were also observed following a regimen of intraperitoneal administration--thus, systemic delivery. These results indicate that selective inhibition of STAT3, while not directly toxic to tumor cells, leads to impaired VEGF signaling, and thereby, to inhibition of tumor angiogenesis [56, 57].

CPA-7

CPA-7, originally identified as a platinum (IV) complex with DNA-alkylating activity, was reported to disrupt STAT3 signaling and its downstream sequela [58] (Fig. **6**). STAT3 activity (phosphorylation) was inhibited at low uM levels of CPA-7, and growth inhibition and apoptotic induction were observed in cells that had constitutively activated STAT3. This original study also established the anti-tumor activity of CPA-7 in a mouse CT26 colon tumor model.

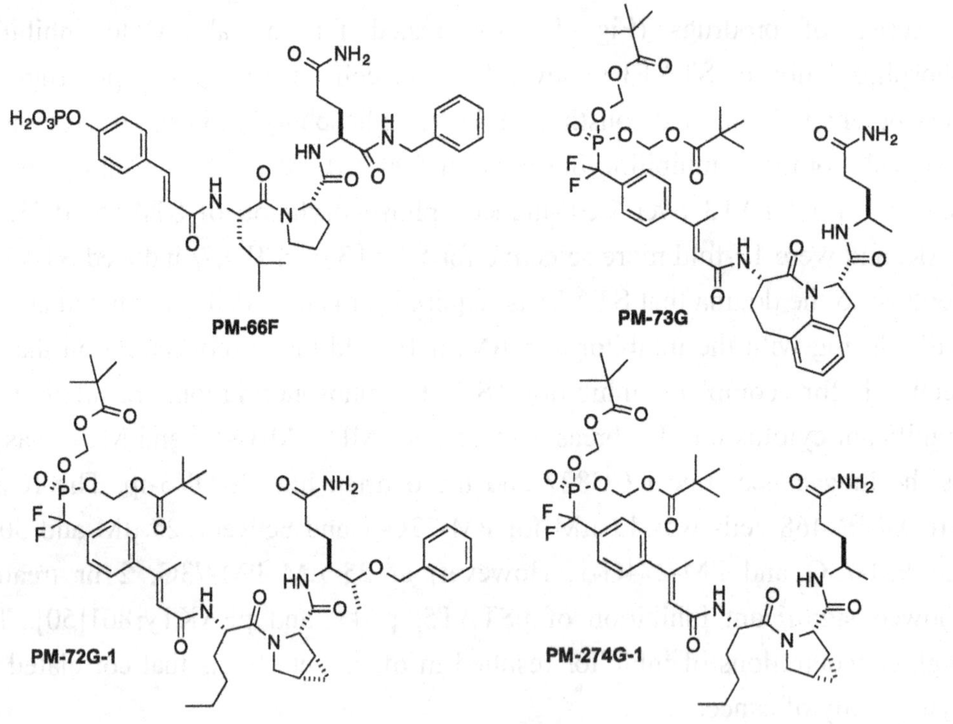

PM-66F

PM-73G

PM-72G-1

PM-274G-1

Figure 5: Phosphopeptide prodrugs derived from gp130 Tyr904 targeting the SH2 domain of STAT3.

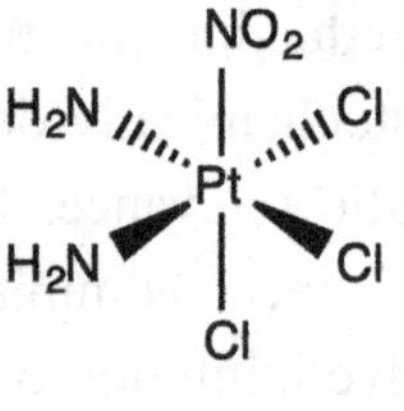

CPA-7

Figure 6: The structure of CPA-7.

In a subsequent study in the murine GL261 glioma model [59], using medium conditioned by these tumor cells, increased STAT3 activity in microglial cells treated *in vitro* by such media was observed, as well as down-regulation of the pro-inflammatory cytokine, IL1-β. In intracranial GL261 tumors, siRNA-mediated inhibition of STAT3 caused tumor growth inhibition. Further development of CPA-7 and characterization of its anti-tumor or anti-angiogenic effects *in vivo* have not been reported.

Inhibition of STAT3-Mediated Angiogenesis with JAK Kinase Inhibitors

AZD-1480

A logical route to the inhibition of STAT3 would be to inhibit the upstream JAK kinase with small molecule inhibitors. Although JAK kinase inhibitors are FDA approved for myelodysplastic neoplasia (Ruxolotinib) and rheumatoid arthritis (Tofacinib), little has been published on the effects of JAK kinase inhibition on angiogenesis in solid tumors. At concentrations that knocked down Tyr705 phosphorylation, the JAK2 inhibitor AZD1480 [60] did not inhibit the proliferation of cancer cell lines *in vitro* suggesting that inhibition of Stat3 phosphorylation is not cytotoxic to solid tumor cells [61, 62]. However, in syngeneic Renca renal mouse tumors, AZD 1480 inhibited tumor growth which was accompanied by a 3-fold reduction in microvessel formation (CD-31 staining) and reduced VEGF and MMP9 expression [55, 60]. Inhibition of tumor infiltration of myeloid cells was also noted. AZD1480 also inhibited tumor growth and angiogenesis in 786-O human renal cell carcinoma xenografts.

Although JAK inhibition appears to inhibit tumor growth and angiogenesis *in vivo*, off-target effects and toxicity are of major concern. Adverse events reported for Ruxolotinib include anemia, thrombocytopenia, gastrointestinal disturbances, metabolic abnormalities, and peripheral neuropathy [57, 62] Tofacinib has been associated with liver test elevation, neutropenia, lipid and creatinine elevation and, importantly, increased incidence of infections, including tuberculosis [58, 63]. Indeed, AZD1480 was withdrawn from clinical trials due to neurotoxicities [63] and the JAK2 inhibitor, fedratinib, was recently withdrawn from clinical trials due to the development of Wernicke's encephalopathy in patients. This is disappointing since JAK inhibition is a simple route to inhibiting Stat3 phosphorylation.

Selective Inhibition of STAT3 Tyr705 Phosphorylation is Not Cytotoxic to Tumor Cells

As mentioned, early work suggested that STAT3 is required for tumor cell growth and proliferation and this has been accepted as dogma [2, 64]. Examination of reports on the effects of JAK inhibitors reveals that inhibition of Tyr705 phosphorylation is not cytotoxic to cancer cells. Pyridone P6 [65] is a pan-JAK inhibitor. Treatment of a panel of melanoma cell lines with this compound brought pSTAT3 levels to below detectable levels, but this had no effect on proliferation *in vitro* [66]. Pyridone P6, like virtually all kinase inhibitors, inhibits other kinases, notably PDK1 [67], but in this case the off-target effects are not cytotoxic. AZD1480 [60] preferentially inhibits JAK2 over the other JAKS [62]. At concentrations that completely inhibited STAT3 phosphorylation, this material had no effect on the proliferation of MDA-MB-468 (breast), DU145 (prostate), and MDAH2774 (ovarian) cancer cells *in vitro* [62]. This finding was recapitulated in subsequent publications from this group [61, 68] and others [69]. The JAK1/2 inhibitor Ruxolitinib, FDA approved for myelodysplastic disorders, reduces pSTAT3 levels, but did not affect growth of lung cancer cell lines *in vitro* [70]. Taken together with the data on the phosphopeptide mimics targeting the SH2 domain (Fig. **4**) [50], the conclusion is that inhibiting the phosphorylation of Tyr705 of STAT3 is not cytotoxic to cancer cells. This is supported by a report showing that treatment of ovarian cancer cells with siltuximab, the anti-IL-6 monoclonal antibody, resulted in pSTAT3 inhibition with no effect on proliferation [71]. By inference, if a compound inhibits STAT3 phosphorylation and it kills cells, it is likely doing the latter by off-target effects.

Recent reports show that there are functions of STAT3 that do not rely on phosphorylation of Tyr705. For instance, in addition to pSTAT3, unphosphorylated STAT3 (USTAT3) is found in the nucleus and has been reported to act as a co-transcription factor with NF-κB [72]. Additionally, Ser727-phosphorylated STAT3 (not phosphorylated on Tyr705) has been found in mitochondria and this form supports the growth of RAS transformed cells [73, 74]. Thus, removing total STAT3 protein from cells with siRNA or antisense approaches will remove these two additional functions of this protein that appear to be linked to cancer cell survival.

Natural Product Inhibitors of STAT3 Function

In addition to the synthetic compounds designed to target the SH2 domain of STAT3, nature has provided several compounds that inhibit STAT3 function in cells. Quercetin, Resveratrol, Curcumin and related analogs have received considerable attention for their abilities to block several facets of STAT3 physiology *in vitro* and *in vivo*. These materials have potential as anti-angiogenic agents.

2-Methoxystypandrone

2-Methoxystypandrone (Fig. **7**), originally identified as a protease inhibitor [75], was the most potent among several anthraquinones isolated from Polygonum cuspidatum with activity against STAT3 activation, with an IC_{50} in the low μM range to [76]. It was since found to block the STAT3 pathway upstream at JAK2 and also to have dual activity against IκB kinase [77]. As it is structurally related to Sta-21 and Plumbagin (Fig. **1**), 2-Methoxystypandrone is likely to have pleiotropic effects. Given this profile, it is unlikely to be an ideal candidate for further development as an anti-angiogenic agent.

Quercetin and Related Flavanoids

A mechanistic linkage between Quercetin (Fig. **7**) and STAT3 signaling was initially revealed by observations that Quercetin blocked IL-12-induced tyrosine phosphorylation of JAK2, TYK2, STAT3, and STAT4 [78]. Quercetin was found to block upstream JAK2 phosphorylation evoked by oxidized LDL in EC [79]. Quercetin inhibited hypoxia-induced VEGF expression in NCI-H157 cells in the low uM range, although concomitantly inducing HIF-1alpha expression [80]. In human EC, Quercetin reduced the gene expression of specific factors implicated in local vascular inflammation including IL-1R, Ccl8, IKK, and STAT3 [81].

Quercetin is a potent inhibitor of IL-6-driven STAT3 signaling in T98G and U87 glioblastoma cells [82]. Quercetin reduced GP130, JAK1 and STAT3 activation by IL-6, as well as downstream expression of cyclin D1 and MMP-2 genes, and strongly decreased both the proliferation and the migratory activity of these cells. Similar results were obtained in cholangiocarcinoma cells; JAK/STAT pathway activation by IL-6 and by IFN-γ was suppressed by pretreatment with Quercetin and EGCG, associated with a decrease in phosphorylated-STAT1 and STAT3 protein levels [83]. These agents also blocked up-regulation of inducible nitric oxide synthase (iNOS) and intercellular adhesion molecule-1 (ICAM-1) by these cytokines, and inhibited proliferation and cytokine-induced migration of these cells. Quercetin treatment inhibited proliferation, induced apoptosis, and suppressed migration and invasion properties of melanoma cells [84]. Quercetin inhibited STAT3 signaling by interference with STAT3 phosphorylation and reduction of STAT3 nuclear localization. In turn, this down-regulated STAT3 targeted transcription of the genes Mcl-1, MMP-2, MMP-9 and VEGF, all involved in cell growth, migration and invasion. Constitutively activated STAT3 partially overcame growth inhibition induced by Quercetin. Quercetin was shown to suppress A375 tumor growth and STAT3 activity in mouse xenograft models, and inhibited syngeneic murine B16F10 cells lung metastasis.

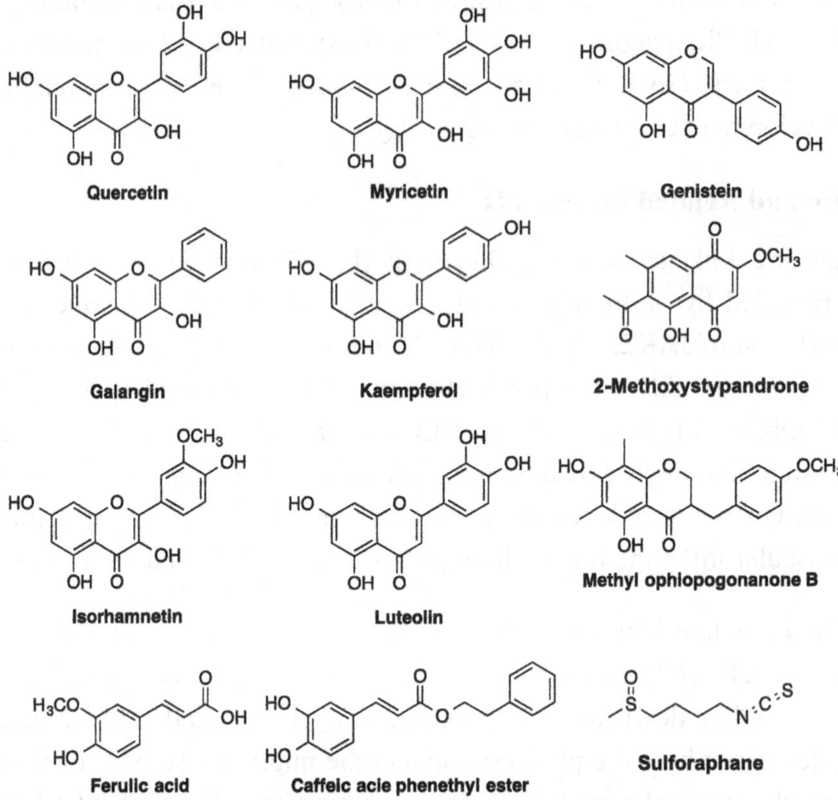

Quercetin Myricetin Genistein

Galangin Kaempferol 2-Methoxystypandrone

Isorhamnetin Luteolin Methyl ophiopogonanone B

Ferulic acid Caffeic acie phenethyl ester Sulforaphane

Figure 7: Flavanoid and other natural product inhibitors of STAT3 function.

The anti-tumor and anti-angiogenic effects of Quercetin have been the subjects of considerable investigation over the last decade and a half. The anti-tumor effects of Quercetin were first described about a decade and a half ago, possibly attributable to immune stimulation, free radical scavenging, alteration of the mitotic cycle, gene regulation, anti-angiogenic activity, and apoptotic induction

[85]. Balb/c mice bearing Colon-25 tumors were orally administered daily doses of Quercetin [86], and tumor size was reduced at day 20 in Quercetin groups with two different dose levels compared to controls.

Quercetin and Resveratrol both inhibited the growth and the migration of bovine aorta endothelial (BAE) cells in *in vitro* angiogenesis assays [87]. Quercetin significantly inhibited tube formation of vascular EC in a 3-D culture model.

HIF-1α translocates to the nucleus where it facilitate transcription of target genes, including those involved in angiogenesis. Quercetin was found to also activate HIF-1α in a manner similar to hypoxia [88]. Quercetin, an inhibitor of Ser/Thr kinases, stabilized HIF-1α and caused its nuclear localization of the protein in a transcriptionally active state.

Quercetin was found to inhibit multiple steps of angiogenesis in human microvascular dermal EC including proliferation, migration, and tube formation [89]. Quercetin displayed an anti-angiogenic effect in the chicken chorioallantoic membrane assay. Quercetin caused a decrease in the expression and activity of MMP-2: involved in the angiogenic process of migration, invasion, and tube formation. Several flavonoids and homoisoflavonoids, including Quercetin, were screened using a reporter under the control of the hypoxia-response element (HRE) in order to identify those that inhibit the activation of HRE under hypoxic conditions [90]. Isorhamnetin, luteolin, Quercetin, and methyl ophiopogonanone B (Fig. 7) were effective at 3-9 ug/ml range in inhibiting the reporter activity.

SCID mice implanted with CWR22 prostate tumor cells were treated with either tamoxifen, Quercetin or their combination for 28 days [91]. Quercetin affected the expression and phosphorylation of cdc-2 and cyclin B1, and strongly inhibited the Ki-67 index. Combined tamoxifen-Quercetin delayed the appearance of tumors, and markedly inhibited the final tumor volume and reduced the endpoint tumor weight. The Ki-67 index, VEGF121 and VEGF165 mRNA and MVD were strongly decreased by the combined treatment.

The flavonols, myricetin, Quercetin, kaempferol and galangin (Fig. 7), were tested for antioxidant and anti-proloferative activity against HUVECs [92]. When tested at about half of their LD_{50} concentrations, a rank order of myricetin > Quercetin > kaempferol > galangin was observed in suppression of VEGF-stimulated HUVEC tubular structure formation: thus, not correlating with the -OH moieties present in their structures. However, the extent of suppression of activated U937 monocytic cell adhesion to HUVEC did show association with the number of -OH moieties.

Quercetin and its main circulating conjugates in man (quercetin-3'-sulphate (Q3'S) and quercetin-3-glucuronide (Q3G)) had opposing effects on angiogenesis [93]. Strikingly, Quercetin and Q3G inhibited VEGF-induced EC functions and angiogenesis, whereas Q3'S promoted EC proliferation and angiogenesis. The effect of Q3G was linked to ERK1/2 phosphorylation elicited by VEGF; the activation of EC by Q3'S was associated with stimulation of VEGFR-2 and its downstream signaling (PI3 kinase/Akt and NO synthase pathways), leading to ERK1/2 phosphorylation. Q3G and Q3'S are the two major Quercetin conjugates in plasma, but their ratio may be dependent on nuances of metabolism *in vivo*.

Quercetin was shown to inhibit angiogenesis *via* suppression of endothelial NO synthase (eNOS) and by inducing early M-phase cell cycle arrest [94]. Quercetin treatment of BAE cells resulted in eNOS phosphorylation at Ser 617 and Akt phosphorylation at Ser 473 phosphorylation both being quickly reduced; early M-phase arrest and disruption of mitotic microtubule polymerization was noted with persistent exposure. Administration of Quercetin to mice bearing both syngeneic EMT-6 ectopic mammary tumors and Matrigel implants suppressed angiogenesis.

Clinical trials of Quercetin have been hampered by its poor pharmacological profile, particularly its extreme water insolubility. A liposomal formulation of Quercetin was developed to ascertain effects on its biodistribution and antitumor efficacy [95]. Liposomal Quercetin could be readily dissolved prior to i.v. injection and accumulated in tumor tissues. Its plasma half-time of clearance was 2 hr. Liposomal Quercetin induced tumor cell apoptosis *in vitro* and significantly inhibited tumor growth *in vivo*. Immunohistochemistry analysis showed that tumor angiogenesis was reduced, as assessed by CD31 staining.

Another formulation of Quercetin, using sulfobutyl ether-7-β-cyclodextrin, was evaluated for angiogenesis inhibition *in vivo* in addition to *in vitro* human tumor cell growth inhibition [96]. The cyclodextrin complex of Quercetin was more potent for inhibiting cell proliferation in human erythroleukaemia and cervical cancer cells *in vitro*. Decreased tumor MVD in a mouse melanoma model was observed following oral Quercetin administration; the Quercetin/cyclodectrin complex showed significantly improved anti-cancer activity at much lower concentration than the free drug.

The effects of physiologically-relevant (~100 nM) concentrations of several polyphenols from diverse classes on gene expression were investigated in HUVEC, using both microarray and quantitative RT-PCR assays [97]. Treatment with ferulic acid (Fig. 7), Quercetin or Resveratrol resulted in significant (>2-fold) down-regulation of ~3.6% of 10,000 genes in the microarray, and significant (>2-fold) up-regulation of ~2.3% of genes in the array. Quantitative RT-PCR studies indicated that Resveratrol significantly increased the expression of eNOS, whereas both Resveratrol and Quercetin decreased expression of endothelin-1, a vasoconstrictor.

Several components derived from propolis, including Quercetin, were evaluated in *in vitro* models of angiogenesis using HUVECs [98]. Two such components, Quercetin and caffeic acid phenethyl ester (Fig. 7), were found to possess strong inhibitory effects on tube formation and on EC proliferation.

The effect of Quercetin on *in vitro* and *in vivo* models of pancreatic cancer stem cells (CSC) demonstrated diminished ALDH1 activity and reduced apoptotic resistance in response to treatment [99]. The combination of Quercetin with sulforaphane, an isothiocyanate enriched in broccoli, revealed synergy. Quercetin caused enhanced binding of NF-kB, but co-incubation with sulforaphane blocked this. Quercetin prevented expression of proteins involved in EMT, which was enhanced by sulforaphane. Quercetin inhibited growth of CSC-enriched xenografts associated with reduced proliferation, angiogenesis, CSC-marker expression and apoptotic induction.

VEGF gene transcription in MCF-7 cells was inhibited by Quercetin [100]. Reporter gene assays containing hypoxia response elements and activator protein-1 (AP-1) elements revealed that the activities of HIF-1α and AP-1, both transcription factors for VEGF gene transcription, were suppressed by Quercetin.

Nuclear levels of c-Jun and HIF-1α in Tamoxifen-resistant-MCF-7 cells were reduced by Quercetin, which also inhibited the enhanced VEGF secretion in these cells. Similar effects were observed in chick chorioallantoic membrane assays.

Quercetin and sulforaphane (Fig. 7) in combination inhibited the proliferation and migration of B16F10 melanoma cells as well as tumor growth *in vivo* more effectively than either agent alone [101]; the effectiveness of the combination effect was predominantly due to decreased MMP-9 expression in the tumors.

Intra-tumoral administration of Quercetin into Dalton's lymphoma ascites-induced solid tumors in Swiss albino mice decreased tumor volume [102]. Quercetin promoted tumor cell apoptosis by down-regulating Hsp90 and Hsp70; it also down-regulated the expression of HIF-1α and VEGF. Selected flavonoids were docked *in silico* onto pro-angiogenic peptides including human VEGF, HIF-1α, and VEGFR-2. Genistein, kaempferol, and Quercetin were found to be effective inhibitors of angiogenesis in the choriallantoic membrane model [103].

Quercetin suppressed both COX-2 mRNA and its protein expression, prostaglandin PGE(2) production, and COX-2 promoter activation in MDA-MB-231 and MCF-7 human breast cancer cells, as well as blocking COX-2-mediated angiogenesis in human EC [104]. Quercetin inhibited the binding of the transactivators CREB2, C-Jun, C/EBPβ and NF-κB, and blocked the recruitment of the coactivator p300, to the COX-2 promoter. It also inhibited p300 histone acetyltransferase (HAT) activity, thereby abrogating the p300-mediated acetylation of NF-κB. Conversely, genetic re-constitution with active p300 was shown to overcome Quercetin-mediated inhibition of endogenous HAT activity.

Hypericum attenuatum Choisy contains a number of flavonoids, including Quercetin-4'-O-β-D-glucopyranoside (QODG), and the latter was used to treat HUVECs, and resultant effects on cell viability, cell migration, tube formation and apoptosis were determined, as well as those on VEGFR-2-mediated signaling [105]. QODG inhibited angiogenesis in HUVECs *in vitro via* suppression of VEGF-induced phosphorylation of VEGFR-2. c-Src, FAK, ERK, AKT, mTOR and S6K also impacted this signaling.

EC were pre-incubated with Quercetin before stimulation with PMA and showed reduced tube formation on matrigel and migration in wound healing assays [106]. Quercetin-mediated reduction of angiogenesis was associated with inhibition of PMA-induced COX-2 protein expression and prostanoid production, as well as MMP-9 protein release and gelatinolytic activity, associated with reduced intracellular ROS levels and activation of NF-κB.

Quercetin encapsulated into biodegradable monomethoxy poly(ethylene glycol)-poly(ε-caprolactone) micelles inhibited the growth of A2780S ovarian cancer cells *in vitro*, and *in vivo* administration of these micelles suppressed the growth of A2780S xenografts *via* increased apoptosis and reduced angiogenesis [107]. Quercetin treatment induced apoptosis of A2780S cells through the mitochondrial apoptotic pathway, and decreased phosphorylated p44/42 MAPK and phosphorylated Akt.

Rat aortic ring assays were used to demonstrate that Quercetin inhibited microvessel sprouting and the proliferation, migration, invasion and tube formation of EC [108]. Quercetin treatment inhibited *ex vivo* angiogenesis in the chicken egg chorioallantoic membrane assay and matrigel plug assays, and suppressed VEGF induced phosphorylation of VEGFR-2 and the downstream protein kinases AKT, mTOR, and ribosomal protein S6 kinase in HUVECs.

PEGylated liposomal Quercetin was evaluated in both CDDP-sensitive (A2780s) and -resistant (A2780cp) human ovarian cancer models [109]. The anti-tumor activity of this formulation was verified in both models, compared to free Quercetin or empty liposomes; immunohistochemical and immunofluorescence assays demonstrated that this was accompanied by inhibited proliferation and increased apoptosis of tumor cells, and decreased MVD. Quercetin-loaded micelles demonstrated improved cellular uptake and apoptotic induction, and stronger inhibitory effects on proliferation, migration, and invasion of 4T1 breast carcinoma cells compared to free Quercetin, and suppressed proliferation, migration, invasion, and tube formation of HUVECs [110]. In s.c. 4T1 tumors, Quercetin micelles were more effective than free drug in suppressing tumor growth and spontaneous pulmonary metastasis, and in prolonging host survival;

tumors in the micelle-treated group had increased apoptosis, less proliferation and fewer microvessels.

Quercetin interfered with the formation of intersegmental vessels, the dorsal aorta and the posterior cardinal vein in transgenic zebrafish embryos [111]; in HUVECs, Quercetin inhibited cell viability, expression of VEGFR-2 and tube formation, possibly *via* suppression of the ERK pathway.

Curcumin and Analogues

Treatment of activated T cells with Curcumin (Fig. **8**) was found to inhibit IL-12-induced tyrosine phosphorylation of JAK2, STAT3, tyrosine kinase 2, and STAT4 [112]. Curcumin inhibited both constitutive and IL-6-induced STAT3 phosphorylation and consequent STAT3 nuclear translocation in multiple myeloma cells [113]. Curcumin was without effect on STAT5 phosphorylation, but inhibited IFN-alpha-induced STAT1 phosphorylation. Inhibition was both rapid and reversible. HTLV-I-transformed T cell leukemias, MT-2, HuT-102, and SLB-1, express constitutively phosphorylated JAK3, TYK2, STAT3, and STAT5; *in vitro* treatment with Curcumin induced a dose-dependent decrease in JAK and STAT phosphorylation, concomitant with growth-arrest and apoptosis [114]. In K562 chronic erythroleukemia cells, Curcumin treatment decreased nuclear STAT3, STAT5a and STAT5b, without affecting STAT1 or the phosphorylation state of these three STAT proteins [115]. The decrease in nuclear STAT5a and STAT5b occurred concomitantly with an increase in truncated STAT5 isoforms—thus indicating that Curcumin induced STAT5 cleavage into dominant negative variants lacking the STAT5 C-terminal region. IFNγ treatment increased nuclear STAT1 and STAT3 levels, as well as their phosphorylated isoforms, the latter blocked by Curcumin. Curcumin also inhibited expression of JAK2 mRNA as well as cyclin D1 and v-src gene expression in these cells.

Treatment of microglial cells with Curcumin increased phosphorylation and association of SH2 domain-containing protein tyrosine phosphatases, SHP-2, with JAK1/2--thereby inhibiting the initiation of JAK-STAT signaling in activated microglia [116]. Curcumin strongly inhibited phosphorylation of STAT1 and STAT3 as well as JAK1 and 2 in microglia activated with gangliosides, LPS, or

IFN-gamma. Curcumin blocked the expression of inflammation-associated genes, including ICAM-1 and MCP 1, whose promoters contain STAT-binding elements. Treatment of ovarian and endometrial cancer cell lines with Curcumin decreased both constitutive and IL-6-induced STAT-3 phosphorylation, which was associated with reduced cell viability and increased cleavage of caspase-3, resulting in its activation [117]. Inhibition by Curcumin was reversible and pSTAT-3 levels were normalized again 24 h after Curcumin removal. Basal Suppressors Of Cytokine Signaling Protein 3 (SOCS-3) expression was higher in cancer cell lines compared to normal cell lines, and Curcumin treatment decreased this expression. Overexpression of SOCS-3 in Curcumin-treated cells increased both pSTAT3 levels as well as cell viability. Normal ovarian and endometrial cells exhibited high expression of Protein Inhibitors of Activated STAT3 (PIAS-3) protein, whereas basal expression in cancer cells was lower, but increased following Curcumin exposure. siRNA-mediated knockdown of PIAS-3 blocked these inhibitory effects of Curcumin on STAT-3 phosphorylation and on cell viability.

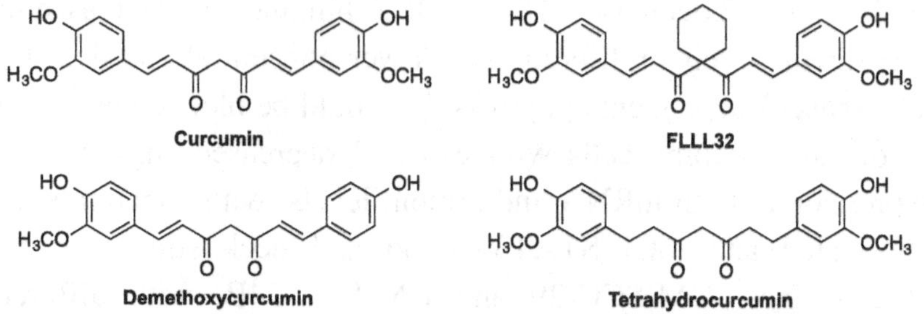

Figure 8: Curcumin and analogs.

Curcumin inhibited JAK1,2/STAT3 tyrosine-phosphorylation in murine glioma cell lines [118]. Curcumin induced G2/M arrest and down-regulated transcription of the STAT3 target genes c-Myc, MMP-9, Snail, and Twist, and of Ki67. Curcumin reduced glioma cell migratory and invasive behavior, which could be overcome by constitutively active STAT3C. *In vivo*, Curcumin reduced the growth rate and midline crossing of intracranially implanted gliomas, resulting in significantly improved long-term host survival. Assays of mitochondrial dehydrogenase activity, phosphatidylserine externalization, esterase staining, caspase activation and poly-adenosine diphosphate ribose polymerase cleavage indicated that Curcumin suppressed phosphorylation of STAT3 at both Tyr705 and Ser727, and JNK-1 phosphorylation in cholangiocarcinoma cell lines KKU100, KKU-M156 and KKU-M213 [119].

Polycystin-1 is encoded by the Pkd1 gene, and Pkd1-deletion mice were treated with Curcumin, which has previously been reported to affect several pathways modulated in autosomal dominant polycystic kidney disease (ADPKD) [120]. mTOR signaling was elevated in cystic kidneys after such conditional inactivation of Pkd1, and activation of STAT3 signaling was correlated with cyst progression. Both the mTOR and STAT3 pathways were inhibited by Curcumin *in vitro,* and Pkd1-deletion mice treated with Curcumin showed improved renal histology and reductions in STAT3 activation upon sacrifice, as well as in proliferation and cystic indices, and in kidney weight/body weight ratios; renal failure was also significantly delayed.

Using structure-based design, a Curcumin analogue, FLLL32 (Fig. **8**), was generated which had superior biochemical properties and more specificity for STAT3 [121]. FLLL32 decreased STAT3 DNA binding and decreased human and canine osteosarcoma cell proliferation at lower concentrations than Curcumin--leading to caspase-3-dependent apoptosis that could be blocked by Z-VAD-FMK. Treatment of osteosarcoma cells with FLLL32 decreased survivin, VEGF, and MMP2 expression at both mRNA and protein levels, with concomitant decreases in phosphorylated and total STAT3. Head and neck squamous cell cancer (HNSCC) cell lines, UM-SCC-29 and UM-SCC-74B, with different CDDP sensitivity showed expression of pSTAT3 [122]. CDDP-resistant UM-SCC-29 cells and CDDP-sensitive UM-SCC-74B cells both express STAT3 and pSTAT3. FLLL32 down-regulated pSTAT3 in both HNSCC cell lines and induced a potent antitumor effect; FLLL32, alone or with CDDP, increased the proportion of apoptotic cells. FLLL32 sensitized CDDP-resistant UM-SCC-29 cells ~ four-fold.

Curcumin inhibited small cell lung cancer (SCLC) cell proliferation, cell cycling, migration, invasion and angiogenesis through suppression of STAT3 signaling and its downstream targets [123]. Curcumin suppressed expression of proliferative proteins, Survivin, Bcl-X(L) and Cyclin B1, and invasive proteins: VEGF, MMP-2, MMP-7 and ICAM-1. STAT3 knockdown by siRNA was able to induce anti-invasive effects, whereas activation by IL-6 increased cell proliferation, cell survival, angiogenesis, invasion, migration and tumor growth.

Computational modeling of STAT3 dimerization implicated Lys591, Arg609, Ser611, Glu612, Ser613, Ser636 and Val637 in the binding of Curcumin and analogues with the SH2 domain of monomeric STAT3 [124]. Demethoxycurcumin and hexahydrocurcuminol were predicted to be the most potent inhibitors of all Curcumin derivatives and known inhibitors: FLLL32, STA21, and Stattic. The Curcumin-proline conjugate (1,7-Bis(4-O-L-prolinoyl-3-methoxyphenyl)-1,4,6-heptatriene-5-ol-3-one) was predicted to be the most potent inhibitor of STAT3 dimerization of Curcumin-amino acid conjugates.

The potential anti-angiogenic activity of demethoxycurcumin (DC) (Fig. **8**), a structural analog of Curcumin, was investigated in cultured HUVECs using cDNA microarray analysis [125]. Of 1024 human genes in the cancer-focused microarray, 187 were up-regulated and 72 down-regulated at least two-fold. Nine angiogenesis-related genes were down-regulated > 5-fold in response to DC, suggesting that genetic reprogramming was involved in its anti-angiogenic properties. Matrix metalloproteinase-9 (MMP-9), the product of one of the strongly down-regulated angiogenesis-related genes down-regulated over 5-fold by DC, was assayed *via* gelatin zymography; DC was found to strongly inhibit MMP-9 expression, without a direct effect on its activity.

A panel of synthetic Curcumin analogs was prepared and screened for *in vitro* anti-cancer and anti-angiogenesis activities; some analogs performed with high potency in the NCI *in vitro* anti-cancer cell line cytotoxicity screening assay and were superior to CDDP by these criteria [126]. Analogs that were most active in the cell line screens were also effective in *in vitro* anti-angiogenesis assays, one being almost on par with the anti-angiogenic drug, TNP-470. A lead compound was further evaluated *in vivo* and was shown to effectively reduce the volume of human breast tumor xenografts and did so with minimal toxicity.

The effects of Curcumin and tetrahydrocurcumin (THC, Fig. **8**) on tumor angiogenesis in human HepG2 hepatocellular carcinoma models were evaluated [127]. HepG2 xenografts were innoculated into a dorsal skin-fold chamber in male nude mice. Daily oral administration of Curcumin or THC was initiated a day later. The tumor microvasculature was observed using fluorescence videomicroscopy and capillary vascularity (CV) at weekly intervals thereafter.

Multiple pathological angiogenic features, including microvascular dilatation, tortuosity, and hyper-permeability were markedly reduced by treatment with either drug; treatments also caused significant decreases in the CV.

Validation *in silico* of wet laboratory data for Curcumin, aeroplysinin-1, and halofuginone has been reported [128]. The inhibition patterns of vascular EC differentiation and of capillary tube formation mediated by anti-angiogenic growth factors from treatment with these compounds aligned well with *in silico* results for several angiogenic targets, *e.g.*, bFGF and VEGFR-1.

Resveratrol

Resveratrol (Fig. **9**), a component of red wine, inhibited both TNFα- and IL-6-induced Tyr705 phosphorylation of STAT3 in IL-6-treated ECs [129]. Resveratrol induced eNOS activity and thereby NO production; treatment of ECs with the NO donor, SNAP, reduced IL-6-induced STAT3 phosphorylation, whereas exposure of ECs to a NO Synthase inhibitor reversed these effects of Resveratrol. ECs transfected with constitutively active Rac1 showed increased ICAM-1 promoter activity, intracellular ROS levels and also STAT3 phosphorylation: all blocked by Resveratrol.

Resveratrol-treated cells with constitutively-active STAT3 underwent irreversible G0-G1 arrest, observed for v-Src-transformed mouse NIH3T3 fibroblasts, MDA-MB-231 human breast, Panc-1 pancreatic, and DU145 prostate carcinoma cells, whereas arrest was at the S phase in MDA-MB-468 human breast and Colo-357 pancreatic cancer cells [130]. In contrast, cells treated with Resveratrol and lacking aberrant STAT3 activity showed reversible growth arrest and only slight loss of viability. In DU145 cells and v-Src-transformed mouse fibroblasts, Resveratrol down-regulated STAT3-regulated cyclin D1, Bcl-xL and Mcl-1 genes.

Resveratrol decreased STAT3 activation in DR5-positive melanomas [131]. In another melanoma model, Resveratrol inhibited STAT3 and NF-kB-dependent transcription, thereby suppressing cFLIP and Bcl-xL expression, while activating the MAPK- and the ATM-Chk2-p53 pathways [132]. Resveratrol also upregulated TRAIL promoter activity and induced TRAIL surface expression in melanoma cell lines, resulting in rapid apoptosis; however, for melanoma lines with suppressed cell surface translocation of TRAIL, a necrotic mechanism was predominant following radiation.

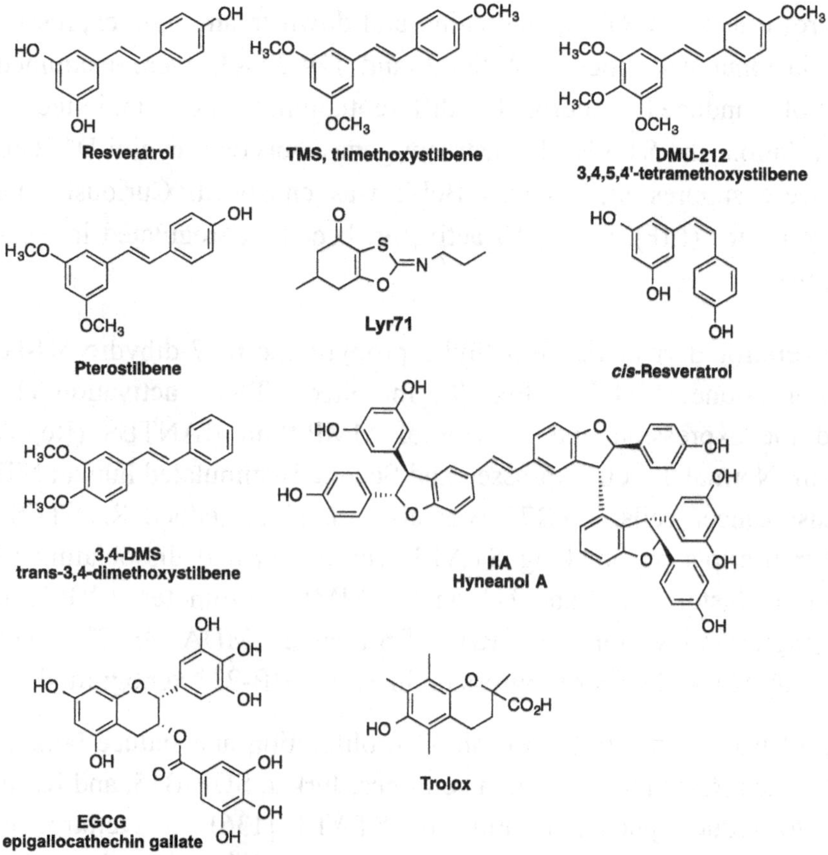

Figure 9: Resveratrol, related stilbene derivatives, and co-administered natural product inhibitors of STAT3 function.

Resveratrol and an analog, 3,4,5,4'-tetramethoxystilbene (DMU-212, Fig. **9**), were investigated for effects on cell viability, cell cycle traverse, STAT3 activation, and microtubule dynamics in MDA-MB-435 and MCF-7 human breast cancer cells [133]. DMU-212 exerted higher levels of growth inhibition against both cell lines compared to Resveratrol, and acted *via* different mechanisms. DMU-212 induced predominantly G2/M arrest, whereas Resveratrol induced G0/G1 arrest; Resveratrol induced more prominent changes in Cyclin D1 expression compared to DMU-212. DMU-212 induced apoptosis and reduced the expression of multiple anti-apoptotic proteins more so than Resveratrol. Although both agents inhibited STAT3 phosphorylation, treatments of DMU-212 also led to increased tubulin polymerization.

Resveratrol effects on STAT3 signaling and downstream gene expression in the medulloblastoma cell lines, UW228-2 and UW228-3, were examined [134]. Resveratrol induced neuronal differentiation, and inhibited STAT3 phosphorylation; the STAT3 downstream genes, survivin, cyclin D1, Cox-2, and c-Myc, were suppressed, whereas Bcl-2 was enhanced. Curiously, Leukemia Inhibitory Factor (LIF), a STAT3 activator, became upregulated in Resveratrol-treated cells.

The Resveratrol derivative, 6-methyl-2-propylimino-6, 7-dihydro-5H-benzo [1, 3]-oxathiol- 4-one (LYR71) (Fig. **9**), inhibited STAT3 activation [135], and inhibited the expression and activity of MMP-9 in RANTES (Regulated on Activation, Normal T Cell Expressed and Secreted)-stimulated human MDA-MB-231 breast cancer cells. LYR71 was also found to reduce RANTES-induced MMP-9 transcripts by blocking STAT3 recruitment and dissociating p300 and deacetylating histones H3 and H4 on the MMP-9 promoter. LYR71 inhibited tumor migration/invasion in RANTES-treated MDA-MB-231 cells and consequently blocked tumor progression in MDA-MB-231 xenografted mice.

Resveratrol was demonstrated to inhibit proliferation and induce both apoptosis and cell cycle arrest in the leukemia cell lines, Jurkat, SUP-B15, and Kasumi-1, as well as to reduce phosphorylation of STAT3 [136]. Resveratrol treatment improved the survival of Kasumi-1-bearing mice, and down-regulated the STAT3 pathway *in vivo*. Resveratrol was also investigated for its anti-tumor effects against the NK cell lines KHYG-1, NKL, NK-92 and NK-YS, representative of NK cell leukemias and lymphomas [137]. Resveratrol induced G0/G1 arrest, suppressed proliferation and induced apoptosis. It also suppressed constitutively active STAT3 and inhibited JAK2 phosphorylation, but had no effect on other upstream mediators of STAT3 activation, such as PTEN, TYK2, and JAK1. Resveratrol, as expected, induced down-regulation of the anti-apoptotic proteins MCL1 and survivin, downstream effectors of STAT3; combined treatment with 5-fluorouracil (5FU) and Resveratrol sensitized HT-29 and SW-620 colorectal carcinoma cell lines to 5FU, inducing increased oxidative stress and inhibition of AKT and STAT3 pathways [138].

Trimethoxyl stilbene (TMS, Fig. **9**), a derivative of Resveratrol, inhibited proliferation and induced apoptosis of A549 NSCLC cells [139], associated with up-regulation/activation of caspase-3 and IκB, and down-regulation of NFκB, STAT3, STAT5b, and JAK2 signal transduction.

Primary glioblastoma multiforme (GBM)-CD133(+)-derived radioresistant tumor initiating cells (TIC) were highly tumorigenic and had high levels of phosphorylated STAT3 [140]. Treatment with either shRNA-STAT3 or the STAT3 inhibitor, AG490, significantly inhibited the cancer stem cell (CSC)-like properties and the radioresistance of the GBM-CD133(+) TIC, both *in vitro* and *in vivo*; treatment with Resveratrol induced TIC apoptosis and enhanced their radiosensitivity by suppressing STAT3 signaling.

Tumors harbor elevated levels of STAT3 with Lysine acetylation, and the consequences of altering such acetylation in human MDA-MB-468 triple-negative breast cancer (TNBC) and HCT116 colon cancer cells, as well as in A2058 and M223 melanomas were investigated. Genetically altering STAT3 at Lys685 reduced tumor growth, which was accompanied by the demethylation and reactivation of tumor-suppressor genes [141]. Such mutation of STAT3 at Lys685 disrupted DNA methyltransferase 1/STAT3 interactions, which paralleled observations following Resveratrol treatment, acting as an acetylation inhibitor. Of note, reduction of acetylated STAT3 in TNBC cells caused demethylation and activation of the estrogen receptor (ER)-α gene, thereby sensitizing the tumor cells to anti-estrogens.

Pterostilbene (PTE), a dimethylated analog of Resveratrol with higher bioavailability, was evaluated in human osteosarcoma cells to determine its impact on JAK2/STAT3 and apoptosis-related signaling pathways [142]. PTE exhibited strong antitumor activity, as evidenced by reductions in tumor cell viability, adhesion, migration and mitochondrial membrane potential and increases in apoptotic index and ROS production. It also inhibited the phosphorylation of JAK2 at Tyr 1007 and thus the downstream activation of STAT3 and its target genes, including Bcl-xL and Mcl-1, leading to the up-regulation of mitochondrial apoptosis pathway-related proteins and cyclin-dependent kinase inhibitors: p21 and p27.

Resveratrol was found to inhibit lung metastasis from orthotopic syngeneic 4T1 breast tumors in mice [143], involving inactivation of STAT3, and preventing the generation and function of tumor-evoked regulatory B cells (tBregs), and also negating TGF-β-dependent conversion of metastasis-promoting Foxp3(+) regulatory T cells (Tregs).

The effects of the methylated analog of Resveratrol, 3,4,5,4'-trans-tetramethoxystilbene, DMU-212 (Fig. **9**) on the activation of the transcription factors NF-κB, AP-1, and STAT3 on rat hepatocarcinogenesis were evaluated [144]. Activation of NF-κB and of STAT3 induced by N-nitrosodiethylamine/phenobarbital treatment was suppressed by DMU-212, as was the level of iNOS protein, downstream of NF-κB; treatment with DMU-212 increased constitutive levels of the AP-1 subunits, c-Jun and c-Fos, and the binding of c-Jun to TRE consensus site.

Human EJ transitional cell carcinoma (TCC) cells were briefly exposed to Resveratrol to attempt to imitate pulsatile *in vivo* drug exposure in the bladder from intravesical instillation; in addition, an orthotopic TCC nude mouse xenograft model was established by injecting EJ cells into the sub-urothelial layer of the bladder and treating the tumor-implanted mice with short-term intravesical Resveratrol instillation [145]. Resveratrol treatment caused S phase arrest and apoptosis *in vitro,* associated with attenuated phosphorylation, nuclear translocation and transcription of STAT3, and down-regulation of survivin, cyclinD1, c-Myc and VEGF: all STAT3 downstream genes. Resveratrol treatment of the mouse tumor xenograft model caused tumor growth suppression, apoptosis and STAT3 inactivation without affecting normal urothelium.

Resveratrol inhibited the growth and migration of BAE cells and inhibited their tube formation [87]. Resveratrol inhibited cellular proliferation, cell-cycle progression, and matrix metalloproteinase (MMP)-9 expression in TNF-α-induced human vascular smooth muscle cells (VSMC) [146]. Resveratrol inhibited TNF-α-stimulated DNA synthesis in VSMC, which was associated with reduced levels of ERK 1/2 activity and cell-cycle arrest at the G(1) phase; it also down-regulated the expression of cyclins and their cyclin-dependent kinases (CDKs), while upregulating the expression of the CDK1 inhibitor, p21/WAF1.

Zymographic and immunoblot analyses both revealed that Resveratrol suppressed the TNF-α-induced expression of MMP-9 in VSMC.

Cis- and *trans*-Resveratrol stereoisomers were compared for their effects on angiogenesis and EC αv-β3 integrin function [147]. The *trans* isomer inhibited EC proliferation and the repair of mechanically wounded EC monolayers, as well as preventing EC sprouting in fibrin gel, collagen gel invasion, and morphogenesis on Matrigel. *Trans*-Resveratrol inhibited vascularization of the chick embryo vasculosa and also murine melanoma B16 tumor growth and neovascularization. In contrast, the *cis* isomer had null or limited effects. *Trans-* but not *cis*-Resveratrol inhibited αv-β3 integrin-dependent EC adhesion and the recruitment of enhanced GFP-tagged beta3 integrin in focal adhesion contacts.

The mechanism of anti-angiogenic activity of Resveratrol and its trans-3,5,4'-trimethoxystilbene analogue in HUVECs and in transgenic zebrafish was investigated [148]. The analogue had more potent anti-angiogenic activity than Resveratrol in both gene expression analyses of the VEGF receptor and cell-cycle assays. In zebrafish, trans-3,5,4'-trimethoxystilbene caused intersegmental vessel regression and down-regulated VEGFR2 mRNA expression, and also induced G2/M cell-cycle arrest, predominantly in EC of zebrafish embryos. Resveratrol reduced EC tube formation on matrigel and EC migration in wound healing assays [106]. This effect was associated with inhibition of PMA-induced COX-2 protein expression and prostanoid production, as well as MMP-9 protein release and gelatinolytic activity, which were accompanied by a significant reduction in the stimulated intracellular ROS levels and in activation of redox-sensitive NF-κB. Trans-3,4-dimethoxystilbene (3,4-DMS, Fig. **9**) inhibited proliferation, migration, tube formation, and endogenous neovascularization of EC [149]. The anti-angiogenic effect was proposed to occur *via* induction of EC apoptosis; 3,4-DMS also induced macroautophagy in EC through activation of AMPK and the resultant downstream inhibition of mTOR. Combination of 3,4-DMS with 3-methyladenine (3-MA), an inhibitor of autophagy, or with an siRNA specific for the autophagy-related gene (ATG), potentiated both pro-apoptotic and anti-angiogenic effects.

Resveratrol demonstrated cytotoxic effects and induction of rat RT-2 glioma cell apoptosis [150]. Resveratrol exerted antitumor effects on s.c. (ectopic) tumors, including slower tumor growth rate and improved host survival; Resveratrol was less effective against intracerebral (orthotopic) tumors. The expression of VEGF in RT-2 cells and the proliferation of ECV304 HUVEC were both inhibited by Resveratrol, and immunohistochemical analyses indicated that treatment reduced MVD in the s.c. gliomas. The effects of Resveratrol on angiogenesis in s.c. rat RT-2 gliomas were investigated by color Doppler ultrasound [151]. Rats treated with Resveratrol had slower tumor growth rates than controls, and the reduced color Doppler vascularity index, MVD and tumor size were highly correlated.

Resveratrol was found to reduce primary tumor volume and weight and lung metastasis in mice bearing the Lewis lung carcinoma (LLC) [152]. Resveratrol inhibited tumor-induced neovascularization *in vivo* and inhibited the formation of capillary-like tube formation from HUVEC *in vitro*. Resveratrol inhibited the binding of VEGF to HUVEC. These Investigators proposed that the anti-tumor and anti-metastatic activities of Resveratrol could be attributed to the inhibition of DNA synthesis in LLC cells and the inhibition of LLC-induced angiogenesis. The anti-tumor activity of heyneanol A (HA, Fig. **9**), a tetramer of Resveratrol, was assessed in the C57BL/6 mouse s.c. LLC model [153]. HA decreased tumor growth, accompanied by marked tumor cell apoptosis--defined by cleaved caspase-3 and TUNEL assays--decreased tumor cell proliferation, as well as MVD. Treatment of LLC cells *in vitro* with HA or Resveratrol significantly increased apoptosis; both agents induced cleavage of caspase-9, caspase-3 and PARP, which were blocked by the pan-caspase inhibitor, Z-VAD-FMK. HA and Resveratrol suppressed the bFGF-induced proliferation and capillary differentiation of HUVEC, and inhibited the binding of bFGF to its receptor.

Resveratrol demonstrated potent inhibition of FGF2-induced angiogenesis in EC and inhibited platelet/fibrin clot-promoted human colon and fibrosarcoma tumor growth in the chick chorioallantoic membrane tumor model [154]. Resveratrol promoted apoptosis in FGF2-stimulated EC by increasing p53 protein production.

Resveratrol-treated mice had significantly lower tumor growth, decreased angiogenesis, and increased apoptotic index in their MDA-MB-231 xenografted tumors [155]; Resveratrol-treated MDA-MB-231 cells had reduced extracellular levels of VEGF *in vitro*.

The effect of Resveratrol on HPV-16 E6- and E7-induced HIF-1α and VEGF gene expression in cervical carcimoma cells, C-33A and HeLa, was examined [156]. HPV-16 E6- and E7-transfected cells expressed increased HIF-1α protein and VEGF. These increases were blocked by cotransfection with either HIF-1α siRNA or by treatment with Resveratrol. Blocking ERK 1/2 and PI3-kinase by PD98059 and LY294002, respectively, abolished 16 E6- and E7-induced HIF-1α and VEGF expression. HPV-16 E6- and E7-transfected cervical cancer cells stimulated *in vitro* capillary or tubule formation, which could be abolished either by cotransfection with HIF-1α siRNA or with Resveratrol treatment.

Cell proliferation, migration and differentiation of HUVECs was strongly increased by coculture with RPMI 8226 multiple myeloma cells, and Resveratrol inhibited all three effects [157]. Resveratrol treatment of RPMI 8226 cells caused decreased MMP-2 and MMP-9 activity, and inhibited VEGF and bFGF protein levels. Decreased levels of VEGF, bFGF, MMP-2 and MMP-9 mRNA indicated Reservatrol-mediated anti-angiogenic control at the level of gene expression.

Resveratrol modulated HIF-1α and VEGF expression in human ovarian cancer cells A2780/CP70 and OVCAR-3 [158]. It reduced both basal-level and growth factor-induced HIF-1α protein levels and inhibited VEGF expression. Resveratrol inhibited AKT and MAPK activation and inhibited insulin-like growth factor 1 (IGF-1)-induced HIF-1α expression *via* inhibition of ribosomal protein S6 kinase 1, S6 ribosomal protein, eukaryotic initiation factor 4E-binding protein 1, and eukaryotic initiation factor 4E. Resveratrol also induced HIF-1α protein degradation through the proteasome pathway. The effects of Lysophosphatidic acid (LPA) and Resveratrol on cell migration and HIF-1α and VEGF expression in human ovarian cancer cells was evaluated [159]. LPA treatment under hypoxia increased HIF-1α protein levels and thus increased expression of VEGF protein and mRNA; these increases in HIF-1α and VEGF expression were markedly blunted by Resveratrol. This is mechanistically associated with inactivation of both p42/p44 MAPK and p70S6K and enhanced degradation of HIF-1α protein—with a net decrease in VEGF expression and cell migration.

Resvertarol was observed to inhibit both pro-MMP-9 production and VEGF-induced angiogenesis in HUVECs [160]. Inhibition of pro-MMP-9 production in Colon 26 cells and VEGF-induced angiogenesis by three dihydroxystilbenes selected from a panel of 21 stilbenes was greater than those achieved by Resveratrol. These dihydroxystilbenes inhibited tumor growth in Colon 26-bearing mice and inhibited neovascularization in a Colon 26 chamber-bearing mouse model, and also inhibited VEGF-induced VEGFR-2 phosphorylation. However, they had no effect on either VEGFR-1 or VEGFR-2 expression, nor on VEGF-induced VEGFR-1 phosphorylation in HUVECs. These Investigators suggested that inhibition of tumor-induced neovascularization was attributable to inhibition of VEGF-induced EC migration and VEGF-induced angiogenesis mediated *via* inhibition of VEGF-induced VEGFR-2 phosphorylation in EC and pro-MMP-9 expression in Colon 26 tumor cells.

The effects of Resveratrol and EGCG on VEGF secretion by endometrial cancer cells were determined [161]. VEGF secretion from both hysterectomy-derived surgical samples and immortalized endometrial cell lines was detectable, but treatment with either Resveratrol or EGCG significantly reduced this.

Treatment of A549 lung adenocarcinoma cells with Resveratrol inhibited migratory and invasive activities [162]. Resveratrol also inhibited Heme oxygenase-1 (HO-1)-mediated MMP-9 and MMP-2 expression; HO-1 inhibition and silencing suppressed MMPs and invasion of lung cancer cells.

Vasculogenic mimicry, the ability of highly invasive tumor cells to form capillary-like structures and repetitive, matrix-rich networks in 3D culture that mimic embryonic counterparts, was studied for the possible effects of Resveratrol, EGCG, N-acetyl-cysteine (NAC) and Trolox on this process in a melanoma model [163]. The levels of VEGF, VEGFRs and active caspase-3 were strongly decreased in cells treated with these anti-oxidants, and the formation of capillary-like structures was strongly related to ROS levels. The effects of Resveratrol on tumor cell release of anti-angiogenic Thrombospondin-1 (TSP-1) and VEGF into the ECM and on vascular EC apoptosis were established [164]. Melanoma cell lines co-cultured with monolayers of vascular EC or in 3D spheroids were used as models. Resveratrol effects were environment-dependent: it stimulated vascular EC proliferation in monolayer cultures, whereas it caused EC growth inhibition when they were in 3D co-cultures with melanoma cells. The responses correlated with increased expression of p53 and TSP1 by the tumor cells, as well as with decreased hypoxia-driven expression of HIF-1α and inhibition of VEGF production.

Human LNCaP prostate cancer cells treated with Resveratrol demonstrated growth inhibition [165]. Subsequent gene expression analyses identified androgen-responsive genes affected by Resveratrol through inhibition of both androgen- and estrogen-mediated transcription. In LNCaP xenografts, Resveratrol treatment delayed tumor growth; however, contrary to expectations, treatment led to increased MVD and inhibition of apoptosis. Resveratrol and TRAIL each inhibited growth of PC-3 prostate cancer xenografts in nude mice by inhibiting tumor cell proliferation and inducing apoptosis [166]. Their combination was more effective in inhibiting tumor growth than either agent alone. In PC-3 xenografts, treatment with Resveratrol and TRAIL each inhibited angiogenesis (reduced MVD and VEGF and VEGFR2(+) cells) and MMP-2 and MMP-9; their combination was more effective than either agent alone.

The effect of Resveratrol on VEGF expression and angiogenesis in hepatocellular carcinoma (HCC) was investigated [167]. VEGF protein and mRNA expressions in HCC cells treated with Resveratrol were significantly decreased, as was activation of NF-kB. Tumor growth in nude mice and MVD was decreased following Resveratrol treatment.

Treatments of intradermally-implanted Ehrlich ascites carcinoma (EAC)-bearing mice with Resveratrol or its combination with CDDP caused reduction in MVD [168]. Plasma levels of VEGF were also significantly reduced following treatment with Resveratrol or when combined with CDDP. There was a strong association between plasma VEGF levels and MVD. Treatment with Resveratrol reduced the level of intra-tumoral VEGF receptor type-2 (Flk-1) in tumors, and co-administration with CDDP caused further reduction.

Inhibition of STAT3-Mediated Angiogenesis with Promiscuous Inhibitors of Kinases and Other Target Enzymes

Sorafenib

Sorafenib/Nexavar/Bay43-9006 (Fig. **10**), originally described as an orally-active selective inhibitor of Raf-1, a protein kinase downstream from Ras in the mitogen-activated protein kinase (MAPK) pathway [169], has since been established to be much broader in its inhibitory profile, including wild-type B-Raf, mutant b-raf V599E, VEGFR-2, VEGFR-3, PDGFR-β, Flt3, c-KIT and p38α as targets. As such, it has combined anti-proliferative and anti-angiogenic properties, classifying it as a so-called triple angiokinase inhibitor [170].

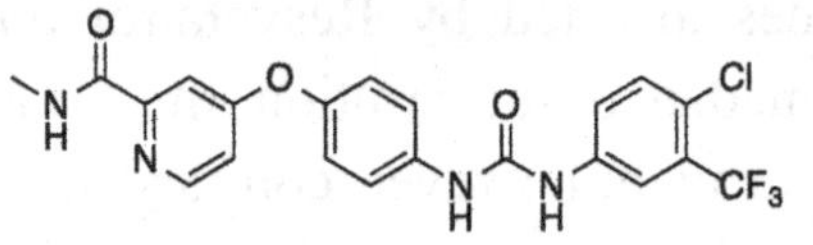

Sorafenib

Figure 10: Structure of sorafenib.

To add to this complex picture of interacting targets of Sorafenib, it was recognized fairly early in its development that it also blocked activation of the STAT3 pathway *in vitro* [171, 172]. Subsequent mechanistic studies rather rapidly accounted for this interaction, and identified activation of a phosphatase that de-phosphorylated pSTAT3 as a target of Sorafenib: specifically, the phosphatase Shatterproof 2 [173, 174].

In the decade since the initial clinical evaluation of Sorafenib, most of the clinical trials have been in vascularized tumors: metastatic renal cell carcinoma, grain tumors including malignant glioma, and non-small cell carcinoma [170, 175-180]. In renal cell carcinoma, Bevacizumab, the humanized anti-VEGF monoclonal antibody, plus IFN produced superior progression-free survival (PFS) and overall response rate (ORR) compared to IFN, but it also yielded a marked increase for the risk of serious side effects. The overall survival (OS) was extended by Sorafenib (17.8 months) and Sunitinib (26.4 months) as compared with IFN (13 months). It was concluded that compared with IFN therapy, VEGF pathway-targeted therapies improved PFS and achieved significant therapeutic benefits, but the net impact on the risk to benefit ratio remains to be established.

Some of these side effects focus on renal physiology; the mechanism for anti-VEGF therapy-induced hypertension is not well understood; however, NO pathway inhibition, rarefaction (thinning), and oxidative stress may be important in its pathogenesis [181]. Glomerular injury may develop from loss of VEGF effects on maintaining the filtration barrier.

In malignant glioma, clinical trials using Bevacizumab as well as small-molecule tyrosine kinase inhibitors (TKIs) that target different VEGF receptors, such as Sorafenib, either as monotherapy or in combination with other drugs, tended to indicate that glioma developed resistance to anti-angiogenic treatments [175]. In fact, prolonged anti-angiogenic therapy targeting only the VEGF-VEGFR axis might activate mechanisms involving other pro-angiogenic factors: *e.g.*, basic fibroblast growth factor (bFGF), stromal derived factor 1 (SDF-1) and Tie-2. These could stimulate angiogenesis by mobilizing bone marrow derived endothelial progenitor cells that promote angiogenesis.

TKIs that block VEGFRs are still under investigation in NSCLC. Their lack of target specificity likely will lead to unexpected toxicities. In early trials, Sorafenib monotherapy has shown no clinical benefit in pre-treated NSCLC patients, although combination with Erlotinib or with cytotoxics has demonstrated some optimism [176]. However, in more advanced clinical testing of Sorafenib and a variety of agents that inhibit VEGF in addition to either platelet derived growth factor (PDGF) or fibroblast derived growth factor (FGF), none even in combination with chemotherapy have resulted in improvements in OS for advanced NSCLC [170].

More recently, drugs against EGFR, *e.g.*, the TKIs Erlotinib and Gefitinib, combined with Bevacizumab, have been found to provide clinical benefit in advanced NSCLC, and are currently approved. The most compelling evidence has been derived from the combination of either Erlotinib plus Bevacizumab, or Erlotinib plus either Sunitinib or Sorafenib. Phase III trials have typically yielded improvements in PFS, but an impact on OS has yet to be demonstrated [177].

SUMMARY

Inhibition of angiogenesis has for some time been heralded as a major target for cancer treatment, a view accompanied by an ever-increasing understanding of underlying tumor angiogenic mechanisms that has been achieved over the last few decades. Discovery and development of anti-angiogenic agents is a major challenge, and the FDA approval of Bevacizumab, a humanized monoclonal antibody directed against VEGF, has been a milestone in that quest. Initially encouraging results prompted the development of other agents targeting angiogenesis, with some now approved for a number of cancers, usually in combination treatment settings. Although the clinical experience has typically been that PFS was increased, a favorable impact on OS has been more elusive, with most patients unfortunately relapsing due to acquired resistance and, in some cases, with even more aggressive disease. Thus, there continues to be a strong need for new anti-angiogenic candidates and an understanding of their optimal utilization.

This Review has emphasized approaches under development that act on the STAT3/VEGF/VEGFR axis, including small molecules, peptidomimetics and natural products (Fig. **11**). Clearly, the disappointing toxicities observed clinically with AZD1480, acting upstream of STAT3 at JAK2, serve as a caution that this approach will also be met with considerable and unanticipated challenges. Thus, the existence of a multitude of approaches in various stages of the developmental pipeline should be viewed as a positive, from which we hope that agents of real therapeutic benefit to cancer patients will in time emerge.

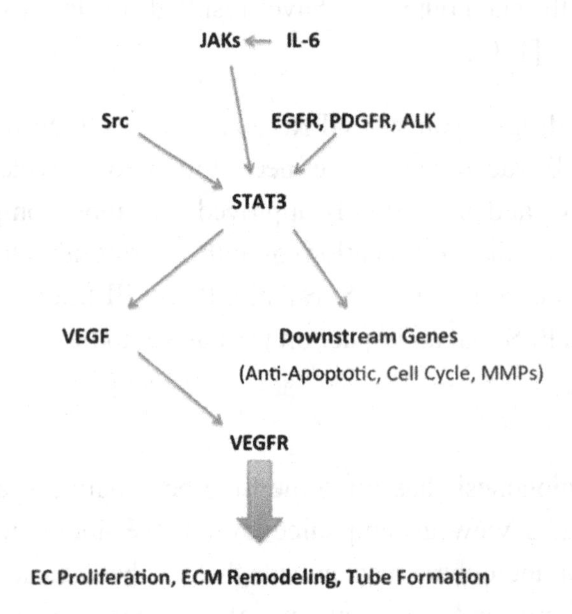

Figure 11: Cellular and Molecular Pathways Engaged in STAT3 Regulation of Angiogenesis.

ACKNOWLEDGEMENTS

Declared None.

CONFLICT OF INTEREST

The authors confirm that this chapter contents have no conflict of interest.

REFERENCES

[1] Huang S. Regulation of metastases by signal transducer and activator of transcription 3 signaling pathway: clinical implications. Clinical cancer research : an official journal of the American Association for Cancer Research. 2007; 13(5):1362-6.

[2] Yu H, Jove R. The STATs of cancer--new molecular targets come of age. Nat Rev Cancer. 2004; 4(2):97-105.

[3] Yu H, Pardoll D, Jove R. STATs in cancer inflammation and immunity: a leading role for STAT3. Nat Rev Cancer 2009; 9(11):798-809.

[4] Dave B, Landis MD, Tweardy DJ, Chang JC, Dobrolecki LE, Wu MF, *et al.* Selective small molecule Stat3 inhibitor reduces breast cancer tumor-initiating cells and improves recurrence free survival in a human-xenograft model. PLoS One 2012; 7(8):e30207.

[5] Haftchenary S, Avadisian M, Gunning PT. Inhibiting aberrant Stat3 function with molecular therapeutics: a progress report. Anticancer Drugs. 2011; 22(2):115-27.

[6] Lavecchia A, Di Giovanni C, Novellino E. STAT-3 inhibitors: state of the art and new horizons for cancer treatment. Curr Med Chem. 2011; 18(16):2359-75.

[7] Miklossy G, Hilliard TS, Turkson J. Therapeutic modulators of STAT signalling for human diseases. Nat Rev Drug Discovery. 2013; 12(4_testjo):611-29.

[8] Niu G, Wright KL, Huang M, Song L, Haura E, Turkson J, *et al.* Constitutive Stat3 activity up-regulates VEGF expression and tumor angiogenesis. Oncogene. 2002; 21(13):2000-8.

[9] Masuda M, Suzui M, Lim JTE, Deguchi A, Soh J-W, Weinstein IB. Epigallocatechin-3-gallate decreases VEGF production in head and neck and breast carcinoma cells by inhibiting EGFR-related pathways of signal transduction. J Exp Ther Oncol. 2002; 2(6):350-9.

[10] Song H, Wang R, Wang S, Lin J. A low-molecular-weight compound discovered through virtual database screening inhibits Stat3 function in breast cancer cells. Proc Natl Acad Sci U S A. 2005; 102(13):4700-5.

[11] Wei C-C, Ball S, Lin L, Liu A, Fuchs JR, Li P-K, *et al.* Two small molecule compounds, LLL12 and FLLL32, exhibit potent inhibitory activity on STAT3 in human rhabdomyosarcoma cells. Int J Oncol. 2011; 38(1):279-85.

[12] Bhasin D, Cisek K, Pandharkar T, Regan N, Li C, Pandit B, *et al.* Design, synthesis, and studies of small molecule STAT3 inhibitors. Bioorg Med Chem Lett. 2008; 18(1):391-5.

[13] O'Brien PJ. Molecular mechanisms of quinone cytotoxicity. Chem-Biol Interact. 1991; 80(1):1-41.

[14] Padhye S, Dandawate P, Yusufi M, Ahmad A, Sarkar FH. Perspectives on medicinal properties of plumbagin and its analogs. Med Res Rev,. 2012; 32(6):1131-58.

[15] Bid HK, Oswald D, Li C, London CA, Lin J, Houghton PJ. Anti-angiogenic activity of a small molecule STAT3 inhibitor LLL12. PLoS One. 2012; 7(4):e35513.

[16] Matsuno K, Masuda Y, Uehara Y, Sato H, Muroya A, Takahashi O, *et al.* Identification of a New Series of STAT3 Inhibitors by Virtual Screening. ACS Med Chem Lett. 2010; 1(8):371-5.

[17] Ashizawa T, Miyata H, Ishii H, Oshita C, Matsuno K, Masuda Y, *et al.* Antitumor activity of a novel small molecule STAT3 inhibitor against a human lymphoma cell line with high STAT3 activation. International journal of oncology. 2011; 38(5):1245-52.

[18] Ashizawa T, Miyata H, Iizuka A, Komiyama M, Oshita C, Kume A, *et al.* Effect of the STAT3 inhibitor STX-0119 on the proliferation of cancer stem-like cells derived from recurrent glioblastoma. International journal of oncology. 2013; 43(1):219-27.

[19] Assi HH, Paran C, VanderVeen N, Savakus J, Doherty R, Petruzzella E, *et al.* Preclinical characterization of signal transducer and activator of transcription 3 small molecule inhibitors for primary and metastatic brain cancer therapy. The Journal of pharmacology and experimental therapeutics. 2014; 349(3):458-69.

[20] Iwamaru A, Szymanski S, Iwado E, Aoki H, Yokoyama T, Fokt I, *et al.* A novel inhibitor of the STAT3 pathway induces apoptosis in malignant glioma cells both *in vitro* and *in vivo*. Oncogene. 2007; 26(17):2435-44.

[21] Shin DS, Kim HN, Shin KD, Yoon YJ, Kim SJ, Han DC, *et al.* Cryptotanshinone inhibits constitutive signal transducer and activator of transcription 3 function through blocking the dimerization in DU145 prostate cancer cells. Cancer research. 2009; 69(1):193-202.

[22] Lu L, Li C, Li D, Wang Y, Zhou C, Shao W, *et al.* Cryptotanshinone inhibits human glioma cell proliferation by suppressing STAT3 signaling. Molecular and cellular biochemistry. 2013; 381(1-2):273-82.

[23] Jin YC, Kim CW, Kim YM, Nizamutdinova IT, Ha YM, Kim HJ, *et al.* Cryptotanshinone, a lipophilic compound of Salvia miltiorrriza root, inhibits TNF-alpha-induced expression of adhesion molecules in HUVEC and attenuates rat myocardial ischemia/reperfusion injury *in vivo*. Eur J Pharmacol. 2009; 614(1-3):91-7.

[24] Zhang F, Zheng W, Pi R, Mei Z, Bao Y, Gao J, *et al.* Cryptotanshinone protects primary rat cortical neurons from glutamate-induced neurotoxicity *via* the activation of the phosphatidylinositol 3-kinase/Akt signaling pathway. Exp Brain Res. 2009; 193(1):109-18.

[25] Dat NT, Jin X, Lee JH, Lee D, Hong YS, Lee K, *et al.* Abietane diterpenes from Salvia miltiorrhiza inhibit the activation of hypoxia-inducible factor-1. Journal of natural products. 2007; 70(7):1093-7.

[26] Gong Y, Li Y, Lu Y, Li L, Abdolmaleky H, Blackburn GL, *et al.* Bioactive tanshinones in Salvia miltiorrhiza inhibit the growth of prostate cancer cells *in vitro* and in mice. International journal of cancer Journal international du cancer. 2011; 129(5):1042-52.

[27] Luo Y, Chen W, Zhou H, Liu L, Shen T, Alexander JS, *et al.* Cryptotanshinone inhibits lymphatic endothelial cell tube formation by suppressing VEGFR-3/ERK and small GTPase pathways. Cancer prevention research. 2011; 4(12):2083-91.

[28] Lee HJ, Jung DB, Sohn EJ, Kim HH, Park MN, Lew JH, *et al.* Inhibition of Hypoxia Inducible Factor Alpha and Astrocyte-Elevated Gene-1 Mediates Cryptotanshinone Exerted Antitumor Activity in Hypoxic PC-3 Cells. Evidence-based complementary and alternative medicine : eCAM. 2012; 2012:390957.

[29] Li Y, Gong Y, Li L, Abdolmaleky HM, Zhou J-R. Bioactive tanshinone I inhibits the growth of lung cancer in part *via* downregulation of Aurora A function. Mol Carcinog. 2013; 52(7):535-43.
[30] Schust J, Sperl B, Hollis A, Mayer TU, Berg T. Stattic: a small-molecule inhibitor of STAT3 activation and dimerization. Chem Biol. 2006; 13(11):1235-42.
[31] Szelag M, Sikorski K, Czerwoniec A, Szatkowska K, Wesoly J, Bluyssen HA. *In silico* simulations of STAT1 and STAT3 inhibitors predict SH2 domain cross-binding specificity. European journal of pharmacology. 2013; 720(1-3):38-48.
[32] Kubota T, Fujiwara H, Amaike H, Takashima K, Inada S, Atsuji K, *et al.* Reduced HGF expression in subcutaneous CT26 tumor genetically modified to secrete NK4 and its possible relation with antitumor effects. Cancer science. 2004; 95(4):321-7.
[33] Kuba K, Matsumoto K, Date K, Shimura H, Tanaka M, Nakamura T. HGF/NK4, a four-kringle antagonist of hepatocyte growth factor, is an angiogenesis inhibitor that suppresses tumor growth and metastasis in mice. Cancer research. 2000; 60(23):6737-43.
[34] Matsumura A, Kubota T, Taiyoh H, Fujiwara H, Okamoto K, Ichikawa D, *et al.* HGF regulates VEGF expression *via* the c-Met receptor downstream pathways, PI3K/Akt, MAPK and STAT3, in CT26 murine cells. Int J Oncol. 2013; 42(2):535-42.
[35] Xu X, Kasembeli MM, Jiang X, Tweardy BJ, Tweardy DJ. Chemical probes that competitively and selectively inhibit Stat3 activation. PloS one. 2009; 4(3):e4783.
[36] Becker S, Groner B, Muller CW. Three-dimensional structure of the Stat3beta homodimer bound to DNA. Nature. 1998; 394(6689):145-51.
[37] Turkson J, Ryan D, Kim JS, Zhang Y, Chen Z, Haura E, *et al.* Phosphotyrosyl peptides block Stat3-mediated DNA binding activity, gene regulation, and cell transformation. The Journal of biological chemistry. 2001; 276(48):45443-55.
[38] Turkson J, Kim JS, Zhang S, Yuan J, Huang M, Glenn M, *et al.* Novel peptidomimetic inhibitors of signal transducer and activator of transcription 3 dimerization and biological activity. Mol Cancer Ther. 2004; 3(3):261-9.
[39] Siddiquee KA, Gunning PT, Glenn M, Katt WP, Zhang S, Schrock C, *et al.* An oxazole-based small-molecule Stat3 inhibitor modulates Stat3 stability and processing and induces antitumor cell effects. ACS Chem Biol. 2007; 2(12):787-98.
[40] Siddiquee K, Zhang S, Guida WC, Blaskovich MA, Greedy B, Lawrence HR, *et al.* Selective chemical probe inhibitor of Stat3, identified through structure-based virtual screening, induces antitumor activity. Proceedings of the National Academy of Sciences of the United States of America. 2007; 104(18):7391-6.
[41] Zhang X, Yue P, Page BD, Li T, Zhao W, Namanja AT, *et al.* Orally bioavailable small-molecule inhibitor of transcription factor Stat3 regresses human breast and lung cancer xenografts. Proc Natl Acad Sci USA. 2012; 109(24):9623-8.
[42] Urlam MK, Pireddu R, Ge Y, Zhang X, Sun Y, Lawrence HR, *et al.* Development of new N-arylbenzamides as STAT3 dimerization inhibitors. MedChemComm. 2013; 4(6):932-41.
[43] Zhang X, Sun Y, Pireddu R, Yang H, Urlam MK, Lawrence HR, *et al.* A Novel Inhibitor of STAT3 Homodimerization Selectively Suppresses STAT3 Activity and Malignant Transformation. Cancer Res. 2013; 73(6):1922-33.
[44] Ren Z, Cabell LA, Schaefer TS, McMurray JS. Identification of a high-affinity phosphopeptide inhibitor of Stat3. Bioorg Med Chem Lett. 2003; 13(4):633-6.
[45] Stahl N, Farruggella TJ, Boulton TG, Zhong Z, Darnell JE, Jr., Yancopoulos GD. Choice of STATs and other substrates specified by modular tyrosine-based motifs in cytokine receptors. Science. 1995; 267(5202):1349-53.

[46] Gerhartz C, Heesel B, Sasse J, Hemmann U, Landgraf C, Schneider-Mergener J, *et al.* Differential activation of acute phase response factor/STAT3 and STAT1 *via* the cytoplasmic domain of the interleukin 6 signal transducer gp130. I. Definition of a novel phosphotyrosine motif mediating STAT1 activation. The Journal of biological chemistry. 1996; 271(22):12991-8.

[47] Wiederkehr-Adam M, Ernst P, Muller K, Bieck E, Gombert FO, Ottl J, *et al.* Characterization of phosphopeptide motifs specific for the Src homology 2 domains of signal transducer and activator of transcription 1 (STAT1) and STAT3. The Journal of biological chemistry. 2003; 278(18):16117-28.

[48] Coleman DRIV, Kaluarachchi K, Ren Z, Chen X, McMurray JS. Solid phase synthesis of phosphopeptides incorporating 2,2-dimethyloxazolidine pseudoproline analogs: evidence for trans Leu-Pro peptide bonds in Stat3 inhibitors. Int J Pept Res Ther 2008; 14(1):1-9.

[49] Coleman DRt, Ren Z, Mandal PK, Cameron AG, Dyer GA, Muranjan S, *et al.* Investigation of the binding determinants of phosphopeptides targeted to the SRC homology 2 domain of the signal transducer and activator of transcription 3. Development of a high-affinity peptide inhibitor. J Med Chem. 2005; 48(21):6661-70.

[50] Mandal PK, Gao F, Lu Z, Ren Z, Ramesh R, Birtwistle JS, *et al.* Potent and Selective Phosphopeptide Mimetic Prodrugs Targeted to the Src Homology 2 (SH2) Domain of Signal Transducer and Activator of Transcription 3. J Med Chem. 2011; 54(10):3549-5463.

[51] Mandal PK, Heard PA, Ren Z, Chen X, McMurray JS. Solid-phase synthesis of Stat3 inhibitors incorporating O-carbamoylserine and O-carbamoylthreonine as glutamine mimics. Bioorg Med Chem Lett. 2007; 17(3):654-6.

[52] Mandal PK, Limbrick D, Coleman DR, Dyer GA, Ren Z, Birtwistle JS, *et al.* Conformationally constrained peptidomimetic inhibitors of signal transducer and activator of transcription 3: evaluation and molecular modeling. J Med Chem. 2009; 52(8):2429-42.

[53] Mandal PK, Ren Z, Chen X, Kaluarachchi K, Liao WSL, McMurray JS. Structure-Activity Studies of Phosphopeptidomimetic Prodrugs Targeting the Src Homology 2 (SH2) Domain of Signal Transducer and Activator of Transcription 3 (Stat3). Int J Pept Res Ther. 2012:Ahead of Print.

[54] Mandal PK, Ren Z, Chen X, Xiong C, McMurray JS. Structure-affinity relationships of glutamine mimics incorporated into phosphopeptides targeted to the SH2 domain of signal transducer and activator of transcription 3. J Med Chem. 2009; 52(19):6126-41.

[55] Mandal PK, Liao WS, McMurray JS. Synthesis of phosphatase-stable, cell-permeable peptidomimetic prodrugs that target the SH2 domain of Stat3. Organic letters. 2009; 11(15):3394-7.

[56] Auzenne EJ, Klostergaard J, Mandal PK, Liao WS, Lu Z, Gao F, *et al.* A phosphopeptide mimetic prodrug targeting the SH2 domain of STAT3 inhibits tumor growth and angiogenesis. J Exp Ther Oncol. 2012; 10(2):155-62.

[57] McMurray JS, Mandal PK, Liao WS, Klostergaard J, Robertson FM. The consequences of selective inhibition of signal transducer and activator of transcription 3 (STAT3) tyrosine705 phosphorylation by phosphopeptide mimetic prodrugs targeting the Src homology 2 (SH2) domain. JAK-STAT. 2012; 1(4):263-73.

[58] Turkson J, Zhang S, Palmer J, Kay H, Stanko J, Mora LB, *et al.* Inhibition of constitutive signal transducer and activator of transcription 3 activation by novel platinum complexes with potent antitumor activity. Molecular cancer therapeutics. 2004; 3(12):1533-42.

[59] Zhang L, Alizadeh D, Van Handel M, Kortylewski M, Yu H, Badie B. Stat3 inhibition activates tumor macrophages and abrogates glioma growth in mice. Glia. 2009; 57(13):1458-67.

[60] Ioannidis S, Lamb ML, Wang T, Almeida L, Block MH, Davies AM, *et al.* Discovery of 5-Chloro-N(2)-[(1S)-1-(5-fluoropyrimidin-2-yl)ethyl]-N(4)-(5-methyl-1H-pyr azol-3-yl)pyrimidine-2,4-diamine (AZD1480) as a Novel Inhibitor of the Jak/Stat Pathway. J Med Chem. 2011; 54(1):262-76.

[61] Xin H, Herrmann A, Reckamp K, Zhang W, Pal S, Hedvat M, *et al.* Antiangiogenic and antimetastatic activity of JAK inhibitor AZD1480. Cancer Res. 2011; 71(21):6601-10.

[62] Hedvat M, Huszar D, Herrmann A, Gozgit JM, Schroeder A, Sheehy A, *et al.* The JAK2 inhibitor AZD1480 potently blocks Stat3 signaling and oncogenesis in solid tumors. Cancer Cell. 2009; 16(6):487-97.

[63] Plimack ER, Lorusso PM, McCoon P, Tang W, Krebs AD, Curt G, *et al.* AZD1480: a phase I study of a novel JAK2 inhibitor in solid tumors. Oncologist. 2013; 18(7):819-20.

[64] Costantino L, Barlocco D. STAT 3 as a Target for Cancer Drug Discovery. Curr Med Chem. 2008; 15:834-43.

[65] Thompson JE, Cubbon RM, Cummings RT, Wicker LS, Frankshun R, Cunningham BR, *et al.* Photochemical preparation of a pyridone containing tetracycle: a Jak protein kinase inhibitor. Bioorg Med Chem Lett. 2002; 12(8):1219-23.

[66] Kreis S, Munz GA, Haan S, Heinrich PC, Behrmann I. Cell density dependent increase of constitutive signal transducers and activators of transcription 3 activity in melanoma cells is mediated by Janus kinases. Mol Cancer Res. 2007; 5(12):1331-41.

[67] Bobkova EV, Weber MJ, Xu Z, Zhang YL, Jung J, Blume-Jensen P, *et al.* Discovery of PDK1 kinase inhibitors with a novel mechanism of action by ultrahigh throughput screening. The Journal of biological chemistry. 2010; 285(24):18838-46.

[68] Scuto A, Krejci P, Popplewell L, Wu J, Wang Y, Kujawski M, *et al.* The novel JAK inhibitor AZD1480 blocks STAT3 and FGFR3 signaling, resulting in suppression of human myeloma cell growth and survival. Leukemia. 2011; 25(3):538-50.

[69] Loveless ME, Lawson D, Collins M, Prasad Nadella MV, Reimer C, Huszar D, *et al.* Comparisons of the Efficacy of a Jak1/2 Inhibitor (AZD1480) with a VEGF Signaling Inhibitor (Cediranib) and Sham Treatments in Mouse Tumors Using DCE-MRI, DW-MRI, and Histology. Neoplasia. 2012; 14(1):54-64.

[70] Looyenga BD, Hutchings D, Cherni I, Kingsley C, Weiss GJ, Mackeigan JP. STAT3 Is Activated by JAK2 Independent of Key Oncogenic Driver Mutations in Non-Small Cell Lung Carcinoma. PLoS One. 2012; 7(2):e30820.

[71] Coward J, Kulbe H, Chakravarty P, Leader D, Vassileva V, Leinster DA, *et al.* Interleukin-6 as a therapeutic target in human ovarian cancer. Clin Cancer Res. 2011; 17(18):6083-96.

[72] Yang J, Stark GR. Roles of unphosphorylated STATs in signaling. Cell Res. 2008; 18(4):443-51.

[73] Gough DJ, Corlett A, Schlessinger K, Wegrzyn J, Larner AC, Levy DE. Mitochondrial STAT3 supports Ras-dependent oncogenic transformation. Science. 2009; 324(5935):1713-6.

[74] Wegrzyn J, Potla R, Chwae YJ, Sepuri NB, Zhang Q, Koeck T, *et al.* Function of mitochondrial Stat3 in cellular respiration. Science. 2009; 323(5915):793-7.

[75] Singh SB, Graham PL, Reamer RA, Cordingley MG. Discovery, total synthesis, HRV 3C-protease inhibitory activity, and structure-activity relationships of 2-methoxystypandrone and its analogues. Bioorganic & medicinal chemistry letters. 2001; 11(24):3143-6.

[76] Liu J, Zhang Q, Chen K, Liu J, Kuang S, Chen W, *et al.* Small-molecule STAT3 signaling pathway modulators from Polygonum cuspidatum. Planta medica. 2012; 78(14):1568-70.

[77] Kuang S, Qi C, Liu J, Sun X, Zhang Q, Sima Z, *et al.* 2-Methoxystypandrone inhibits signal transducer and activator of transcription 3 and nuclear factor-kappaB signaling by inhibiting Janus kinase 2 and IkappaB kinase. Cancer science. 2014; 105(4):473-80.

[78] Muthian G, Bright JJ. Quercetin, a flavonoid phytoestrogen, ameliorates experimental allergic encephalomyelitis by blocking IL-12 signaling through JAK-STAT pathway in T lymphocyte. Journal of clinical immunology. 2004; 24(5):542-52.

[79] Choi JS, Kang SW, Li J, Kim JL, Bae JY, Kim DS, *et al.* Blockade of oxidized LDL-triggered endothelial apoptosis by quercetin and rutin through differential signaling pathways involving JAK2. Journal of agricultural and food chemistry. 2009; 57(5):2079-86.

[80] Anso E, Zuazo A, Irigoyen M, Urdaci MC, Rouzaut A, Martinez-Irujo JJ. Flavonoids inhibit hypoxia-induced vascular endothelial growth factor expression by a HIF-1 independent mechanism. Biochemical pharmacology. 2010; 79(11):1600-9.

[81] Kleemann R, Verschuren L, Morrison M, Zadelaar S, van Erk MJ, Wielinga PY, *et al.* Anti-inflammatory, anti-proliferative and anti-atherosclerotic effects of quercetin in human *in vitro* and *in vivo* models. Atherosclerosis. 2011; 218(1):44-52.

[82] Michaud-Levesque J, Bousquet-Gagnon N, Beliveau R. Quercetin abrogates IL-6/STAT3 signaling and inhibits glioblastoma cell line growth and migration. Exp Cell Res. 2012; 318(8):925-35.

[83] Senggunprai L, Kukongviriyapan V, Prawan A, Kukongviriyapan U. Quercetin and EGCG Exhibit Chemopreventive Effects in Cholangiocarcinoma Cells *via* Suppression of JAK/STAT Signaling Pathway. Phytotherapy research : PTR. 2013.

[84] Cao HH, Tse AK, Kwan HY, Yu H, Cheng CY, Su T, *et al.* Quercetin exerts anti-melanoma activities and inhibits STAT3 signaling. Biochemical pharmacology. 2014; 87(3):424-34.

[85] Lamson DW, Brignall MS. Antioxidants and cancer, part 3: quercetin. Alternative medicine review : a journal of clinical therapeutic. 2000; 5(3):196-208.

[86] Hayashi A, Gillen AC, Lott JR. Effects of daily oral administration of quercetin chalcone and modified citrus pectin on implanted colon-25 tumor growth in Balb-c mice. Alternative medicine review : a journal of clinical therapeutic. 2000; 5(6):546-52.

[87] Igura K, Ohta T, Kuroda Y, Kaji K. Resveratrol and quercetin inhibit angiogenesis *in vitro*. Cancer letters. 2001; 171(1):11-6.

[88] Wilson WJ, Poellinger L. The dietary flavonoid quercetin modulates HIF-1 alpha activity in endothelial cells. Biochemical and biophysical research communications. 2002; 293(1):446-50.

[89] Tan WF, Lin LP, Li MH, Zhang YX, Tong YG, Xiao D, *et al.* Quercetin, a dietary-derived flavonoid, possesses antiangiogenic potential. European journal of pharmacology. 2003; 459(2-3):255-62.

[90] Hasebe Y, Egawa K, Yamazaki Y, Kunimoto S, Hirai Y, Ida Y, *et al.* Specific inhibition of hypoxia-inducible factor (HIF)-1 alpha activation and of vascular endothelial growth factor

(VEGF) production by flavonoids. Biological & pharmaceutical bulletin. 2003; 26(10):1379-83.

[91] Ma ZS, Huynh TH, Ng CP, Do PT, Nguyen TH, Huynh H. Reduction of CWR22 prostate tumor xenograft growth by combined tamoxifen-quercetin treatment is associated with inhibition of angiogenesis and cellular proliferation. International journal of oncology. 2004; 24(5):1297-304.

[92] Kim JD, Liu L, Guo W, Meydani M. Chemical structure of flavonols in relation to modulation of angiogenesis and immune-endothelial cell adhesion. The Journal of nutritional biochemistry. 2006; 17(3):165-76.

[93] Donnini S, Finetti F, Lusini L, Morbidelli L, Cheynier V, Barron D, *et al.* Divergent effects of quercetin conjugates on angiogenesis. The British journal of nutrition. 2006; 95(5):1016-23.

[94] Jackson SJ, Venema RC. Quercetin inhibits eNOS, microtubule polymerization, and mitotic progression in bovine aortic endothelial cells. The Journal of nutrition. 2006; 136(5):1178-84.

[95] Yuan ZP, Chen LJ, Fan LY, Tang MH, Yang GL, Yang HS, *et al.* Liposomal quercetin efficiently suppresses growth of solid tumors in murine models. Clinical cancer research : an official journal of the American Association for Cancer Research. 2006; 12(10):3193-9.

[96] Kale R, Saraf M, Juvekar A, Tayade P. Decreased B16F10 melanoma growth and impaired tumour vascularization in BDF1 mice with quercetin-cyclodextrin binary system. The Journal of pharmacy and pharmacology. 2006; 58(10):1351-8.

[97] Nicholson SK, Tucker GA, Brameld JM. Effects of dietary polyphenols on gene expression in human vascular endothelial cells. The Proceedings of the Nutrition Society. 2008; 67(1):42-7.

[98] Ahn MR, Kunimasa K, Kumazawa S, Nakayama T, Kaji K, Uto Y, *et al.* Correlation between antiangiogenic activity and antioxidant activity of various components from propolis. Molecular nutrition & food research. 2009; 53(5):643-51.

[99] Zhou W, Kallifatidis G, Baumann B, Rausch V, Mattern J, Gladkich J, *et al.* Dietary polyphenol quercetin targets pancreatic cancer stem cells. International journal of oncology. 2010; 37(3):551-61.

[100] Oh SJ, Kim O, Lee JS, Kim JA, Kim MR, Choi HS, *et al.* Inhibition of angiogenesis by quercetin in tamoxifen-resistant breast cancer cells. Food and chemical toxicology : an international journal published for the British Industrial Biological Research Association. 2010; 48(11):3227-34.

[101] Pradhan SJ, Mishra R, Sharma P, Kundu GC. Quercetin and sulforaphane in combination suppress the progression of melanoma through the down-regulation of matrix metalloproteinase-9. Experimental and therapeutic medicine. 2010; 1(6):915-20.

[102] Anand K, Asthana P, Kumar A, Ambasta RK, Kumar P. Quercetin mediated reduction of angiogenic markers and chaperones in DLA-induced solid tumours. Asian Pacific journal of cancer prevention : APJCP. 2011; 12(11):2829-35.

[103] Gacche RN, Shegokar HD, Gond DS, Yang Z, Jadhav AD. Evaluation of selected flavonoids as antiangiogenic, anticancer, and radical scavenging agents: an experimental and *in silico* analysis. Cell biochemistry and biophysics. 2011; 61(3):651-63.

[104] Xiao X, Shi D, Liu L, Wang J, Xie X, Kang T, *et al.* Quercetin suppresses cyclooxygenase-2 expression and angiogenesis through inactivation of P300 signaling. PLoS One. 2011; 6(8):e22934.

[105] Lin.C, Wu M, Dong J. Quercetin-4'-O-β--glucopyranoside (QODG) inhibits angiogenesis by suppressing VEGFR2-mediated signaling in zebrafish and endothelial cells. PLoS One. 2012; 7(2):e31708.

[106] Scoditti E, Calabriso N, Massaro M, Pellegrino M, Storelli C, Martines G, *et al.* Mediterranean diet polyphenols reduce inflammatory angiogenesis through MMP-9 and COX-2 inhibition in human vascular endothelial cells: a potentially protective mechanism in atherosclerotic vascular disease and cancer. Archives of biochemistry and biophysics. 2012; 527(2):81-9.

[107] Gao X, Wang B, Wei X, Men K, Zheng F, Zhou Y, *et al.* Anticancer effect and mechanism of polymer micelle-encapsulated quercetin on ovarian cancer. Nanoscale. 2012; 4(22):7021-30.

[108] Poyil P, Budhraja A, Son Y-O, Wang X, Zhang Z, Ding S, *et al.* Quercetin inhibits angiogenesis mediated human prostate tumor growth by targeting VEGFR-2 regulated AKT/mTOR/P70S6K signaling pathways. PLoS One. 2012; 7(10):e47516.

[109] Long Q, Xiel Y, Huang Y, Wu Q, Zhang H, Xiong S, *et al.* Induction of apoptosis and inhibition of angiogenesis by PEGylated liposomal quercetin in both cisplatin-sensitive and cisplatin-resistant ovarian cancers. Journal of biomedical nanotechnology. 2013; 9(6):965-75.

[110] Wu Q, Deng S, Li L, Sun L, Yang X, Liu X, *et al.* Biodegradable polymeric micelle-encapsulated quercetin suppresses tumor growth and metastasis in both transgenic zebrafish and mouse models. Nanoscale. 2013; 5(24):12480-93.

[111] Zhao D, Qin C, Fan X, Li Y, Gu B. Inhibitory effects of quercetin on angiogenesis in larval zebrafish and human umbilical vein endothelial cells. European journal of pharmacology. 2014; 723:360-7.

[112] Natarajan C, Bright JJ. Curcumin inhibits experimental allergic encephalomyelitis by blocking IL-12 signaling through Janus kinase-STAT pathway in T lymphocytes. Journal of immunology. 2002; 168(12):6506-13.

[113] Bharti AC, Donato N, Aggarwal BB. Curcumin (diferuloylmethane) inhibits constitutive and IL-6-inducible STAT3 phosphorylation in human multiple myeloma cells. Journal of immunology. 2003; 171(7):3863-71.

[114] Rajasingh J, Raikwar HP, Muthian G, Johnson C, Bright JJ. Curcumin induces growth-arrest and apoptosis in association with the inhibition of constitutively active JAK-STAT pathway in T cell leukemia. Biochemical and biophysical research communications. 2006; 340(2):359-68.

[115] Blasius R, Reuter S, Henry E, Dicato M, Diederich M. Curcumin regulates signal transducer and activator of transcription (STAT) expression in K562 cells. Biochemical pharmacology. 2006; 72(11):1547-54.

[116] Kim HY, Park EJ, Joe EH, Jou I. Curcumin suppresses Janus kinase-STAT inflammatory signaling through activation of Src homology 2 domain-containing tyrosine phosphatase 2 in brain microglia. Journal of immunology. 2003; 171(11):6072-9.

[117] Saydmohammed M, Joseph D, Syed V. Curcumin suppresses constitutive activation of STAT-3 by up-regulating protein inhibitor of activated STAT-3 (PIAS-3) in ovarian and endometrial cancer cells. Journal of cellular biochemistry. 2010; 110(2):447-56.

[118] Weissenberger J, Priester M, Bernreuther C, Rakel S, Glatzel M, Seifert V, *et al.* Dietary curcumin attenuates glioma growth in a syngeneic mouse model by inhibition of the

JAK1,2/STAT3 signaling pathway. Clinical cancer research : an official journal of the American Association for Cancer Research. 2010; 16(23):5781-95.

[119] Prakobwong S, Gupta SC, Kim JH, Sung B, Pinlaor P, Hiraku Y, *et al.* Curcumin suppresses proliferation and induces apoptosis in human biliary cancer cells through modulation of multiple cell signaling pathways. Carcinogenesis. 2011; 32(9):1372-80.

[120] Leonhard WN, van der Wal A, Novalic Z, Kunnen SJ, Gansevoort RT, Breuning MH, *et al.* Curcumin inhibits cystogenesis by simultaneous interference of multiple signaling pathways: *in vivo* evidence from a Pkd1-deletion model. Am J Physiol. 2011; 300(5, Pt. 2):F1193-F202.

[121] Fossey SL, Bear MD, Lin J, Li C, Schwartz EB, Li P-K, *et al.* The novel curcumin analog FLLL32 decreases STAT3 DNA binding activity and expression, and induces apoptosis in osteosarcoma cell lines. BMC Cancer. 2011; 11:112.

[122] Abuzeid WM, Davis S, Tang AL, Saunders L, Brenner JC, Lin J, *et al.* Sensitization of head and neck cancer to cisplatin through the use of a novel curcumin analog. Archives of otolaryngology--head & neck surgery. 2011; 137(5):499-507.

[123] Yang C-L, Liu Y-Y, Ma Y-G, Xue Y-X, Liu D-G, Ren Y, *et al.* Curcumin blocks small cell lung cancer cells migration, invasion, angiogenesis, cell cycle and neoplasia through Janus Kinase-STAT3 signalling pathway. PLoS One. 2012; 7(5):e37960.

[124] Kumar A, Bora U. Molecular docking studies on inhibition of Stat3 dimerization by curcumin natural derivatives and its conjugates with amino acids. Bioinformation. 2012; 8(20):988-93.

[125] Kim JH, Shim JS, Lee SK, Kim KW, Rha SY, Chung HC, *et al.* Microarray-based analysis of anti-angiogenic activity of demethoxycurcumin on human umbilical vein endothelial cells: crucial involvement of the down-regulation of matrix metalloproteinase. Japanese journal of cancer research : Gann. 2002; 93(12):1378-85.

[126] Adams BK, Ferstl EM, Davis MC, Herold M, Kurtkaya S, Camalier RF, *et al.* Synthesis and biological evaluation of novel curcumin analogs as anti-cancer and anti-angiogenesis agents. Bioorganic & medicinal chemistry. 2004; 12(14):3871-83.

[127] Yoysungnoen P, Wirachwong P, Changtam C, Suksamrarn A, Patumraj S. Anti-cancer and anti-angiogenic effects of curcumin and tetrahydrocurcumin on implanted hepatocellular carcinoma in nude mice. World journal of gastroenterology : WJG. 2008; 14(13):2003-9.

[128] Arif JM, Siddiqui MH, Akhtar S, Al-Sagair OA. Exploitation of *in silico* potential in prediction, validation and elucidation of mechanism of anti-angiogenesis by novel compounds: comparative correlation between wet lab and *in silico* data. Int J Bioinf Res Appl. 2013; 9(4):336-48.

[129] Wung BS, Hsu MC, Wu CC, Hsieh CW. Resveratrol suppresses IL-6-induced ICAM-1 gene expression in endothelial cells: effects on the inhibition of STAT3 phosphorylation. Life sciences. 2005; 78(4):389-97.

[130] Kotha A, Sekharam M, Cilenti L, Siddiquee K, Khaled A, Zervos AS, *et al.* Resveratrol inhibits Src and Stat3 signaling and induces the apoptosis of malignant cells containing activated Stat3 protein. Molecular cancer therapeutics. 2006; 5(3):621-9.

[131] Ivanov VN, Partridge MA, Johnson GE, Huang SXL, Zhou H, Hei TK. Resveratrol sensitizes melanomas to TRAIL through modulation of antiapoptotic gene expression. Exp Cell Res. 2008; 314(5):1163-76.

[132] Johnson GE, Ivanov VN, Hei TK. Radiosensitization of melanoma cells through combined inhibition of protein regulators of cell survival. Apoptosis. 2008; 13(6):790-802.

[133] Ma Z, Molavi O, Haddadi A, Lai R, Gossage RA, Lavasanifar A. Resveratrol analog trans 3,4,5,4'-tetramethoxystilbene (DMU-212) mediates anti-tumor effects *via* mechanism different from that of resveratrol. Cancer Chemother Pharmacol. 2008; 63(1):27-35.

[134] Yu LJ, Wu ML, Li H, Chen XY, Wang Q, Sun Y, *et al.* Inhibition of STAT3 expression and signaling in resveratrol-differentiated medulloblastoma cells. Neoplasia. 2008; 10(7):736-44.

[135] Kim JE, Kim HS, Shin YJ, Lee CS, Won C, Lee SA, *et al.* LYR71, a derivative of trimeric resveratrol, inhibits tumorigenesis by blocking STAT3-mediated matrix metalloproteinase 9 expression. Experimental & molecular medicine. 2008; 40(5):514-22.

[136] Li T, Wang W, Chen H, Li T, Ye L. Evaluation of anti-leukemia effect of resveratrol by modulating STAT3 signaling. Int Immunopharmacol. 2010; 10(1):18-25.

[137] Trung LQ, Espinoza JL, Takami A, Nakao S. Resveratrol induces cell cycle arrest and apoptosis in malignant NK cells *via* JAK2/STAT3 pathway inhibition. PLoS One. 2013; 8(1):e55183.

[138] Santandreu FM, Valle A, Oliver J, Roca P. Resveratrol Potentiates the Cytotoxic Oxidative Stress Induced by Chemotherapy in Human Colon Cancer Cells. Cell Physiol Biochem. 2011; 28(2):219-28.

[139] Liu P, Wang X, Hu C, Hu T. Inhibition of proliferation and induction of apoptosis by trimethoxyl stilbene (TMS) in a lung cancer cell line. Asian Pacific journal of cancer prevention : APJCP. 2011; 12(9):2263-9.

[140] Yang Y-P, Chang Y-L, Huang P-I, Chiou G-Y, Tseng L-M, Chiou S-H, *et al.* Resveratrol suppresses tumorigenicity and enhances radiosensitivity in primary glioblastoma tumor initiating cells by inhibiting the STAT3 axis. J Cell Physiol. 2012; 227(3):976-93.

[141] Lee H, Zhang P, Herrmann A, Yang C, Xin H, Wang Z, *et al.* Acetylated STAT3 is crucial for methylation of tumor-suppressor gene promoters and inhibition by resveratrol results in demethylation. Proc Natl Acad Sci U S A. 2012; 109(20):7765-9, S/1-S/5.

[142] Liu Y, Wang L, Wu Y, Lv C, Li X, Cao X, *et al.* Pterostilbene exerts antitumor activity against human osteosarcoma cells by inhibiting the JAK2/STAT3 signaling pathway. Toxicology. 2013; 304:120-31.

[143] Lee-Chang C, Bodogai M, Martin-Montalvo A, Wejksza K, Sanghvi M, Moaddel R, *et al.* Inhibition of Breast Cancer Metastasis by Resveratrol-Mediated Inactivation of Tumor-Evoked Regulatory B Cells. J Immunol. 2013; 191(8):4141-51.

[144] M C, W B-D, M W, M M, J J-L. 3,4,5,4'-trans-tetramethoxystilbene (DMU-212) modulates the activation of NF-κB, AP-1, and STAT3 transcription factors in rat liver carcinogenesis induced by initiation-promotion regimen. Mol Cell Biochem. 2014; Feb 13. [Epub ahead of print].

[145] Wu ML, Li H, Yu LJ, Chen XY, Kong QY, Song X, *et al.* Short-term resveratrol exposure causes *in vitro* and *in vivo* growth inhibition and apoptosis of bladder cancer cells. PLoS One. 2014; 9(2):e89806.

[146] Lee B, Moon SK. Resveratrol inhibits TNF-alpha-induced proliferation and matrix metalloproteinase expression in human vascular smooth muscle cells. The Journal of nutrition. 2005; 135(12):2767-73.

[147] Belleri M, Ribatti D, Savio M, Stivala LA, Forti L, Tanghetti E, *et al.* αvβ3 Integrin-dependent antiangiogenic activity of resveratrol stereoisomers. Mol Cancer Ther. 2008; 7(12):3761-70.

[148] Alex D, Leong EC, Zhang Z-J, Yan GTH, Cheng S-H, Leong C-W, *et al.* Resveratrol derivative, trans-3,5,4'-trimethoxystilbene, exerts antiangiogenic and vascular-disrupting

effects in zebrafish through the downregulation of VEGFR2 and cell-cycle modulation. J Cell Biochem. 2010; 109(2):339-46.
[149] Zhang L, Jing HJ, Cui LQ, Li HQ, Zhou B, Zhou GZ, *et al.* 3,4-dimethoxystilbene, a resveratrol derivative with anti-angiogenic effect, induces both macroautophagy and apoptosis in endothelial cells. J Cell Biochem. 2013; 114(3):697-707.
[150] Tseng SH, Lin SM, Chen JC, Su YH, Huang HY, Chen CK, *et al.* Resveratrol suppresses the angiogenesis and tumor growth of gliomas in rats. Clinical cancer research : an official journal of the American Association for Cancer Research. 2004; 10(6):2190-202.
[151] Chen JC, Chen Y, Lin JH, Wu JM, Tseng SH. Resveratrol suppresses angiogenesis in gliomas: evaluation by color Doppler ultrasound. Anticancer research. 2006; 26(2A):1237-45.
[152] Kimura Y, Okuda H. Resveratrol isolated from Polygonum cuspidatum root prevents tumor growth and metastasis to lung and tumor-induced neovascularization in Lewis lung carcinoma-bearing mice. The Journal of nutrition. 2001; 131(6):1844-9.
[153] Lee EO, Lee HJ, Hwang HS, Ahn KS, Chae C, Kang KS, *et al.* Potent inhibition of Lewis lung cancer growth by heyneanol A from the roots of Vitis amurensis through apoptotic and anti-angiogenic activities. Carcinogenesis. 2006; 27(10):2059-69.
[154] Mousa SS, Mousa SS, Mousa SA. Effect of resveratrol on angiogenesis and platelet/fibrin-accelerated tumor growth in the chick chorioallantoic membrane model. Nutrition and cancer. 2005; 52(1):59-65.
[155] Garvin S, Ollinger K, Dabrosin C. Resveratrol induces apoptosis and inhibits angiogenesis in human breast cancer xenografts *in vivo*. Cancer letters. 2006; 231(1):113-22.
[156] Tang X, Zhang Q, Nishitani J, Brown J, Shi S, Le AD. Overexpression of human papillomavirus type 16 oncoproteins enhances hypoxia-inducible factor 1 alpha protein accumulation and vascular endothelial growth factor expression in human cervical carcinoma cells. Clinical cancer research : an official journal of the American Association for Cancer Research. 2007; 13(9):2568-76.
[157] Hu Y, Sun CY, Huang J, Hong L, Zhang L, Chu ZB. Antimyeloma effects of resveratrol through inhibition of angiogenesis. Chinese medical journal. 2007; 120(19):1672-7.
[158] Cao Z, Fang J, Xia C, Shi X, Jiang BH. trans-3,4,5'-Trihydroxystibene inhibits hypoxia-inducible factor 1alpha and vascular endothelial growth factor expression in human ovarian cancer cells. Clinical cancer research : an official journal of the American Association for Cancer Research. 2004; 10(15):5253-63.
[159] Park SY, Jeong KJ, Lee J, Yoon DS, Choi WS, Kim YK, *et al.* Hypoxia enhances LPA-induced HIF-1alpha and VEGF expression: their inhibition by resveratrol. Cancer letters. 2007; 258(1):63-9.
[160] Kimura Y, Sumiyoshi M, Baba K. Antitumor activities of synthetic and natural stilbenes through antiangiogenic action. Cancer science. 2008; 99(10):2083-96.
[161] Dann JM, Sykes PH, Mason DR, Evans JJ. Regulation of Vascular Endothelial Growth Factor in endometrial tumour cells by resveratrol and EGCG. Gynecol Oncol. 2009; 113(3):374-8.
[162] Liu P-L, Tsai J-R, Charles AL, Hwang J-J, Chou S-H, Ping Y-H, *et al.* Resveratrol inhibits human lung adenocarcinoma cell metastasis by suppressing heme oxygenase 1-mediated nuclear factor-κB pathway and subsequently downregulating expression of matrix metalloproteinases. Mol Nutr Food Res. 2010; 54(Suppl. 2):S196-S204.

[163] Vartanian AA, Burova OS, Stepanova EV, Baryshnikov AY, Lichinitser MR. Melanoma vasculogenic mimicry is strongly related to reactive oxygen species level. Melanoma research. 2007; 17(6):370-9.

[164] Trapp V, Parmakhtiar B, Papazian V, Willmott L, Fruehauf JP. Anti-angiogenic effects of resveratrol mediated by decreased VEGF and increased TSP1 expression in melanoma-endothelial cell co-culture. Angiogenesis. 2010; 13(4):305-15.

[165] Wang TTY, Hudson TS, Wang T-C, Remsberg CM, Davies NM, Takahashi Y, *et al.* Differential effects of resveratrol on androgen-responsive LNCaP human prostate cancer cells *in vitro* and *in vivo*. Carcinogenesis. 2008; 29(10):2001-10.

[166] Ganapathy S, Chen Q, Singh KP, Shankar S, Srivastava RK. Resveratrol enhances antitumor activity of TRAIL in prostate cancer xenografts through activation of FOXO transcription factor. PLoS One. 2010; 5(12):e15627.

[167] Yu HB, Zhang HF, Zhang X, Li DY, Xue HZ, Pan CE, *et al.* Resveratrol inhibits VEGF expression of human hepatocellular carcinoma cells through a NF-kappa B-mediated mechanism. Hepato-gastroenterology. 2010; 57(102-103):1241-6.

[168] El-Azab M, Hishe H, Moustafa Y, El-Awady el S. Anti-angiogenic effect of resveratrol or curcumin in Ehrlich ascites carcinoma-bearing mice. European journal of pharmacology. 2011; 652(1-3):7-14.

[169] Hotte SJ, Hirte HW. BAY 43-9006: early clinical data in patients with advanced solid malignancies. Current pharmaceutical design. 2002; 8(25):2249-53.

[170] Ellis PM, Al-Saleh K. Multitargeted anti-angiogenic agents and NSCLC: clinical update and future directions. Critical reviews in oncology/hematology. 2012; 84(1):47-58.

[171] Delgado JS, Mustafi R, Yee J, Cerda S, Chumsangsri A, Dougherty U, *et al.* Sorafenib triggers antiproliferative and pro-apoptotic signals in human esophageal adenocarcinoma cells. Digestive diseases and sciences. 2008; 53(12):3055-64.

[172] Yang F, Van Meter TE, Buettner R, Hedvat M, Liang W, Kowolik CM, *et al.* Sorafenib inhibits signal transducer and activator of transcription 3 signaling associated with growth arrest and apoptosis of medulloblastomas. Molecular cancer therapeutics. 2008; 7(11):3519-26.

[173] Blechacz BRA, Smoot RL, Bronk SF, Werneburg NW, Sirica AE, Gores GJ. Sorafenib inhibits signal transducer and activator of transcription-3 signaling in cholangiocarcinoma cells by activating the phosphatase shatterproof 2. Hepatology (Hoboken, NJ, U S). 2009; 50(6):1861-70.

[174] Yang F, Brown C, Buettner R, Hedvat M, Starr R, Scuto A, *et al.* Sorafenib induces growth arrest and apoptosis of human glioblastoma cells through the dephosphorylation of signal transducers and activators of transcription 3. Molecular cancer therapeutics. 2010; 9(4):953-62.

[175] Arbab AS. Activation of alternative pathways of angiogenesis and involvement of stem cells following anti-angiogenesis treatment in glioma. Histology and histopathology. 2012; 27(5):549-57.

[176] Cabebe E, Wakelee H. Role of anti-angiogenesis agents in treating NSCLC: focus on bevacizumab and VEGFR tyrosine kinase inhibitors. Current treatment options in oncology. 2007; 8(1):15-27.

[177] Di Maio M, Morabito A, Piccirillo MC, Daniele G, Giordano P, Costanzo R, *et al.* Combining anti-Epidermal Growth Factor Receptor (EGFR) and Anti-Angiogenic

Strategies in Advanced NSCLC: We Should have Known Better. Current pharmaceutical design. 2013.
[178] Grimm M-O, Wolff I, Zastrow S, Froehner M, Wirth M. Advances in renal cell carcinoma treatment. Ther Adv Urol. 2010; 2(1):11-7.
[179] Liu F, Chen X, Peng E, Guan W, Li Y, Hu Z, *et al*. VEGF pathway-targeted therapy for advanced renal cell carcinoma: A meta-analysis of randomized controlled trials. J Huazhong Univ Sci Technol, Med Sci. 2011; 31(6):799-806.
[180] Newton HB. Small-molecule and antibody approaches to molecular chemotherapy of primary brain tumors. Current opinion in investigational drugs. 2007; 8(12):1009-21.
[181] Gurevich F, Perazella MA. Renal Effects of Anti-angiogenesis Therapy: Update for the Internist. Am J Med. 2009; 122(4):322-8.

CHAPTER 8

Anti-Angiogenic Therapy and Cardiovascular Diseases: Current Strategies and Future Perspectives

Vasiliki K. Katsi[1,*], Costas T. Psarros[2], Marios G. Krokidis[2], Georgia D. Vamvakou[3], Dimitris Tousoulis[2], Christodoulos I. Stefanadis[2] and Ioannis E. Kallikazaros[1]

[1]Cardiology Department, Hippokration Hospital, Athens, Greece; [2]1st Cardiology Department, Athens University Medical School, Greece and [3]Second Department of Cardiology, University of Athens, Attikon Hospital, Chaidari, Greece

Abstract: The process involving new blood vessel sprouting from already existing ones is regulated by a physiological complex mechanism, known as angiogenesis. It plays a key role in wound healing but is also present in pathophysiological conditions such as cancer and cardiovascular disease, which have the highest rates of morbidity and mortality worldwide. It is stimulated mechanically or chemically, with the latter involving several signaling pathways and proteins widely known as growth factors. Anti-angiogenesis has always been an appealing target for cardiovascular related diseases, such as atherosclerosis, with its role still eluding our grasp. In this chapter we focus on the latest trends in anti-angiogenic therapy and drug discovery as well as highlight the distinct pathways underlying it. Therapies can range from use of peptides, proteins as well as well-defined chemically synthesized molecules. Latest trends involve gene therapy related approaches, with delivery of anti-angiogenic factors to target areas. Furthermore, toxicity issues arising from the use of anti-angiogenic drugs are discussed and highlighted as many of the drugs employed can cause serious side effects, while others may not achieve maximum therapeutic effect. Anti-angiogenic therapy is a very dynamic field and will continue to evolve and improve in the future. A very interesting addition to the anti-angiogenesis drug arsenal can be achieved with the aim of nanotechnology, a novel but promising scientific field. It is certain that in the future new, more potent drugs will be discovered, posing greater therapeutic potential and lower side effects, providing a much needed boost in this continuously evolving scientific field.

Keywords: Angiogenesis, anti-angiogenic drugs, atherosclerosis, cardiovascular disease, endothelial cells, fibroblast growth factor, gene therapy, hypoxia-inducible factor, interleukins, liposomes, macrophages, matrix metalloproteinases, microRNAs, nanomedicine, nicotinamide adenine dinucleotide phosphate-oxidase,

***Corresponding author Vasiliki K. Katsi:** Cardiology Department, Hippokration Hospital, 10 Lefkados street, Kifisia, Athens, Greece, BOX 14562; Tel: 00306934364281; E-mail: vkkatsi@yahoo.gr

Atta-ur-Rahman and Muhammad Iqbal Choudhary (Eds)
Copyright © 2014 Bentham Science Publishers Ltd. Published by Elsevier Inc. All rights reserved.
10.1016/B978-0-12-803963-2.50008-9

polyphenols, reactive oxygen species, statins, transforming growth factor-beta, vascular endothelial growth factor, vascular smooth muscle cells.

INTRODUCTION

Cardiovascular disease (CVD) is the pandemic of the 21st century, affecting not only developed countries but developing ones as well. Although there are various causes underlying CVD progression, atherosclerosis is the prevalent one. It affects all vascular beds and can manifest in the form of cardiac, cerebral, visceral, or peripheral vascular diseases. Atherosclerosis can be defined as a chronic low grade inflammation affecting the vascular wall, which over time and through multiple processes results to the development of atherosclerotic plaques (Fig. **1**) [1].

While atherosclerosis was initially considered a lipid storage disease, later insights revealed that development and progression of atherosclerosis is a multi-variable process with many factors involved, namely hormones, cytokines, adhesion molecules, bacterial products and inflammatory cells [2]. In response to the endothelial injury triggered from the above inflammatory cell infiltration occurs. In addition, two major events of the atherogenic process are the deposition of low density lipoprotein (LDL) and sub-endothelial matrix remodeling. Reactive oxygen species (ROS) as well as several other biochemical mediators and enzymes participate in the oxidation of LDL and the formation of ox-LDL, with the latter being recognized by multiple receptors such as lectin-like oxLDL receptor-1 (LOX-1), scavenger receptor SR-B1, Toll-like receptors (TLRs) and CD205 [3]. Ox-LDL induced the over-expression of several inflammatory mediators, such as monocyte chemoattractant protein-1 (MCP-1), tumor necrosis factor alpha (TNF-α) and interleukin-1β (IL-1β) as well as adhesion molecules like vascular cell adhesion molecule-1 (VCAM-1) and intercellular adhesion molecule-1 (ICAM-1), which in turn favour macrophage and other inflammatory cell recruitment into the sub-endothelial area [4]. Subsequently ox-LDL is accumulated by macrophages, resulting in the formation of foam cells [5]. However, there are reports that ox-LDL may exert anti-inflammatory effects in macrophages through, activation of the peroxisome proliferator-activated receptor gamma (PPARγ) pathway [6-8]. All the above conclude to the fact that ox-LDL is a pluripotent mediator, affecting several pathways [9]. Furthermore, apart from the classic factors, there has been a documented link between the immune system and atherosclerosis, with both innate and adaptive immunity playing key roles in atherosclerosis progression [10]. T helper 1 (Th1) cells have been implicated in the pathophysiology of atherosclerosis as they produce high levels of interferon-γ (IFN-γ) encountered in atherosclerotic lesions [11]. Studies with mice lacking the Th1 cell transcription factor T-bet, IFN-γ, or the IFN-γ receptor have reported resistance to high fat diet-induced atherosclerosis [12, 13]. Interlukin-12 (IL-12) is produced mainly by macrophages, dendritic cells (DCs) and B cells [14]. It can stimulate INF-γ production from T and natural killer (NK) cells in th1 responses [14]. In mice treatment with IL-12 stimulated INF-γ production and increased lesion size, mice lacking ApoE and p40, a subunit of IL-12 demonstrated smaller atherosclerotic lesions [15]. However, given the fact that p40 is also a subunit of another interleukin, IL-23, reduction of atherosclerotic size might occur due to combined deficiency of both IL-12 and IL-23 [15]. There are also recent studies in mice and humans claiming that IL-17-producing CD4+ T (Th17) cells are encountered in atherosclerotic lesions but their role is still debatable and needs to be clarified [16-18]. All these findings pinpoint the critical role of innate and adaptive immunity in atherosclerosis. On the other hand interleukin-10 (IL-10), protects early stage atherosclerosis development possibly by inhibiting INF-γ production, thus exhibiting strong anti-inflammatory action [19].

In addition, in early stage atherosclerosis smooth muscle cell migration from the media to the tunica is observed. Over the course of time, plaques are formed from mature lesions, which in turn can undergo calcification further reducing vessel elasticity [20]. Pro-inflammatory cytokines and proteinases eventually render the plaque unstable, making it prone to rupture. In the event of a plaque rupture, the pro-coagulant exposed lipid core results in the occlusion of the vessel lumen [20]. Depending on the area occluded, these events clinically manifest in the form of acute vascular syndromes, like acute myocardial infarction (AMI) or stroke [21]. The key mechanisms involved in atherosclerosis are displayed in Fig. **2**.

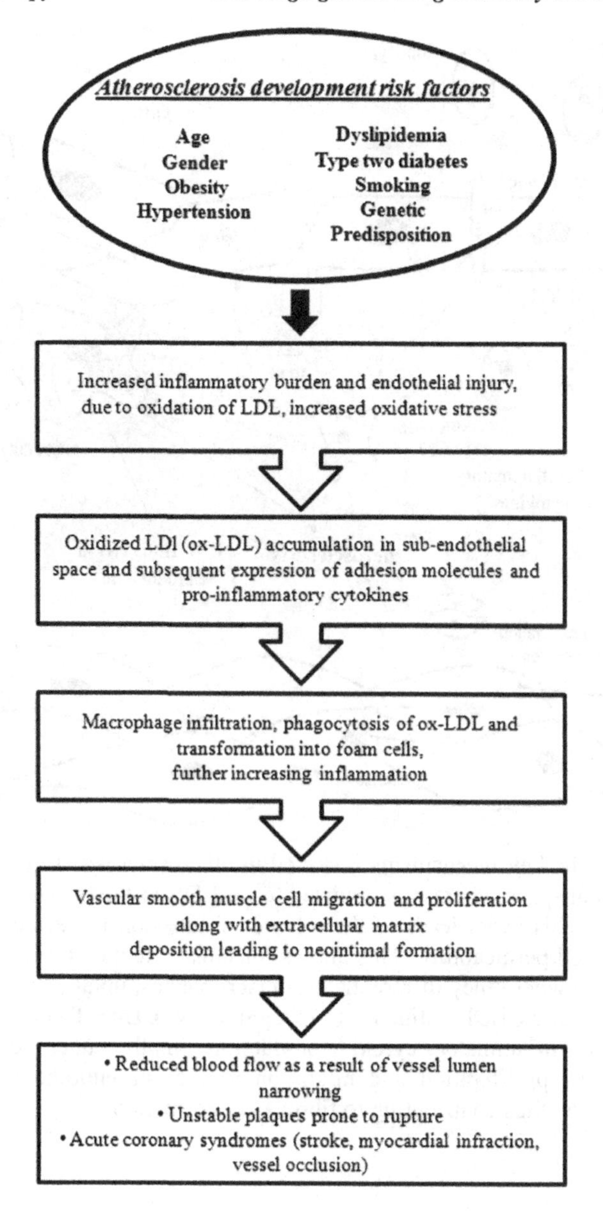

Figure 1: A diagrammatic overview of the development and progression of atherosclerosis. Classic risk factors for the development of atherosclerosis, such as obesity, smoking, hypertension *etc.* can cause oxidative stress, oxidation of LDL to ox-LDL and subsequently promote inflammation. Ox-LDL accumulation in the sub-endothelial space along with secretion of cytokines and adhesion molecules, activate the endothelium, promoting macrophage infiltration. Macrophages accumulate ox-LDL and transform into foam cells further increasing the inflammatory burden. It later steps cytokines release causes VSMC migration and proliferation into the arterial wall, leading to neointimal formation. As a result vessels narrow, reducing blood flow and the atherosclerotic plaques formed may rupture, resulting to acute coronary syndromes.

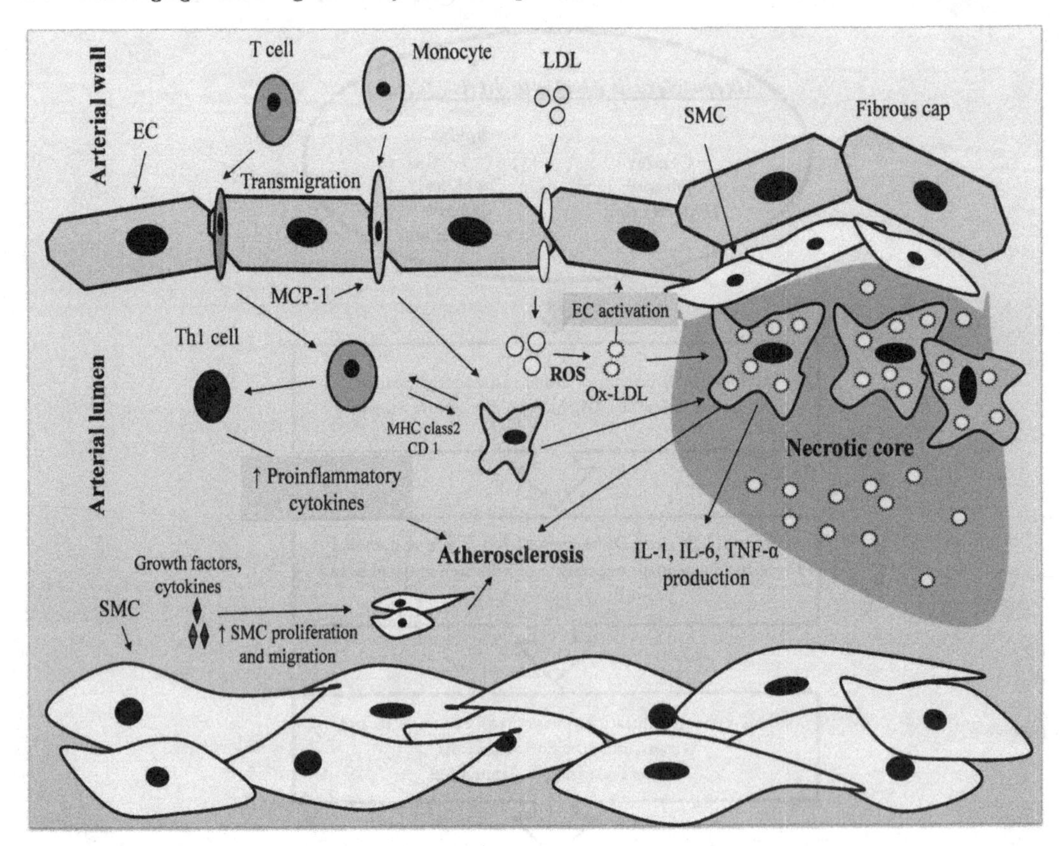

Figure 2: Overview of the key mechanisms involved in atherosclerosis. Low density lipoprotein enters the sub-endothelial space where it is oxidized to ox-LDL in the presence of ROS and other biochemical stimulants. This event leads to the release of adhesion molecules and a subsequent increase in endothelial cell permeability. This allows for leukocyte and T-cell transmigration into the sub-endothelial space where they differentiate to macrophages, uptake ox-LDL and transform to foam cells. Furthermore, ox-LDL antigens are recognized by CD4+ T cells which differentiate to Th1 cells, favoring pro-inflammatory cytokine production. Finally, under the effect of cytokines and growth factors SMC proliferation and migration to the sub-endothelial space occurs, in response to vascular injury, thus contributing to fibrous cap formation.

ANGIOGENESIS

Vascular endothelial cells which along with pericytes and the basal membrane have a pivotal role in orchestrating the angiogenic process. Under normal circumstances endothelial cells (ECs) maintain their function by autocrine signaling of VEGF, NOTCH, angiopoietin-1 and fibroblast growth factor (FGF) [22]. When the physiological conditions are disrupted by low oxygen, inflammation, tumor development or wound healing, endothelial cells starter expressing hypoxia-inducible factors such as HIF-2α and prolyl hydroxylase domain 2 (PHD2) [22] Angiogenesis occurs in several subsequent steps starting with the degradation of the basal membrane followed by reduced adhesion of endothelial cells. Afterwards, vascular sprouting occurs *via* the migration of endothelial leading (tip) and trailing (stalking) cells. Subsequently new lumen is formed by endothelial cells while pericytes are attracted to the newly formed area. Finally, vascular stabilization occurs by tightening of the basal membrane and of cell junctions [22]. On a protein level there are several molecules that regulate angiogenic responses by acting on several distinct signaling cascades. The most important ones are vascular endothelial growth factor (VEGF), fibroblast growth factor-2 (FGF2), transforming growth factor-beta (TGF-β), Notch ligands jagged 1 (JAG1) and Delta like ligand 4 (DLL4) and angiopoietins (Ang-1 and Ang-2). VEGF consists of a family of seven members all sharing a common core region of cysteine knot motif. VEGF is a very specific mitogen of endothelial cells. Following binding to tyrosine receptors, a variety of responses can occur, leading to "endothelial cell proliferation, migration and formation" of new vessels [23]. Furthermore, VEGF interacts with vascular endothelial growth factor receptor-2 (VEGFR2) activating endothelial nitric oxide synthase (eNOS), SRC, RAS-ERK and PI3K-AKT pathways inducing vascular permeability in addition to endothelial proliferation migration and survival [24, 25]. Ang-1 and Ang-2 are the best characterized angiopoietins. They bind to Tie-2 tyrosine kinase receptor, expressed in vascular endothelial cells and in some macrophage subtypes involved in angiogenesis [26]. Angiopoietins are key molecules in the angiogenic process as they control angiogenic switch. Ang-1 is pivotal for endothelial cell proliferation, migration and survival, while Ang-2 disrupts endothelial and perivascular cell connections, thus leading to cell death and vascular regression [26]. Interestingly under the presence of VEGF, Ang-2 exerts pro-angiogenic effects [26]. As with VEFG, FGFs are a family of structurally similar polypeptides, with nine distinct members. FGF-1 and FGF-2, also termed as acidic and basic FGFs respectively have been well-characterized as key modulators of angiogenesis [27].

Delta-Notch signaling is also a critical part of angiogenesis as it regulates sprout formation. In this cell-cell signaling system, VEGF-A induced DLL4 production in tip cells induces Notch receptor activation in stalk cells [28].Apart from direct involving in angiogenesis VEGF, FGF, Notch and TGF-β signaling cascades crosstalk with other pathways such as the canonical WNT and Hedgehog pathways that regulate embryonic and stem cell responses [29, 30]. TGF- exerts diverse cell actions by binding and activating type I and II serine and threonine kinase receptors. TGF-β signaling is without dispute a critical element of angiogenesis and vascular remodeling [31]. TGF-β1 is the key molecule in the formation of the primary vascular structure and the subsequent creation of a more complex network [31].

ASSOCIATION OF ANGIOGENESIS WITH ATHEROSCLEROSIS

Since delivery of nutrients from the lumen to nearby cells is limited, many larger human arteries form a microvasculature at their outmost (adventitial) layers termed as vasa vasorum. Artery branching at common intervals that run longitudinally parallel to the vessel consists of first-order vasa vasorum, while artery arching from the first-order vasa vasorum around the perimeters coronary lumen is termed as second order vasa vasorum [32]. Association of intimal neovascularization and atherosclerosis was first mentioned at the end of the 19th century by Koester, while in the following years several similar observations were made. Almost a century later, it was hypothesized that atherosclerotic plaque progression beyond a critical point, was due to the proliferation of coronary vasculature, by supplementation of oxygen and nutrients [33]. Later on, they proposed that the neovascular network of coronary atherosclerotic plaques can be more rupture-prone, causing plaque destabilization thus promoting acute myocardial infarction development [34]. Since then, there has been ample evidence linking plaque neovascularization with the progression atherosclerosis both *in vivo and in vitro* [35]. Furthermore, in human atherosclerotic lesions where angiogenesis occurs, several proinflammatory cytokines are expresses further establishing the link between these two processes [35]. However, in order to fully elucidate the role of angiogenesis in atherosclerosis, further elucidation of the mechanisms and pathways is required. In the field of angiogenesis there is still much progress to be made.

ANGIOGENESIS AND NEOINTIMAL GROWTH

Following arterial stenting, angioplasty and venous bypass graft, aberrant neovascularization has been observed [36-39]. This can be justified by the hypoxic environment at the sites of intimal hyperplasia, which not only as aforementioned switches to an angiogenic profile but also leads to the up-regulation of several growth factors (mainly VEGF, FGF) and cytokines which in turn regulate vascular smooth muscle cell (VSMC) migration and proliferation [40]. The role of VEGF in several models of intima formation has yielded contradictory results [35]. This can be explained by the use of different animal models and the fact that VEGF effects are concentration dependent [41]. Lower concentrations of VEGF are atheroprotective accompanied with very low angiogenic response. On the other hand, at higher concentrations, the protective effect is nullified, along with an increase of angiogenesis [35, 41]. In even greater concentrations VEGF can exhibit proatherogenic action as demonstrated by intimal thickening, having a detrimental effect in the atherosclerotic progress [35]. There is also evidence that endothelial progenitor cells are a source of VSMCs encountered in atherosclerotic plaques [42, 43]. However, their role in angiogenesis and tissue revascularization still remains to be clarified.

HYPOXIA AND INTRAPLAQUE NEOVASCULARIZATION

Recent evidence emerging from cancer models suggesting that hypoxia is a critical factor for tumor growth, has provided insight on whether it can affect atherosclerotic plaque neovascularization [44]. Theoretically, when vessel thickness exceeds a threshold, as a result of injury or lipid accumulation, oxygen and nutrient supply to the tissues will decrease as the distance between the lumen or the vasa vasorum grows. When this distance surpasses 100μm [45], it forms a hypoxic environment, providing the stimulus for hypoxia-inducible transcription factors (HIF). More specifically HIF-1α induces the expression of VEGF and other pro-angiogenic factor expression, commencing the angiogenic process [46]. Supporting this hypothesis, in atherosclerotic lesions, HIF-1α, FGF and VEGF levels were found elevated [47]. Given the fact that microvessels deliver oxygen and nutrients to both plaques and inflammatory cells, atherosclerosis continues perpetually [48]. However, it should be noted that angiogenesis only promotes plaque formation but does not initiate it. Interestingly, angiogenic responses can be induced by a hypoxia-independent mechanism that of oxidative stress [48]. Overexpression of p22phox, a key subunit of β-nicotinamide adenine dinucleotide phosphate (NADPH) oxidase in a transgenic mouse model demonstrated increased arterial lesion formation [49], suggesting that ROS might have a significant effect in triggering angiogenic responses.

INFLAMMATORY CELLS AND NEOVASCULARIZATION

In atherosclerotic plaques and especially in vulnerable ones, there is a higher concentration of macrophages [50]. These along with VSMCs secrete several cytokines and proinflammatory molecules, greatly contributing to the progression and development of the disease. Furthermore, they secrete a plethora of angiogenic factors [51]. Rupture prone areas consisting of higher numbers of inflammatory cells capable of secreting matrix metalloproteinases (MMPs) can contribute to plaque instability [52]. Comparison of normal coronary arteries with atheromatous ones in humans, revealed that interleukin-18 (IL-18), produced by macrophages is almost exclusively produced in the atheromatous human coronary artery [53]. IL-18 has been described to have similar actions to that of VEGF and FGF-2 [54]. In plaques, expression ICAM-1, E-selectin and VCAM-1 promotes even more inflammatory cell accumulation, thereby accelerating atherosclerosis [55]. A summary of the association between angiogenesis and cardiovascular disease is displayed in Fig. **3**.

ANTIOXIDANTS IN ANGIOGENESIS AND ATHEROSCLEROSIS

All organisms are constantly exposed to free radicals and oxidants, which are either produced as a result of physiological processes or derived from exogenous sources [56]. Free radicals exert both beneficial and hazardous effects, creating a delicate oxidative balance [57]. Once this balance is disrupted, oxidative stress occurs, which has been associated with a multitude of disorders including atherosclerosis and cardiovascular disease [58]. In order to maintain this balance, organisms employ the use of antioxidants [57]. Antioxidants are created *in situ* (endogenous) and can be further classified as enzymatic or non-enzymatic or can

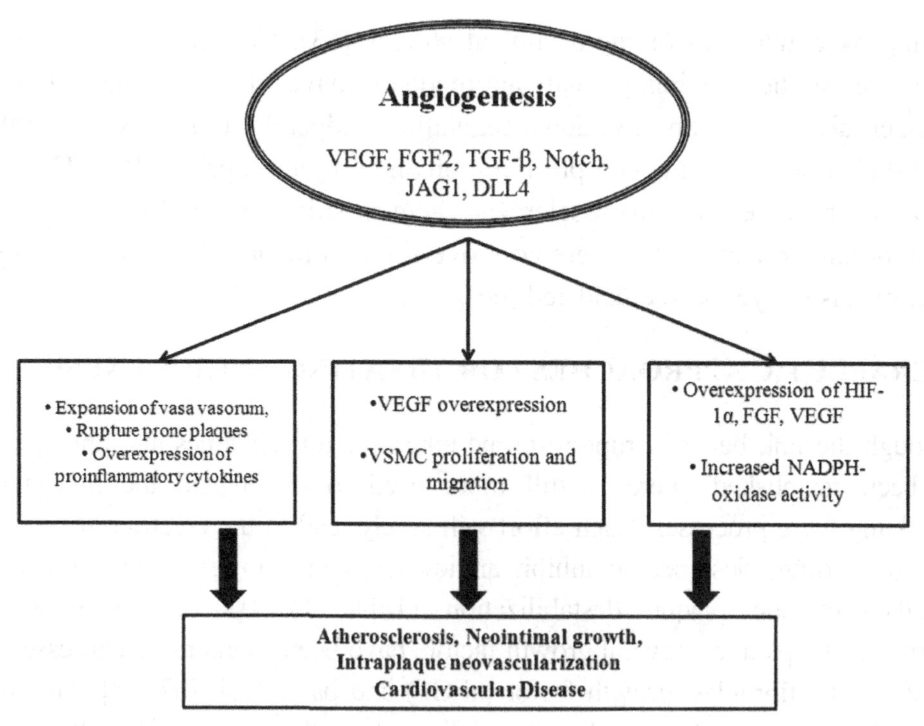

Figure 3: Schematic representation relating angiogenesis with cardiovascular disease. Key molecules in the angiogenic process such as VEGF, FGF2, TGF-β *etc.* lead to the expansion of vasa vasorum, which makes rupture-prone atherosclerotic plaques more vulnerable while promoting proinflammatory cytokine over-expression. Furthermore, in angiogenesis VEGF over-expression facilitates VSMC migration and promotes their proliferation. In addition, hypoxia inducible factors like HIF-1α combined with increased NADPH-oxidase activity are a major source of oxidative stress. All the above factors are major contributors for the progression of the atherosclerotic process, neointimal growth and intraplaque neovascularization, leading to the development of CVD.

be obtained through diet (exogenous) [59]. Enzymatic antioxidants include: catalase, superoxide dismutase (SOD), glutathione peroxidase (GPx) and glutathione reductase (GRx). Examples of endogenous non-enzymatic antioxidants are glutathione, coenzyme Q10, melatonin, transferrin and bilirubin, while exogenous antioxidants are vitamin C and E, flavonoids, polyphenols and carotenoids [59, 60]. Since atherosclerosis progression is mediated through ROS-induced oxidation of LDL, studies have demonstrated that antioxidants can inhibit atherosclerosis by preventing LDL oxidation and subsequent formation of ox-LDL [61, 62]. However, there are some studies claiming that prevention of atherosclerosis by dietary antioxidant vitamin supplementation still needs further proving by conduction of more clinical studies [63]. Concerning angiogenesis, there are studies claiming that antioxidants have a favourable effect in angiogenesis prevention, by down-regulating inducible nitric oxide synthase (iNOS) [64] or by altering cell proliferation and migration profile [65]. However, just as in the case with atherosclerosis, there is still a debate on the favorable action of antioxidants as there are controversial reports and the exact mechanism of action has not yet been elucidated [66].

THERAPEUTIC APPROACHES FOR TREATING ANGIOGENESIS

Although the link between rupture prone atherosclerotic plaques and angiogenesis has been established, there is still more need to understand the mechanisms underlying these processes. Such effort will surely lead to the creation of new anti-angiogenic drugs designed to inhibit angiogenesis thus lowering the progression of atherosclerotic plaque destabilization (Table **1**). As aforementioned, in atherosclerotic plaques, several growth factors have been found to be expressed like VEGF, acidic fibroblast growth factor (aFGF) and basic FGF [67, 68]. The major source of nutrients to the vessel wall is delivered *via* the vasa vasorum. It has been observed that in atherosclerotic plaques a much denser network of vasa vasorum is present, possibly promoting the destabilization of atherosclerotic plaques [69]. Stopping the expansion of vasa vasorum is critical to reducing atheroma progression [70]. There is also evidence of microvessels formation in restenotic lesion as well as in the neointimal [71, 72].

Table 1: Overview of anti-angiogenic therapeutic tools employed for treating cardiovascular diseases

Therapy	Target/Therapeutic Factor	Mode of Action	Refs.
Antibody treatment	VEGFR-1 (Flt-1)	Reduction of early and intermediate lesion size at the aortic root Suppression of macrophage infiltration in the adventitia	[39]
	VEGF	Inhibition of neovascularisation without prevention of endothelization	[34]
	FGF2	Inhibition of SMC proliferation	[41]
Synthetic drugs	Paclitaxel	Inhibition of neointimal formation Inhibition of in stent restenosis	[36, 37]
	SU5402	Reduction of atherosclerosis by FGFR inhibition	[45]

Table 1: contd….

	SSR128129E	Novel multi-FGFR inhibitor, demonstrated significant Reduction of neointimal formation	[46, 51]
	TNP-470	Inhibition of intimal hyperplasia Reduction in plaque growth	[52]
	PI-88	Reduction of intimal thickening and VSMC proliferation	[39]
Anti-angiogenic factors	Angiostatin	Inhibition of neointimal formation	[36]
	sFlt-1	Reduction of adventitial thickening	[47]
	Endostatin	Reduction in plaque growth	[52]
	Interleukin-10	Negative regulation of VEGF expression	[54]
	PEDF	Blocking of NADPH oxidase mediated ROS generation in SMCs	[80]
	sFGFR1	Inhibition of SMC proliferation	[42]
Statins	Fluvastatin	Down-regulation of angiogenic molecules (VEF, HIF-1α, phospho-STAT3) Suppression of	[60]
		ICAM-1 expression Prevention of superoxide-induced lipid peroxidation	
	Cerivastatin	Inhibition of MMP-1,-3,-9 expression in SMCs and macrophage foam cells	[62]
	Simvastatin	Down-regulation of PDGF and VEGF expression	[65]
Gene therapy	miR-17-92 cluster	Inhibition of angiogenic activity in ECs Prevention of neovascularization	[105,109-111]
	miR-23-27-24 cluster	Regulation of angiogenesis and postnatal retinal vascular development Repression of angiogenesis sprouting	[95, 113,114]
	miR-208	Regulation of cardiac stress response	[103, 118]
Nanomedicinal	$a_v\beta_3$-targeted paramagnetic nanoparticles loaded with fumagillin	Inhibition of aortic atherosclerotic progression Possibility of MRI imaging for therapy assessment	[125,126]
	Liposomes loaded with PLP	Anti-inflammatory action, imaging with ^{18}F-FDG-PET/CT	[128]

Stent implantation represents a major breakthrough in the treatment of cardiovascular disease. Despite that there are still major issues that need to be addressed such as in-stent restenosis (ISR) and stent thrombosis (IST), with the former accounting for 15-30% of bare metal stent usage in percutaneous coronary interventions [73, 74]. The development of IST is accredited to the fact that during the balloon deployment of the stent, vessel injury occurs, resulting in neointimal hyperplasia and tissue proliferation [75]. Other factors influencing ISR formation are patient-related (diabetic or patients with small vessel diameter) while the stent coating physicochemical characteristics (elastic recoil, shape and surface properties) are also of great importance [76, 77]. Following stent implantation, endothelialization occurs [78]. This is mostly driven through the deposition of platelets and fibrin followed by migration and penetration of leukocytes into the tissue, with several cytokines regulating the process [78]. If the re-endothelization process is rapidly achieved rapidly and completely after intervention the formation of the neointimal can be significantly reduced [75]. One strategy is to inhibit key components of the most important signaling pathways such as VEGF/Ang-1 with the use of antibodies.

In a study using New Zealand rabbits under a three week atherogenic diet, use of phosphorycholine coated stents with an anti-VEGF antibody inhibited neovascularisation without preventing endothelization [79]. Similarly, another group reported the use of a biodegradable polymeric stent coating releasing hirudin and iloprost, successfully inhibiting neointimal formation after coronary stenting in both sheep and pig models [80]. Another group reported that paclitaxel and angiostatin both offered protection against neointimal formation after administration of recombinant human VEGF in rabbits [81]. Furthermore, use of paclitaxel-coated balloon catheters proved beneficial in patients with in-stent restenosis [82].

PLACENTAL GROWTH FACTOR

Placental growth factor (PIGF) is a homolog of VEGF and has been implemented in pathological angiogenesis by acting through its receptor Flt-1 as well as in the development of atherosclerotic plaques and macrophage accumulation in mice [83]. Treatment with ftl-1 antibodies in Apo $E^{-/-}$ and $PIGF^{-/-}$ deficient mice not only reduced atherosclerotic lesion size and number but macrophage accumulation as well, compared to their Apo $E^{-/-}$ counterparts [83]. Interestingly, the number of plaque microvessels and the growth of advanced atherosclerotic lesion remained unaffected [84].

FIBROBLAST GROWTH FACTOR INHIBITION

Fibroblast growth factors are involved in several biological processes, by participating in a plethora of endocrine signaling pathways [85]. There are so far 22 FGFs identified in vertebrates, having a highly conserved gene structure as well as amino acid sequence. FGFs play key roles in both embryonic development and in adult organism functions. Regarding the former, FGFs play "diverse roles in cell proliferation, migration and differentiation", while concerning the latter they regulate functions associated with tissue repair in response to injury [85]. Aberrant FGF expression has also been associated with cancer development and progression [86]. Concerning atherosclerosis, the role of FGFs still remains unclear. Given the fact that FGF is a potent stimulant for smooth muscle cell (SMC) and endothelial cells which both contribute to the stability of atherosclerotic plaques, there were many doubts as to whether inhibition of FGFs or its receptor could be beneficial for treating atherosclerosis. Two older studies demonstrated that SMC proliferation was inhibited after treatment with anti-FGF2 or administration of soluble FGFR1 after balloon injury or aortic transplants, pinpointing the role of FGFs in restenosis in different models of disease [87, 88]. However, despite the fact that FGF1 and 2 as well as their receptors have been identified as components of atherosclerotic plaques [89], the role of FGFRs in early atherosclerotic lesion formation still remains largely unknown. Recently, a study highlighted the effects of FGFR2 in acceleration of atherosclerosis in $ApoE^{-/-}$ mice over-expressing FGF-R2, through promotion of $p21^{Cip1}$-mediated endothelial cell dysfunction as well as platelet-derived growth factor (PDGF) induced VSMC proliferation [90], further complicating the role of FGFRs in atherosclerosis progression. An older study using the compound SU5402, an FGFR inhibitor, demonstrated successful attenuation of atherosclerotic progression in $ApoE^{-/-}$ mice. It should be mentioned that this inhibition was not FGF specific and could be mediated through VEGFR signaling as well [91]. In a recent study, treatment with SSR128129E, a novel multi-FGF inhibitor was beneficial not only in mice undergone vein graft but in ApoE$^{-/-}$ deficient mice as well. The former showed significant reduction of neointimal formation, while the latter reduced lesion size in the aortic sinus [92].

VASCULAR ENDOTHELIAL GROWTH FACTOR RECEPTOR

As aforementioned, VEGF is a key molecule of the angiogenic process. It interacts *via* several receptors but mainly through VEGFR-1 also known as Flt-1. It has been proposed that Flt-1 is involved in pathological angiogenesis so it may be a lucrative target for treating atherosclerotic plaque formation as well as other angiogenesis related diseases [93]. Admission of a monoclonal anti-Flt-1 antibody in a model of ApoE$^{-/-}$ mice for a period of five weeks demonstrated significant reduction of early and intermediate lesion size at the aortic root. Furthermore the Flt-1 antibody suppressed macrophage infiltration in the adventitia, reducing inflammation [83]. An interesting study highlighted the use of sFlt-1 to block VEGF and FGF with a (with a dominant negative form of FGF receptor 1 [FGF-R1DN]) attenuated adventitial thickening. Interestingly, the study concluded that adventitial thickening was not initiated by angiogenesis but only stimulated [94]. In a very interesting study, adenosine (Ado) was found to modulate the balance between soluble and membrane and Flt-1 receptor, switching from an anti-angiogenic to a pro-angiogenic profile respectively in human primary macrophages [95].

VASCULAR ENDOTHELIAL GROWTH FACTOR -C,-D

Although there is ample scientific information about VEGF and PLGF and their interactions with their inhibitors, little is known about other the other isoforms VEGF-C,-D and their receptor Flt-4 and their relation with angiogenesis. A recent study demonstrated that macrophages and monocytes express both these growth factors as well as their receptor in advanced atherosclerotic plaques both *in vivo* and *in vitro* [96]. This new link between angiogenesis and atherosclerosis can be exploited for developing new anti-angiogenic therapies.

Potent inhibitors of angiogenesis include the fumagillin family of natural products. These are employed in combating tumor angiogenesis and metastasis. A synthetic analogue of fumagillin is TNP-470 which is employed in clinical trials as an anticancer drug. In an older study, SMCs that underwent treatment with TNP-470 demonstrated inhibition of DNA synthesis [97]. More recently, in a study examining the effects of TNP-470 in SMCs and found that intimal hyperplasia was inhibited in a dose dependant manner highlighting its application for preventing vascular intimal hyperplasia [98]. As well as this, in ApoE$^{-/-}$ mice, treatment with TNP-470 or endostatin, a natural occurring fragment from type XVIII collagen with anti-angiogenic properties, resulted in significant reduction in plaque growth even when the treatment started after 32 weeks. However, there was no significant effect at the very early stages of atherosclerotic plaque formation [99].

Use of a synthetic polysulfated oligosaccharide, Phosphomannopentaose sulfate (PI-88) attenuated intimal thickening and reduced VSMC proliferation after balloon injury in rats [84]. It has been proposed that PI-88 has a dual mechanism of action: Firstly, by binding to FGF-2 and thus blocking FGF-2 receptor dependant ERK activation and secondarily by inhibiting heparinise activity, an enzyme that degrades heparin sulphate in both the extracellular matrix (ECM) and cell surface [84].

INTERLEUKIN THERAPY

Following the inflammatory reaction, several cytokines are being secreted in an effort to regulate the inflammation. A broad range of cytokines and especially interleukins are critical mediators of the atherogenic process, greatly contributing to plaque build-up [100]. There are some interleukins, such as IL-10, a potent deactivator of macrophages, thus exhibiting a significant anti-inflammatory action [101]. In addition, IL-10 down-regulates the expression of VEGF, TNF-α and MMP-9, preventing tumor growth-related angiogenesis [102]. In a mouse model of ischemia-induced angiogenesis IL-10 proved to have anti-angiogenic effects by negatively regulating VEGF expression [103].

STATINS FOR TREATING PATHOLOGICAL ANGIOGENESIS

3-hydroxy-3-methyl-glutaryl-CoA (HMG-CoA) reductase inhibitors better known as statins are a class of drugs are well known due to their pleiotropic lipid lowering effects. However, it was later proved that the benefit of statins in vascular disease was "independent of their lipid lowering effects" [104]. Statins act by inhibiting the enzyme HMG-CoA reductase which is present at the early stages of the mevalonate pathway which yields several products such as cholesterol, coenzyme Q10, heme-A, and several isoprenylated proteins [105]. There have been several studies which highlight the beneficial use of statins. Statins are very intriguing as they have multiple effects by regulating cytokines, adhesion molecules redox sensitive transcriptional pathways thus expanding their beneficial effects as described by meta-analysis studies [104, 106].

Interestingly there are studies proving that statin treatment can reduce VEGF levels in serum as well as atherosclerotic plaque volume [107, 108]. In a study using statins and more specifically fluvastatin, in a mouse model demonstrated that statin treatment can have beneficial effects by inhibiting the up-regulation of angiogenic molecules such as VEF, HIF-1α and phospho-STAT3. In addition it suppressed expression of adhesion molecule ICAM-1 and prevented superoxide production and lipid peroxidation, suggesting that the anti-angiogenic effects of statins are because of their anti-oxidant and anti-inflammatory properties [109]. Furthermore, the anti-angiogenic effect of statins has been demonstrated due to their inhibitory effect in cyclooxygenase-2 (COX-2) and MMP-9 "expression and activity in endothelial cells", thereby contributing to plaque stabilization [110]. In a study with human saphenous veins and rabbit aortic SMCs and macrophages 50nM cerivastatin dosage inhibited MMP-1,-3,-9 expression in both SMCs and macrophage foam cells [111]. Statin therapy has also been associated with reduced plaque angiogenesis in carotid therapy [112]. Another group reported that daily administration of simvastatin (40mg) can attenuate angiogenesis by down-regulation of PDGF and VEGF levels [113]. There are reports however, of biphasic effects of statins and more specifically of cerivastatin and atorvastatin. For example, low doses of statins can stimulate angiogenesis but higher statin doses possessed anti-angiogenic effects associated with increased endothelial apoptosis and decrease in VEGF expression [114]. Of great interest is the fact that statins are also utilized in other fields (*i.e.* oncology, hematology) due to their "anti-angiogenic and anti-tumor properties" [115]. However, long-term effects of statin treatment still remain under investigation. An interesting study highlighted that chronic statin treatment in patients with AMI was associated with reduced positive remodeling in the culprit lesions [116].

REDOX CONTROL OF ANGIOGENESIS

Reactive oxygen species include a great variety of oxygen containing chemically reactive molecules that upon reaction with key biological molecules (*i.e.* lipids and proteins) can drastically lead to an increase of oxidant production leading to oxidative stress [117]. Oxidative stress is termed as the overproduction of ROS and reactive nitrogen species (RNS) and has been identified as a fundamental mechanism for cell damage [118]. Most notable examples of ROS are superoxide anion, hydroxyl radical and hydrogen peroxide [119]. It is worthwhile mentioning that low concentrations of radicals regulate several cellular functions and most notably cell communication and angiogenesis [120, 121]. From the vast majority of ROS, superoxide radicals have been implicated in several pathological conditions including atherosclerosis, vascular remodeling, myocardial infarction and ischemic stroke [122]. There are several oxidant enzyme systems that have been identified as key players for the production of superoxide. These include xanthine oxidase (XO), uncoupled NOS and cytochrome P450 [123]. These enzymatic sources, cannot account for the bulk amount of superoxide production under physiological and pathophysiological conditions [124]. NADPH oxidase is an enzyme with similar structure of the enzymatic complex found in phagocytic leucocytes. NADPH consists of at least 5 subunits: p47phox and p67phox, found at the cytosol and the membrane bound p22phox gp91phox (better known as Nox2) and Rac as small g protein [101]. Translocation of the cytosolic components to the membrane-bound complex triggers a superoxide production burst from the extracellular part of the membrane [101]. This is achieved *via* the reduction of oxygen and the use of NADPH as the electron donor [125].

In endothelial cells, several pro-angiogenic stimuli such as cytokines and growth factors can activate NADPH-dependent ROS production [126]. This effect can also work the opposite way, as increased ROS concentration may subsequently affect redox sensitive molecules, pathways or transcriptional factors [121]. More specifically, Nox2 and Nox4 isoforms in endothelial cells are responsible for ROS production and knockout of these two isoforms greatly reduced endothelial cell proliferation and survival [127]. A group reported that in human microvascular and lung microvascular endothelial cells, Nox4 is a key mediator of angiogenesis through activation of the ERK signaling pathway [128, 129]. VSMCs treated with thrombin and PDGFAB, increased ROS production, following "activation of a p22-containing NADPH oxidase" [130, 131]. Furthermore, there is growing evidence linking oxidative stress and neointimal formation after angioplasty [132, 133], further associating NADPH oxidase with pathological angiogenesis and CVD. Given the role of NADPH-oxidase in cardiovascular diseases and inflammation, its inhibition may prove a new therapeutic solution for several diseases, by affecting the distinct pathways underlying these disorders.

In a rat carotid artery balloon injury model, treatment with Pigment epithelium-derived factor (PEDF) drastically inhibited neointimal hyperplasia after vascular injury [134]. PEDF is a glycoprotein and belongs to the family of serine protease inhibitors [135]. It is a potent inhibitor of angiogenesis and blocks TNF-α and angiotensin II (Ang II) induced EC activation, due to its anti-oxidant properties [135]. The inhibition observed from PEGF treatment was derived by blocking of NADPH oxidase mediated ROS generation in SMCs [134]. Pharmacological inhibition of NADPH oxidase is also been applied for the treatment of retinopathy, which involves aberrant neovascularization of the retina, and suppression of tumor growth [136].

NATURAL POLYPHENOLS AS ANTI-ANGIOGENIC DRUGS

Polyphenols are a class of natural occurring chemicals found in plants, although some can be synthetic or semisynthetic [137]. Their main characteristic is the presence of multiple phenol rings in their structure. The most important characteristic of polyphenols is their strong anti-oxidant properties [137]. Depending on their structure they can act as chain breakers or free radical scavengers [138]. Polyphenols are believed to inhibit atherosclerotic plaque progression not only to their anti-oxidant abilities but also through inhibition of new blood vessel formation [139]. Apart from these effects, the beneficial actions of polyphenols may expand to prevention of LDL oxidation, inhibition of MCP-1 and tissue factor as well as activation of platelets [139]. Recently there is growing evidence of the anti-angiogenic effect of polyphenols *in vitro* and *in vivo* models.

In smooth muscle cells, red wine polyphenolic compounds (RWPCs) attenuated VEGF expression. It was also highlighted that the RWPCs reduced $PDGF_{AB}$ *via* the redox sensitive p38 MAPK pathway [140]. Polyphenols also inhibit "enzymatic sources of ROS production such as NADPH oxidase and xanthine oxidase in cells", [141, 142] while in the meantime "enhance the activity of antioxidant enzymes such as catalase and glutathione peroxidase" [143]. These last two enzymes play a vital role in maintaining oxidative homeostasis by decomposing hydrogen peroxide to oxygen and water. In addition RWPCs offered sustained inhibition of $PDGF_{AB}$, attenuating VEGF expression. Interestingly other antioxidant compounds like vitamin C, N-acetylcysteine and diphenylene iodonium demonstrated only partial reduction in $PDGF_{AB}$-induced VEGF expression [140].

A major feature during the progression of the atherosclerotic process is the extensive remodeling of the arterial wall [144]. This procedure is tightly regulated by the MMPs, which can degrade components of the extracellular matrix. In vascular tissues MMP-1, -2 and MMP-9 have key roles "in the turnover of type IV collagen", promoting angiogenesis [145] and atherosclerosis [146]. Since the role of MMPs in angiogenesis and atherosclerosis is well established [147], inhibition of MMPs can provide a potent therapeutic target for treating pathological angiogenesis. Polyphenols can inhibit thrombin and membrane bound MT1-MMP activity in VSMCs, resulting in significant reduction in MMP-2 levels [148]. Another hallmark event in angiogenesis, atherosclerosis and restenosis is the aberrant "proliferation and migration of ECs and VSMCs" [149]. Several polyphenols have been associated with inhibition of such cellular events. These effects have been associated with "decreased expression of CREB and ATF-1 transcription factors and subsequent down-regulation of cyclin A gene" [150]. Also the anti-angiogenic action exerted from Polyphenols is due to "specific inhibition of p38 MAPK and PI3-kinase/Akt pathways" [150]. Another polyphenol of interest is honokiol, which was found in pre-clinical models to have significant anti-angiogenic and anti-inflammatory effects with minimal toxicity [151].

Despite the fact that there are several *in vitro* studies for the anti-angiogenic properties of polyphenols, there are few *in vivo* models used to determine their action. "Local application of RWPCs to chick embryo chorioallantoic membrane", had a strong reduction of small blood vessel number and length, marking decreased angiogenesis after treatment for a period of 48 hours [152]. There are also older studies that pinpoint the anti-angiogenic action of several polyphenolic compounds and highlight their potential application in several angiogenesis related disorders [153]. Despite the distinct observed anti-angiogenic effect, there is still much to be elucidated about the mechanism of action of natural polyphenols.

GENE THERAPY

RNA interference (RNAi), represents a post-transcriptional gene regulation process that is conserved in many different organisms. Small non-coding RNAs (ncRNAs) play critical roles in several biological processes and dysregulation of ncRNAs is associated with several diseases including developmental timing, skeletal muscle proliferation, tumor progression, neurogenesis, brain morphogenesis, transposon silencing, viral defence, and many other cellular processes using the same RNA-processing complex to direct silencing [154-156].

Recent studies suggested that atherosclerosis is an angiogenic disease. The formation of microvessels, contributes to the development of plaques making them rupture-prone. Neovascularisation is proposed to greatly contribute to plaque progression and is frequently observed in human coronary arteries [48, 71]. Recent evidence has strongly pinpointed the role of microRNAs (miRNAs) in CVDs as well as other diseases. Mi-RNAs are small non-coding RNAs that "bind to a target mRNA, causing either degradation or translational repression thus regulating gene expression" [36]. Several studies indicate that these crucial regulators of gene expression have great potential as therapeutics, especially in the regulation of the angiogenic process and cardiogenesis. The identification of circulating miRNAs in patients with CVDs renders them as potential biomarkers for clinical diagnosis, leading to novel therapeutic approaches [157, 158].

Generation of miRNAs is mediated by two enzymes, Dicer and Drosha and is achieved in a two-step processing pathway. When cells encounter long double-stranded RNA molecules, Dicer, a ribonuclease III type enzyme, cleaves them into small interfering RNAs (siRNAs) of 21–23 nucleotides [36, 155, 159]. "Dicer is constitutively expressed in endothelial cells, with its expression remaining unaffected by either response to stimuli, such as VEGF, or by cell proliferation status" [160]. "This small RNA attaches to an RNA interference silencing complex (RISC) and is directed to the messenger RNA (mRNA) of interest" [157, 161]. Several studies indicate that inhibition or hindrance of miRNA biogenesis pathway provides new therapeutic opportunities for treating diseases "characterized by aberrant angiogenesis (cancer or macular degeneration) or irregular angiogenesis (myocardial ischemia or peripheral vascular disease)" [162]. "The role of miRNAs in ECs was assessed by specific silencing of Dicer, by use of short interfering (si)RNA in human umbilical endothelial cells", indicating diminished tube formation and cell migration with remarkable effect on several angiogenic regulators, such as TEK/Tie-2, KDR/VEGFR2, Tie-1, angiopoietin-like 4 (ANGPTL4), IL-8 and eNOS [163-166]. The silencing of Drosha, the other nuclear type-III ribonuclease which processes the pri-miRNAs has been dissected, producing less pronounced effects on angiogenesis than Dicer [163-167].

Dysregulation of miRNAs has been widely studied in angiogenesis and several miRNAs included in this function process. "Knockdown of Dicer in ECs is rescued by adding individual miRNAs in the miR-17-92 cluster, a polycistronic miRNA gene categorized into four families (miR-17-, miR-18-, miR-19- and miR-92 family) and characterized as negative regulators of angiogenesis" [164, 168-170]. "Overexpression of miR-92a targets ITGa5 and inhibits angiogenesis in ECs, while administration of antagomir-92a blocks neovascularization in a mouse hindlimb ischemia model and minimizes tissue injury in myocardial infarction" [171]. It has been demonstrated that KLF-2 and its regulated-genes such as eNOS and thrombomodulin (TM), up-regulation by atheroprotective shear flow in primary Sjögren's syndrome and laminar shear stress were "repressed by over-expression of miR-92a in ECs" [171].

"The miR-23-27-24 cluster participates in angiogenesis and endothelial apoptosis in cardiac ischemia and retinal vascular development and miRNAs encoded by the miR-23-27-24 gene clusters are elevated in endothelial cells and highly vascularized tissues. Inhibition of miR-23 and miR-27 function by locked nucleic acid-modified anti-miRNAs represses angiogenesis *in vitro* and *in vivo*" [154, 172]. MiR-23 and miR-27 silencing, represses angiogenesi, consequently up-regulating Sprouty2 and Sema6A proteins and subsequent attenuation of MAPK and VEGFR2 signaling by Raf activation [173]. MiR-27 is involved in early stage atherosclerosis, while miR-27b inhibited thrombospondin-1, a multifunctional protein which "binds to the reelin receptors, ApoER2 and VLDLR, thereby affecting neuronal migration in the rostral migratory stream" [174, 175].

"Human atherosclerotic plaques were compared to non-atherosclerotic left internal thoracic arteries (LITA) concerning their miRNA expression profile, their correlation between miR/mRNA expression profiles and processes in atherosclerosis" [154, 176]. The expression levels of miR-21, -34a, -146a, -146b-5p, and -210 in patients with CVDs were investigated and "predicted targets of these miRNAs" were found to be down-regulated [162]. Additional studies were carried out including "a few highly expressed miRNAs (miR-2, -15b, -16, -20, -21, 181a, -191, -221, -222, -320, let-7, let-7b, and let-7c), with receptors of angiogenic factors (Flt-1, Nrp-2, Fgf-R, c-Met, and c-kit) as putative mRNA targets" [163, 165].

"Furthermore, miR-208 belongs to a cardiac specific miRNA, encoded by an intron in the gene that encodes α-myosin heavy chain and functions within a regulatory network, controlling cardiac stress responses" [162, 177]. Additionally, miR-126, a highly-characterized EC-specific miRNA, is enriched in tissues characterized by a high vascular component, like the heart and lung [162, 178, 179]. "MiR-126 is encoded by intron 7 of the EGF-like domain 7 gene also known as VE-statin, which encodes an EC-specific secreted peptide that acts as a chemoattractant and inhibitor of smooth muscle cell migration" [180, 181]. Finally, there are also indications that anti-angiogenic gene delivery even locally can exhibit beneficial action. In a study using a rabbit model, local gene delivery of sFlt-1 proved beneficial by suppressing plaque formation and angiogenesis within the atheromatic plaque [182].

Although pharmacological manipulation of miRNAs is still at its infancy, much more research is required before the above key players in angiogenesis can be taken into clinical practice, the overall body of evidence indicates that miRNAs might prove to be potent therapeutic tools in the future for controlling vascular inflammation and regulate the progression of atherosclerosis by controlling the angiogenic switch.

NANOMEDICINAL APPROACHES FOR TREATING PATHOLOGICAL ANGIOGENESIS IN CARDIOVASCULAR DISEASES

Over the past few years scientific discoveries in the field of nanotechnology have been achieved with tremendous speed. Nanotechnology is the scientific field of synthesizing materials with distinct compositions, sizes and properties at the nanoscale level (nm) and utilizing them according to their properties. Nanoparticles are molecular assemblies that due to their unique composition and size exhibit extraordinary physicochemical, optical and mechanical properties [183]. Employment of these devices for medicinal applications gave rise to the field of nanomedicine. Although nanomedicine is still at an infant stage, there has been remarkable progress of nanomedicinal applications for almost every type of disease. Lately, efforts are being made to use such nanodevices for treating pathological angiogenesis.

In an effort to create a platform for sustained delivery of anti-angiogenic agents a group used a single injection in a rabbit model of a $a_v\beta_3$-targeted paramagnetic nanoparticle formulation for site-specific delivery of fumagillin successfully inhibiting aortic atherosclerotic progression for a period of 3 weeks [184]. This treatment when combined with oral administration of atorvastatin prolonged its beneficial effects [185]. Furthermore, the paramagnetic nature of the nanoparticles used rendered MRI imaging possible for monitoring and evaluating the progress of the therapy [185]. Also in an effort to successfully monitor the progress of angiogenesis, dendritic biodegradable nanoprobes targeting specific $a_v\beta_3$ integrin were utilized for detection of peripheral artery disease *via* positron emission tomography (PET) [186]. Novel nanomedicinal approaches can also aim to improve the efficiency of already tested anti-inflammatory and anti-angiogenic therapies. Such an example is the construction of a liposomal nano-formulation loaded with glucocorticoid, prednisolone phosphate (PLP). Glucocorticoids are a drug class with significant anti-inflammatory action in several models of atherosclerosis. However, extensive use of this drug class has been avoided due to poor pharmacokinetics and several side effects. In a rabbit model of experimental atherosclerosis, these liposomal formulations containing a mixture of lipids, polyethylene glycol (PEG) and gadolinium with diethylenetriaminepentacetate and bis(stearylamide) (Gd-DTPA-BSA) were successfully delivered in atherosclerotic plaques, as verified by magnetic resonance imaging (MRI) and ^{18}F-fluoro-deoxy-glucose positron emission tomography combined with computed tomography (^{18}F-FDG-PET/CT) by monitoring the uptake of ^{18}F-FDG at the atherosclerotic aortas after injecting the rabbits [187].

ANTI-ANGIOGENESIS DRUG TOXICITY

Several years ago anti-angiogenic drug development and therapy has been proposed in order to combat angiogenesis related diseases (*i.e.* vascular, cancer, rheumatoid) [188]. The excitement from the promising results yielded from *in vivo* and *in vitro* studies soon gave place to disappointment as clinical trials gave very poor results [189]. This was often due to the fact that the mechanisms underlying these processed are often not thoroughly understood. There are still many concerns about the safety of the anti-angiogenic drugs often employed. This is especially important in CVD, given the fact that it is a multivariable disease. Thus, successful treatment of one parameter many have several negative effect to other target cells or organs.

One of the most frequently observed, site-specific side effects of anti-angiogenic therapies is hypertension. There are several studies that describe this phenomenon as a result of anti-angiogenic drug treatment [190, 191]. The mechanism of hypertension development after these treatments still eludes our grasp. There are reports that hypertension is caused due to VEGF related inhibition of nitric oxide (NO) synthesis as well as the rarefaction of the capillary bed [192]. Furthermore, anti-angiogenic therapy has also been associated with increased cardiotoxicity, arterial and venous thrombosis and bleeding [193]. Furthermore, some anti-angiogenic therapies on phase II have exhibited a great increase in the rate of vascular toxicity (26.1%) along with lower but still significant rates of transient ischemic attack (4.3%) and cerebral vascular incidents (4.3%) [194].

Another issue when administering long term anti-angiogenic therapy are potential delayed toxicity issues. Since most of the therapies under development are at an early stage, there is a possibility that toxic effects will not be detected in animal models or early phase clinical trials. There is still much that need to be done in order all potential side effects of anti-angiogenic therapy to be fully assessed before being administered for therapeutic purposes.

EXCITING FUTURE PROSPECTS

The first idea for developing anti-angiogenic drugs was conceived over 25 years ago, mainly for cancer therapies. Despite this initial hesitations and disappointments, anti-angiogenic therapy has evolved, and is now slowly starting to be applied for treating several other pathophysiological conditions associated with aberrant angiogenesis such as atherosclerosis, diabetic retinopathy, age-related macular degeneration and more. Given the fact that the common link between these diseases is angiogenesis, pharmacological advances to one field will surely be beneficial to the others as well. Furthermore, the progress achieved in this exciting scientific field is also verified by the increased number of drugs currently undergoing different stages in clinical trials. These drugs are assessed not only for their chemopreventing role but also aim in optimizing treatment (Table **2**). Given the fact that our knowledge about the molecules involved in the angiogenic processes increases, new drug targets are identified, expanding our anti-angiogenesis drug arsenal. However, it is of the utmost importance that better *in vivo* and *in vitro* models are created in order to fully assess the potential action and side effects these drugs might pose. The advances in gene therapy and nanotechnology are surely considered to be pioneers in the efforts to create new potent anti-angiogenic therapies.

Table 2: List of anti-angiogenic drugs, their targets and stage of development

Drug	Target Molecule(s)	Development
Bevacizumab	VEGF-A	**FDA approved**
Ranibizumab	VEGF-A	
Pegaptanib	$VEGF_{165}$	
Sorafenib	VEGFR-2/3, PDGFR-β, FLT3 and c-kit	
Erlotinib	Epidermal growth factor (EGFR)	
Sunitinib (SU11248)	VEGFR-2/3, PDGFR-β, FLT3 and c-kit	
VEGF-trap	VEGF-A, PlGF	**Phase III**

Table 2: contd....

Combretastatin A-4	Vascular endothelial cells	
Neovastat	VEGF, MMP-2,-9,-12	
BMS 275291	MMP-1, -2, -8, -9, -13, -14	
Pegaptanib	VEGF	**Phase II/III**
cyclic RGD peptide	$\alpha_v\beta_3$ integrin, Endothelial cell proliferation, migration	
PTK787	VEGF	**Phase II**

ACKNOWLEDGEMENTS

Declared None.

CONFLICT OF INTEREST

The authors confirm that this chapter contents have no conflict of interest.

ABBREVIATIONS

^{18}F-FDG-PET/CT = ^{18}F-fluoro-deoxy-glucose positron emission tomography combined with computed tomography

aFGF = Acidic fibroblast growth factor

AMI = Acute myocardial infarction

Ang II = Angiotensin II

Ang-1 = Angiopoietin-1

Ang-2 = Angiopoietin-2

ANGPTL4 = Angiopoietin-like 4

COX-2 = Cyclooxygenase-2

CVD = Cardiovascular disease

DCs = Dendritic cells

DLL4 = Delta like ligand 4

ECM = Extracellular matrix

ECs = Endothelial cells

eNOS = Endothelial nitric oxide synthase

FGF = Fibroblast growth factor

FGF-R1DN = Dominant-negative form of FGF receptor 1

FLT3 = Fms-related tyrosine kinase 3

Gd-DTPA-BSA = Gadolinium with diethylenetriaminepentacetate and bis(stearylamide)

GPx = Glutathione peroxidase

GRx = Glutathione reductase

HIF = Hypoxia-inducible transcription factor

HMG-CoA = 3-hydroxy-3-methyl-glutaryl-CoA

ICAM-1 = Intercellular adhesion molecule-1

IL-10 = Interleukin-10

IL-12 = Interlukin-12

IL-18 = Interleukin-18

IL-1β = Interleukin-1β

INF-γ = Interferon-γ

iNOS	=	Inducible nitric oxide synthase
ISR	=	In-stent restenosis
IST	=	In-stent thrombosis
JAG 1	=	Jagged-1
LDL	=	Low density lipoprotein
LITA	=	Left internal thoracic arteries
LOX-1	=	Lectin-like oxLDL receptor-1
MCP-1	=	Monocyte chemoattractant protein-1
MHC class 2	=	Major histocompatibility complex class 2
miRNAs	=	microRNAs
MMPs	=	Matrix metalloproteinases
MRI	=	Magnetic resonance imaging
mRNA	=	Messenger RNA
NADPH	=	β-nicotinamide adenine dinucleotide phosphate
ncRNA	=	Non-coding RNA
NK cells	=	Natural killer cells
ox-LDL	=	Oxidized-low density lipoprotein
PDGF	=	Platelet-derived growth factor
PDGFR-β	=	platelet-derived growth factor receptor β
PEDF	=	Pigment Epithelium-Derived Factor

PEG	=	Polyethylene glycol
PET	=	Positron emission tomography
PHD2	=	Prolyl hydroxylase domain 2
PI-88	=	Phosphomannopentaose sulfate
PIGF	=	Placental growth factor
PLP	=	Prednisolone phosphate
PPAR-γ	=	Proliferator-activated receptor gamma
RISC	=	RNA interference silencing complex
RNAi	=	RNA interference
RNS	=	Reactive nitrogen species
ROS	=	Reactive oxygen species
RWPCs	=	Red wine polyphenolic compounds
siRNAs	=	Small interfering RNAs
SMC	=	Smooth muscle cell
SOD	=	Superoxide dismutase
TGF-β	=	Transforming growth factor-beta
Th1 cells	=	T helper 1 cells
TLR	=	Toll-like receptor
TM	=	Thrombomodulin
TNF-α	=	Tumor necrosis factor alpha

VCAM-1 = Vascular cell adhesion molecule-1

VEGF = Vascular endothelial growth factor

VEGFR = Vascular endothelial growth factor receptor

VSMC = Vascular smooth muscle cell

XO = Xanthine oxidase

REFERENCES

[1] Libby P, Ridker PM, Hansson GK. Inflammation in atherosclerosis: from pathophysiology to practice. J Am Coll Cardiol 2009; 54(23): 2129-38.
[2] Libby P, Theroux P. Pathophysiology of coronary artery disease. Circulation 2005; 111(25): 3481-8.
[3] Goyal T, Mitra S, Khaidakov M, *et al.* Current Concepts of the Role of Oxidized LDL Receptors in Atherosclerosis. Curr Atheroscler Rep 2012.
[4] Hansson GK, Hermansson A. The immune system in atherosclerosis. Nat Immunol 2011; 12(3): 204-12.
[5] Stocker R, Keaney JF, Jr. Role of oxidative modifications in atherosclerosis. Physiol Rev 2004; 84(4): 1381-478.
[6] Chawla A, Repa JJ, Evans RM, *et al.* Nuclear receptors and lipid physiology: opening the X-files. Science 2001; 294(5548): 1866-70.
[7] Moore KJ, Rosen ED, Fitzgerald ML, *et al.* The role of PPAR-gamma in macrophage differentiation and cholesterol uptake. Nat Med 2001; 7(1): 41-7.
[8] Nagy L, Tontonoz P, Alvarez JG, *et al.* Oxidized LDL regulates macrophage gene expression through ligand activation of PPARgamma. Cell 1998; 93(2): 229-40.
[9] Lim H, Kim YU, Sun H, *et al.* Proatherogenic conditions promote autoimmune T helper 17 cell responses in vivo. Immunity 2014; 40(1): 153-65.
[10] Libby P, Lichtman AH, Hansson GK. Immune effector mechanisms implicated in atherosclerosis: from mice to humans. Immunity 2013; 38(6): 1092-104.
[11] Gotsman I, Lichtman AH. Targeting interferon-gamma to treat atherosclerosis. Circ Res 2007; 101(4): 333-4.
[12] Laurat E, Poirier B, Tupin E, *et al.* In vivo downregulation of T helper cell 1 immune responses reduces atherogenesis in apolipoprotein E-knockout mice. Circulation 2001; 104(2): 197-202.
[13] Tellides G, Tereb DA, Kirkiles-Smith NC, *et al.* Interferon-gamma elicits arteriosclerosis in the absence of leukocytes. Nature 2000; 403(6766): 207-11.
[14] Andersson J, Libby P, Hansson GK. Adaptive immunity and atherosclerosis. Clin Immunol 2010; 134(1): 33-46.
[15] Davenport P, Tipping PG. The role of interleukin-4 and interleukin-12 in the progression of atherosclerosis in apolipoprotein E-deficient mice. Am J Pathol 2003; 163(3): 1117-25.
[16] Danzaki K, Matsui Y, Ikesue M, *et al.* Interleukin-17A deficiency accelerates unstable atherosclerotic plaque formation in apolipoprotein E-deficient mice. Arterioscler Thromb Vasc Biol 2012; 32(2): 273-80.

[17] Eid RE, Rao DA, Zhou J, *et al.* Interleukin-17 and interferon-gamma are produced concomitantly by human coronary artery-infiltrating T cells and act synergistically on vascular smooth muscle cells. Circulation 2009; 119(10): 1424-32.
[18] Erbel C, Chen L, Bea F, *et al.* Inhibition of IL-17A attenuates atherosclerotic lesion development in apoE-deficient mice. J Immunol 2009; 183(12): 8167-75.
[19] Pinderski Oslund LJ, Hedrick CC, Olvera T, *et al.* Interleukin-10 blocks atherosclerotic events in vitro and in vivo. Arterioscler Thromb Vasc Biol 1999; 19(12): 2847-53.
[20] Demer LL, Tintut Y. Vascular calcification: pathobiology of a multifaceted disease. Circulation 2008; 117(22): 2938-48.
[21] Finn AV, Nakano M, Narula J, *et al.* Concept of vulnerable/unstable plaque. Arterioscler Thromb Vasc Biol 2010; 30(7): 1282-92.
[22] Deveza L, Choi J, Yang F. Therapeutic angiogenesis for treating cardiovascular diseases. Theranostics 2012; 2(8): 801-14.
[23] Hoeben A, Landuyt B, Highley MS, *et al.* Vascular endothelial growth factor and angiogenesis. Pharmacol Rev 2004; 56(4): 549-80.
[24] Coultas L, Chawengsaksophak K, Rossant J. Endothelial cells and VEGF in vascular development. Nature 2005; 438(7070): 937-45.
[25] Olsson AK, Dimberg A, Kreuger J, *et al.* VEGF receptor signalling - in control of vascular function. Nat Rev Mol Cell Biol 2006; 7(5): 359-71.
[26] Fagiani E, Christofori G. Angiopoietins in angiogenesis. Cancer Lett 2013; 328(1): 18-26.
[27] Cross MJ, Claesson-Welsh L. FGF and VEGF function in angiogenesis: signalling pathways, biological responses and therapeutic inhibition. Trends Pharmacol Sci 2001; 22(4): 201-7.
[28] Carmeliet P, De Smet F, Loges S, *et al.* Branching morphogenesis and antiangiogenesis candidates: tip cells lead the way. Nat Rev Clin Oncol 2009; 6(6): 315-26.
[29] Katoh M. WNT signaling pathway and stem cell signaling network. Clin Cancer Res 2007; 13(14): 4042-5.
[30] Katoh Y, Katoh M. Hedgehog signaling, epithelial-to-mesenchymal transition and miRNA (review). Int J Mol Med 2008; 22(3): 271-5.
[31] Goumans MJ, Lebrin F, Valdimarsdottir G. Controlling the angiogenic switch: a balance between two distinct TGF-b receptor signaling pathways. Trends Cardiovasc Med 2003; 13(7): 301-7.
[32] Kwon HM, Sangiorgi G, Ritman EL, *et al.* Enhanced coronary vasa vasorum neovascularization in experimental hypercholesterolemia. J Clin Invest 1998; 101(8): 1551-6.
[33] Barger AC, Beeuwkes R, 3rd, Lainey LL, *et al.* Hypothesis: vasa vasorum and neovascularization of human coronary arteries. A possible role in the pathophysiology of atherosclerosis. N Engl J Med 1984; 310(3): 175-7.
[34] Barger AC, Beeuwkes R, 3rd. Rupture of coronary vasa vasorum as a trigger of acute myocardial infarction. Am J Cardiol 1990; 66(16): 41G-3G.
[35] Khurana R, Simons M, Martin JF, *et al.* Role of angiogenesis in cardiovascular disease: a critical appraisal. Circulation 2005; 112(12): 1813-24.
[36] Bartel DP. MicroRNAs: genomics, biogenesis, mechanism, and function. Cell 2004; 116(2): 281-97.
[37] Edelman ER, Nugent MA, Smith LT, *et al.* Basic fibroblast growth factor enhances the coupling of intimal hyperplasia and proliferation of vasa vasorum in injured rat arteries. J Clin Invest 1992; 89(2): 465-73.

[38] Shibata M, Suzuki H, Nakatani M, *et al.* The involvement of vascular endothelial growth factor and flt-1 in the process of neointimal proliferation in pig coronary arteries following stent implantation. Histochem Cell Biol 2001; 116(6): 471-81.

[39] Shigematsu K, Yasuhara H, Shigematsu H. Topical application of antiangiogenic agent AGM-1470 suppresses anastomotic intimal hyperplasia after ePTFE grafting in a rabbit model. Surgery 2001; 129(2): 220-30.

[40] Rudijanto A. The role of vascular smooth muscle cells on the pathogenesis of atherosclerosis. Acta Med Indones 2007; 39(2): 86-93.

[41] Ozawa CR, Banfi A, Glazer NL, *et al.* Microenvironmental VEGF concentration, not total dose, determines a threshold between normal and aberrant angiogenesis. J Clin Invest 2004; 113(4): 516-27.

[42] Hu Y, Davison F, Zhang Z, *et al.* Endothelial replacement and angiogenesis in arteriosclerotic lesions of allografts are contributed by circulating progenitor cells. Circulation 2003; 108(25): 3122-7.

[43] Shimizu K, Sugiyama S, Aikawa M, *et al.* Host bone-marrow cells are a source of donor intimal smooth- muscle-like cells in murine aortic transplant arteriopathy. Nat Med 2001; 7(6): 738-41.

[44] Hulten LM, Levin M. The role of hypoxia in atherosclerosis. Curr Opin Lipidol 2009; 20(5): 409-14.

[45] Torres Filho IP, Leunig M, Yuan F, *et al.* Noninvasive measurement of microvascular and interstitial oxygen profiles in a human tumor in SCID mice. Proc Natl Acad Sci U S A 1994; 91(6): 2081-5.

[46] Pugh CW, Ratcliffe PJ. Regulation of angiogenesis by hypoxia: role of the HIF system. Nat Med 2003; 9(6): 677-84.

[47] Belgore F, Blann A, Neil D, *et al.* Localisation of members of the vascular endothelial growth factor (VEGF) family and their receptors in human atherosclerotic arteries. J Clin Pathol 2004; 57(3): 266-72.

[48] Sluimer JC, Daemen MJ. Novel concepts in atherogenesis: angiogenesis and hypoxia in atherosclerosis. J Pathol 2009; 218(1): 7-29.

[49] Khatri JJ, Johnson C, Magid R, *et al.* Vascular oxidant stress enhances progression and angiogenesis of experimental atheroma. Circulation 2004; 109(4): 520-5.

[50] Potteaux S, Gautier EL, Hutchison SB, *et al.* Suppressed monocyte recruitment drives macrophage removal from atherosclerotic plaques of Apoe-/- mice during disease regression. J Clin Invest 2011; 121(5): 2025-36.

[51] Libby P. Inflammation in atherosclerosis. Nature 2002; 420(6917): 868-74.

[52] Newby AC. Dual role of matrix metalloproteinases (matrixins) in intimal thickening and atherosclerotic plaque rupture. Physiol Rev 2005; 85(1): 1-31.

[53] Simonini A, Moscucci M, Muller DW, *et al.* IL-8 is an angiogenic factor in human coronary atherectomy tissue. Circulation 2000; 101(13): 1519-26.

[54] Amin MA, Mansfield PJ, Pakozdi A, *et al.* Interleukin-18 induces angiogenic factors in rheumatoid arthritis synovial tissue fibroblasts via distinct signaling pathways. Arthritis Rheum 2007; 56(6): 1787-97.

[55] Galkina E, Ley K. Vascular adhesion molecules in atherosclerosis. Arterioscler Thromb Vasc Biol 2007; 27(11): 2292-301.

[56] Rahman K. Studies on free radicals, antioxidants, and co-factors. Clin Interv Aging 2007; 2(2): 219-36.

[57] Valko M, Leibfritz D, Moncol J, *et al.* Free radicals and antioxidants in normal physiological functions and human disease. Int J Biochem Cell Biol 2007; 39(1): 44-84.
[58] Shah AM, Channon KM. Free radicals and redox signalling in cardiovascular disease. Heart 2004; 90(5): 486-7.
[59] Pham-Huy LA, He H, Pham-Huy C. Free radicals, antioxidants in disease and health. Int J Biomed Sci 2008; 4(2): 89-96.
[60] Willcox JK, Ash SL, Catignani GL. Antioxidants and prevention of chronic disease. Crit Rev Food Sci Nutr 2004; 44(4): 275-95.
[61] Kaliora AC, Dedoussis GV, Schmidt H. Dietary antioxidants in preventing atherogenesis. Atherosclerosis 2006; 187(1): 1-17.
[62] Giugliano D. Dietary antioxidants for cardiovascular prevention. Nutr Metab Cardiovasc Dis 2000; 10(1): 38-44.
[63] Lonn E. Do antioxidant vitamins protect against atherosclerosis? The proof is still lacking*. J Am Coll Cardiol 2001; 38(7): 1795-8.
[64] Polytarchou C, Papadimitriou E. Antioxidants inhibit angiogenesis in vivo through down-regulation of nitric oxide synthase expression and activity. Free Radic Res 2004; 38(5): 501-8.
[65] Matsubara K, Kaneyuki T, Miyake T, *et al.* Antiangiogenic activity of nasunin, an antioxidant anthocyanin, in eggplant peels. J Agric Food Chem 2005; 53(16): 6272-5.
[66] Daghini E, Zhu XY, Versari D, *et al.* Antioxidant vitamins induce angiogenesis in the normal pig kidney. Am J Physiol Renal Physiol 2007; 293(1): F371-81.
[67] Hughes SE, Crossman D, Hall PA. Expression of basic and acidic fibroblast growth factors and their receptor in normal and atherosclerotic human arteries. Cardiovasc Res 1993; 27(7): 1214-9.
[68] Inoue M, Itoh H, Ueda M, *et al.* Vascular endothelial growth factor (VEGF) expression in human coronary atherosclerotic lesions: possible pathophysiological significance of VEGF in progression of atherosclerosis. Circulation 1998; 98(20): 2108-16.
[69] Moreno PR, Purushothaman KR, Sirol M, *et al.* Neovascularization in human atherosclerosis. Circulation 2006; 113(18): 2245-52.
[70] Mulligan-Kehoe MJ. The vasa vasorum in diseased and nondiseased arteries. Am J Physiol Heart Circ Physiol 2010; 298(2): H295-305.
[71] Kumamoto M, Nakashima Y, Sueishi K. Intimal neovascularization in human coronary atherosclerosis: its origin and pathophysiological significance. Hum Pathol 1995; 26(4): 450-6.
[72] Kwon HM, Sangiorgi G, Ritman EL, *et al.* Adventitial vasa vasorum in balloon-injured coronary arteries: visualization and quantitation by a microscopic three-dimensional computed tomography technique. J Am Coll Cardiol 1998; 32(7): 2072-9.
[73] Garg S, Serruys PW. Coronary stents: current status. J Am Coll Cardiol 2010; 56(10 Suppl): S1-42.
[74] Morice MC, Serruys PW, Barragan P, *et al.* Long-term clinical outcomes with sirolimus-eluting coronary stents: five-year results of the RAVEL trial. J Am Coll Cardiol 2007; 50(14): 1299-304.
[75] Busch R, Strohbach A, Rethfeldt S, *et al.* New stent surface materials: The impact of polymer-dependent interactions of human endothelial cells, smooth muscle cells, and platelets. Acta Biomater 2014; 10(2): 688-700.

[76] Kennedy KL, Lucas AR, Wan W. Local delivery of therapeutics for percutaneous coronary intervention. Curr Drug Deliv 2011; 8(5): 534-56.

[77] Stenestrand U, James SK, Lindback J, *et al.* Safety and efficacy of drug-eluting vs. bare metal stents in patients with diabetes mellitus: long-term follow-up in the Swedish Coronary Angiography and Angioplasty Registry (SCAAR). Eur Heart J 2010; 31(2): 177-86.

[78] Luscher TF, Steffel J, Eberli FR, *et al.* Drug-eluting stent and coronary thrombosis: biological mechanisms and clinical implications. Circulation 2007; 115(8): 1051-8.

[79] Stefanadis C, Toutouzas K, Stefanadi E, *et al.* Inhibition of plaque neovascularization and intimal hyperplasia by specific targeting vascular endothelial growth factor with bevacizumab-eluting stent: an experimental study. Atherosclerosis 2007; 195(2): 269-76.

[80] Alt E, Haehnel I, Beilharz C, *et al.* Inhibition of neointima formation after experimental coronary artery stenting: a new biodegradable stent coating releasing hirudin and the prostacyclin analogue iloprost. Circulation 2000; 101(12): 1453-8.

[81] Celletti FL, Waugh JM, Amabile PG, *et al.* Inhibition of vascular endothelial growth factor-mediated neointima progression with angiostatin or paclitaxel. J Vasc Interv Radiol 2002; 13(7): 703-7.

[82] Scheller B, Hehrlein C, Bocksch W, *et al.* Treatment of coronary in-stent restenosis with a paclitaxel-coated balloon catheter. N Engl J Med 2006; 355(20): 2113-24.

[83] Luttun A, Tjwa M, Moons L, *et al.* Revascularization of ischemic tissues by PlGF treatment, and inhibition of tumor angiogenesis, arthritis and atherosclerosis by anti-Flt1. Nat Med 2002; 8(8): 831-40.

[84] Francis DJ, Parish CR, McGarry M, *et al.* Blockade of vascular smooth muscle cell proliferation and intimal thickening after balloon injury by the sulfated oligosaccharide PI-88: phosphomannopentaose sulfate directly binds FGF-2, blocks cellular signaling, and inhibits proliferation. Circ Res 2003; 92(8): e70-7.

[85] Turner N, Grose R. Fibroblast growth factor signalling: from development to cancer. Nat Rev Cancer 2010; 10(2): 116-29.

[86] Ornitz DM, Itoh N. Fibroblast growth factors. Genome Biol 2001; 2(3): REVIEWS3005.

[87] Lindner V, Reidy MA. Proliferation of smooth muscle cells after vascular injury is inhibited by an antibody against basic fibroblast growth factor. Proc Natl Acad Sci U S A 1991; 88(9): 3739-43.

[88] Luo W, Liu A, Chen Y, *et al.* Inhibition of accelerated graft arteriosclerosis by gene transfer of soluble fibroblast growth factor receptor-1 in rat aortic transplants. Arterioscler Thromb Vasc Biol 2004; 24(6): 1081-6.

[89] Brogi E, Winkles JA, Underwood R, *et al.* Distinct patterns of expression of fibroblast growth factors and their receptors in human atheroma and nonatherosclerotic arteries. Association of acidic FGF with plaque microvessels and macrophages. J Clin Invest 1993; 92(5): 2408-18.

[90] Che J, Okigaki M, Takahashi T, *et al.* Endothelial FGF receptor signaling accelerates atherosclerosis. Am J Physiol Heart Circ Physiol 2011; 300(1): H154-61.

[91] Mohammadi M, McMahon G, Sun L, *et al.* Structures of the tyrosine kinase domain of fibroblast growth factor receptor in complex with inhibitors. Science 1997; 276(5314): 955-60.

[92] Frédérique Dol-Gleizes, Nathalie Delesque-Touchard, Anne-Marie Marès, *et al.* A New Synthetic FGF Receptor Antagonist Inhibits Arteriosclerosis in a Mouse Vein Graft Model and Atherosclerosis in Apolipoprotein E-Deficient Mice. PLoS One 2013; 8(11): e80027.

[93] Blann AD, Belgore FM, McCollum CN, *et al.* Vascular endothelial growth factor and its receptor, Flt-1, in the plasma of patients with coronary or peripheral atherosclerosis, or Type II diabetes. Clin Sci (Lond) 2002; 102(2): 187-94.

[94] Ohtani K, Egashira K, Hiasa K, *et al.* Blockade of vascular endothelial growth factor suppresses experimental restenosis after intraluminal injury by inhibiting recruitment of monocyte lineage cells. Circulation 2004; 110(16): 2444-52.

[95] Leonard F, Devaux Y, Vausort M, *et al.* Adenosine modifies the balance between membrane and soluble forms of Flt-1. J Leukoc Biol 2011; 90(1): 199-204.

[96] Schmeisser A, Christoph M, Augstein A, *et al.* Apoptosis of human macrophages by Flt-4 signaling: implications for atherosclerotic plaque pathology. Cardiovasc Res 2006; 71(4): 774-84.

[97] Koyama H, Nishizawa Y, Hosoi M, *et al.* The fumagillin analogue TNP-470 inhibits DNA synthesis of vascular smooth muscle cells stimulated by platelet-derived growth factor and insulin-like growth factor-I. Possible involvement of cyclin-dependent kinase 2. Circ Res 1996; 79(4): 757-64.

[98] Ogata T, Kurabayashi M, Maeno T, *et al.* Angiogenesis inhibitor TNP-470 (AGM-1470) suppresses vascular smooth muscle cell proliferation after balloon injury in rats. J Surg Res 2003; 112(2): 117-21.

[99] Moulton KS, Heller E, Konerding MA, *et al.* Angiogenesis inhibitors endostatin or TNP-470 reduce intimal neovascularization and plaque growth in apolipoprotein E-deficient mice. Circulation 1999; 99(13): 1726-32.

[100] Ait-Oufella H, Taleb S, Mallat Z, *et al.* Recent advances on the role of cytokines in atherosclerosis. Arterioscler Thromb Vasc Biol 2011; 31(5): 969-79.

[101] Asadullah K, Sterry W, Volk HD. Interleukin-10 therapy--review of a new approach. Pharmacol Rev 2003; 55(2): 241-69.

[102] Huang S, Ullrich SE, Bar-Eli M. Regulation of tumor growth and metastasis by interleukin-10: the melanoma experience. J Interferon Cytokine Res 1999; 19(7): 697-703.

[103] Silvestre JS, Mallat Z, Duriez M, *et al.* Antiangiogenic effect of interleukin-10 in ischemia-induced angiogenesis in mice hindlimb. Circ Res 2000; 87(6): 448-52.

[104] Robinson JG, Smith B, Maheshwari N, *et al.* Pleiotropic effects of statins: benefit beyond cholesterol reduction? A meta-regression analysis. J Am Coll Cardiol 2005; 46(10): 1855-62.

[105] Buhaescu I, Izzedine H. Mevalonate pathway: a review of clinical and therapeutical implications. Clin Biochem 2007; 40(9-10): 575-84.

[106] Liu T, Li L, Korantzopoulos P, *et al.* Statin use and development of atrial fibrillation: a systematic review and meta-analysis of randomized clinical trials and observational studies. Int J Cardiol 2008; 126(2): 160-70.

[107] Semenova AE, Sergienko IV, Masenko VP, *et al.* The influence of rosuvastatin therapy and percutaneous coronary intervention on angiogenic growth factors in coronary artery disease patients. Acta Cardiol 2009; 64(3): 405-9.

[108] Sergienko IV, Semenova AE, Masenko VP, *et al.* [Effect of statin therapy on dynamics of vascular endothelial growth factor and fibroblast growth factor in patients with ischemic heart disease]. Kardiologiia 2007; 47(8): 4-7.

[109] Bartoli M, Al-Shabrawey M, Labazi M, *et al.* HMG-CoA reductase inhibitors (statin) prevents retinal neovascularization in a model of oxygen-induced retinopathy. Invest Ophthalmol Vis Sci 2009; 50(10): 4934-40.

[110] Massaro M, Zampolli A, Scoditti E, *et al.* Statins inhibit cyclooxygenase-2 and matrix metalloproteinase-9 in human endothelial cells: anti-angiogenic actions possibly contributing to plaque stability. Cardiovasc Res 2010; 86(2): 311-20.
[111] Luan Z, Chase AJ, Newby AC. Statins inhibit secretion of metalloproteinases-1, -2, -3, and -9 from vascular smooth muscle cells and macrophages. Arterioscler Thromb Vasc Biol 2003; 23(5): 769-75.
[112] Koutouzis M, Nomikos A, Nikolidakis S, *et al.* Statin treated patients have reduced intraplaque angiogenesis in carotid endarterectomy specimens. Atherosclerosis 2007; 192(2): 457-63.
[113] Undas A, Celinska-Lowenhoff M, Stepien E, *et al.* Effects of simvastatin on angiogenic growth factors released at the site of microvascular injury. Thromb Haemost 2006; 95(6): 1045-7.
[114] Weis M, Heeschen C, Glassford AJ, *et al.* Statins have biphasic effects on angiogenesis. Circulation 2002; 105(6): 739-45.
[115] Herrmann J, Lerman LO, Mukhopadhyay D, *et al.* Angiogenesis in atherogenesis. Arterioscler Thromb Vasc Biol 2006; 26(9): 1948-57.
[116] Jinnouchi H, Sakakura K, Wada, H, Ishida, K, Arao, K, Kubo, N, Sugawara, Y, Funayama, H, Ako, J, Momomura, S. Effect of Chronic Statin Treatment on Vascular Remodeling Determined by Intravascular Ultrasound in Patients With Acute Myocardial Infarction. The American Journal of Cardiology 2013.
[117] Ray PD, Huang BW, Tsuji Y. Reactive oxygen species (ROS) homeostasis and redox regulation in cellular signaling. Cell Signal 2012; 24(5): 981-90.
[118] Bertram C, Hass R. Cellular responses to reactive oxygen species-induced DNA damage and aging. Biol Chem 2008; 389(3): 211-20.
[119] Orient A, Donko A, Szabo A, *et al.* Novel sources of reactive oxygen species in the human body. Nephrol Dial Transplant 2007; 22(5): 1281-8.
[120] Terada LS. Specificity in reactive oxidant signaling: think globally, act locally. J Cell Biol 2006; 174(5): 615-23.
[121] Ushio-Fukai M, Alexander RW. Reactive oxygen species as mediators of angiogenesis signaling: role of NAD(P)H oxidase. Mol Cell Biochem 2004; 264(1-2): 85-97.
[122] Wattanapitayakul SK, Bauer JA. Oxidative pathways in cardiovascular disease: roles, mechanisms, and therapeutic implications. Pharmacol Ther 2001; 89(2): 187-206.
[123] Dusting GJ, Selemidis S, Jiang F. Mechanisms for suppressing NADPH oxidase in the vascular wall. Mem Inst Oswaldo Cruz 2005; 100 Suppl 1: 97-103.
[124] Munzel T, Hink U, Heitzer T, *et al.* Role for NADPH/NADH oxidase in the modulation of vascular tone. Ann N Y Acad Sci 1999; 874: 386-400.
[125] Selemidis S, Dusting GJ, Peshavariya H, *et al.* Nitric oxide suppresses NADPH oxidase-dependent superoxide production by S-nitrosylation in human endothelial cells. Cardiovasc Res 2007; 75(2): 349-58.
[126] Frey RS, Ushio-Fukai M, Malik AB. NADPH oxidase-dependent signaling in endothelial cells: role in physiology and pathophysiology. Antioxid Redox Signal 2009; 11(4): 791-810.
[127] Peshavariya H, Dusting GJ, Jiang F, *et al.* NADPH oxidase isoform selective regulation of endothelial cell proliferation and survival. Naunyn Schmiedebergs Arch Pharmacol 2009; 380(2): 193-204.

[128] Datla SR, Peshavariya H, Dusting GJ, *et al.* Important role of Nox4 type NADPH oxidase in angiogenic responses in human microvascular endothelial cells in vitro. Arterioscler Thromb Vasc Biol 2007; 27(11): 2319-24.

[129] Pendyala S, Gorshkova IA, Usatyuk PV, *et al.* Role of Nox4 and Nox2 in hyperoxia-induced reactive oxygen species generation and migration of human lung endothelial cells. Antioxid Redox Signal 2009; 11(4): 747-64.

[130] Gorlach A, Brandes RP, Bassus S, *et al.* Oxidative stress and expression of p22phox are involved in the up-regulation of tissue factor in vascular smooth muscle cells in response to activated platelets. FASEB J 2000; 14(11): 1518-28.

[131] Gorlach A, Diebold I, Schini-Kerth VB, *et al.* Thrombin activates the hypoxia-inducible factor-1 signaling pathway in vascular smooth muscle cells: Role of the p22(phox)-containing NADPH oxidase. Circ Res 2001; 89(1): 47-54.

[132] Shi Y, Niculescu R, Wang D, *et al.* Increased NAD(P)H oxidase and reactive oxygen species in coronary arteries after balloon injury. Arterioscler Thromb Vasc Biol 2001; 21(5): 739-45.

[133] Szocs K, Lassegue B, Sorescu D, *et al.* Upregulation of Nox-based NAD(P)H oxidases in restenosis after carotid injury. Arterioscler Thromb Vasc Biol 2002; 22(1): 21-7.

[134] Nakamura K, Yamagishi S, Matsui T, *et al.* Pigment epithelium-derived factor inhibits neointimal hyperplasia after vascular injury by blocking NADPH oxidase-mediated reactive oxygen species generation. Am J Pathol 2007; 170(6): 2159-70.

[135] Rychli K, Huber K, Wojta J. Pigment epithelium-derived factor (PEDF) as a therapeutic target in cardiovascular disease. Expert Opin Ther Targets 2009; 13(11): 1295-302.

[136] Al-Shabrawey M, Bartoli M, El-Remessy AB, *et al.* Inhibition of NAD(P)H oxidase activity blocks vascular endothelial growth factor overexpression and neovascularization during ischemic retinopathy. Am J Pathol 2005; 167(2): 599-607.

[137] Del Rio D, Rodriguez-Mateos A, Spencer JP, *et al.* Dietary (poly)phenolics in human health: structures, bioavailability, and evidence of protective effects against chronic diseases. Antioxid Redox Signal 2013; 18(14): 1818-92.

[138] Rice-Evans C. Flavonoid antioxidants. Curr Med Chem 2001; 8(7): 797-807.

[139] Oak MH, El Bedoui J, Schini-Kerth VB. Antiangiogenic properties of natural polyphenols from red wine and green tea. J Nutr Biochem 2005; 16(1): 1-8.

[140] Oak MH, Chataigneau M, Keravis T, *et al.* Red wine polyphenolic compounds inhibit vascular endothelial growth factor expression in vascular smooth muscle cells by preventing the activation of the p38 mitogen-activated protein kinase pathway. Arterioscler Thromb Vasc Biol 2003; 23(6): 1001-7.

[141] Lin JK, Chen PC, Ho CT, *et al.* Inhibition of xanthine oxidase and suppression of intracellular reactive oxygen species in HL-60 cells by theaflavin-3,3'-digallate, (-)-epigallocatechin-3-gallate, and propyl gallate. J Agric Food Chem 2000; 48(7): 2736-43.

[142] Ying CJ, Xu JW, Ikeda K, *et al.* Tea polyphenols regulate nicotinamide adenine dinucleotide phosphate oxidase subunit expression and ameliorate angiotensin II-induced hyperpermeability in endothelial cells. Hypertens Res 2003; 26(10): 823-8.

[143] Khan SG, Katiyar SK, Agarwal R, *et al.* Enhancement of antioxidant and phase II enzymes by oral feeding of green tea polyphenols in drinking water to SKH-1 hairless mice: possible role in cancer chemoprevention. Cancer Res 1992; 52(14): 4050-2.

[144] Hyafil F, Vucic E, Cornily JC, *et al.* Monitoring of arterial wall remodelling in atherosclerotic rabbits with a magnetic resonance imaging contrast agent binding to matrix metalloproteinases. Eur Heart J 2011; 32(12): 1561-71.

[145] Nguyen M, Arkell J, Jackson CJ. Human endothelial gelatinases and angiogenesis. Int J Biochem Cell Biol 2001; 33(10): 960-70.

[146] Pasterkamp G, Schoneveld AH, Hijnen DJ, *et al.* Atherosclerotic arterial remodeling and the localization of macrophages and matrix metalloproteases 1, 2 and 9 in the human coronary artery. Atherosclerosis 2000; 150(2): 245-53.

[147] Rundhaug JE. Matrix metalloproteinases and angiogenesis. J Cell Mol Med 2005; 9(2): 267-85.

[148] Oak MH, El Bedoui J, Anglard P, *et al.* Red wine polyphenolic compounds strongly inhibit pro-matrix metalloproteinase-2 expression and its activation in response to thrombin via direct inhibition of membrane type 1-matrix metalloproteinase in vascular smooth muscle cells. Circulation 2004; 110(13): 1861-7.

[149] Gerthoffer WT. Mechanisms of vascular smooth muscle cell migration. Circ Res 2007; 100(5): 607-21.

[150] Iijima K, Yoshizumi M, Hashimoto M, *et al.* Red wine polyphenols inhibit vascular smooth muscle cell migration through two distinct signaling pathways. Circulation 2002; 105(20): 2404-10.

[151] Fried LE, Arbiser JL. Honokiol, a multifunctional antiangiogenic and antitumor agent. Antioxid Redox Signal 2009; 11(5): 1139-48.

[152] Maiti TK, Chatterjee J, Dasgupta S. Effect of green tea polyphenols on angiogenesis induced by an angiogenin-like protein. Biochem Biophys Res Commun 2003; 308(1): 64-7.

[153] Cao Y, Cao R. Angiogenesis inhibited by drinking tea. Nature 1999; 398(6726): 381.

[154] Chen CZ, Li L, Lodish HF, *et al.* MicroRNAs modulate hematopoietic lineage differentiation. Science 2004; 303(5654): 83-6.

[155] Hutvagner G, Zamore PD. A microRNA in a multiple-turnover RNAi enzyme complex. Science 2002; 297(5589): 2056-60.

[156] Zhao Y, Samal E, Srivastava D. Serum response factor regulates a muscle-specific microRNA that targets Hand2 during cardiogenesis. Nature 2005; 436(7048): 214-20.

[157] He L, Hannon GJ. MicroRNAs: small RNAs with a big role in gene regulation. Nat Rev Genet 2004; 5(7): 522-31.

[158] Kuehbacher A, Urbich C, Dimmeler S. Targeting microRNA expression to regulate angiogenesis. Trends Pharmacol Sci 2008; 29(1): 12-5.

[159] Krol J, Loedige I, Filipowicz W. The widespread regulation of microRNA biogenesis, function and decay. Nat Rev Genet 2010; 11(9): 597-610.

[160] Suarez Y, Fernandez-Hernando C, Pober JS, *et al.* Dicer dependent microRNAs regulate gene expression and functions in human endothelial cells. Circ Res 2007; 100(8): 1164-73.

[161] Bartel DP. MicroRNAs: target recognition and regulatory functions. Cell 2009; 136(2): 215-33.

[162] Suarez Y, Sessa WC. MicroRNAs as novel regulators of angiogenesis. Circ Res 2009; 104(4): 442-54.

[163] Fish JE, Santoro MM, Morton SU, *et al.* miR-126 regulates angiogenic signaling and vascular integrity. Dev Cell 2008; 15(2): 272-84.

[164] Kuehbacher A, Urbich C, Zeiher AM, *et al.* Role of Dicer and Drosha for endothelial microRNA expression and angiogenesis. Circ Res 2007; 101(1): 59-68.

[165] Poliseno L, Tuccoli A, Mariani L, *et al.* MicroRNAs modulate the angiogenic properties of HUVECs. Blood 2006; 108(9): 3068-71.

[166] Shilo S, Roy S, Khanna S, *et al.* Evidence for the involvement of miRNA in redox regulated angiogenic response of human microvascular endothelial cells. Arterioscler Thromb Vasc Biol 2008; 28(3): 471-7.

[167] Suarez Y, Fernandez-Hernando C, Yu J, *et al.* Dicer-dependent endothelial microRNAs are necessary for postnatal angiogenesis. Proc Natl Acad Sci U S A 2008; 105(37): 14082-7.

[168] He L, Thomson JM, Hemann MT, *et al.* A microRNA polycistron as a potential human oncogene. Nature 2005; 435(7043): 828-33.

[169] Tanzer A, Stadler PF. Molecular evolution of a microRNA cluster. J Mol Biol 2004; 339(2): 327-35.

[170] Yamakuchi M. MicroRNAs in Vascular Biology. Int J Vasc Med 2012; 2012: 794898.

[171] Bonauer A, Carmona G, Iwasaki M, *et al.* MicroRNA-92a controls angiogenesis and functional recovery of ischemic tissues in mice. Science 2009; 324(5935): 1710-3.

[172] Moura R, Tjwa M, Vandervoort P, *et al.* Thrombospondin-1 activates medial smooth muscle cells and triggers neointima formation upon mouse carotid artery ligation. Arterioscler Thromb Vasc Biol 2007; 27(10): 2163-9.

[173] Zhou Q, Gallagher R, Ufret-Vincenty R, *et al.* Regulation of angiogenesis and choroidal neovascularization by members of microRNA-23~27~24 clusters. Proc Natl Acad Sci U S A 2011; 108(20): 8287-92.

[174] Moura R, Tjwa M, Vandervoort P, *et al.* Thrombospondin-1 deficiency accelerates atherosclerotic plaque maturation in ApoE-/- mice. Circ Res 2008; 103(10): 1181-9.

[175] Urbich C, Kaluza D, Fromel T, *et al.* MicroRNA-27a/b controls endothelial cell repulsion and angiogenesis by targeting semaphorin 6A. Blood 2012; 119(6): 1607-16.

[176] Raitoharju E, Lyytikainen LP, Levula M, *et al.* miR-21, miR-210, miR-34a, and miR-146a/b are up-regulated in human atherosclerotic plaques in the Tampere Vascular Study. Atherosclerosis 2011; 219(1): 211-7.

[177] van Rooij E, Sutherland LB, Qi X, *et al.* Control of stress-dependent cardiac growth and gene expression by a microRNA. Science 2007; 316(5824): 575-9.

[178] Lagos-Quintana M, Rauhut R, Yalcin A, *et al.* Identification of tissue-specific microRNAs from mouse. Curr Biol 2002; 12(9): 735-9.

[179] Wienholds E, Kloosterman WP, Miska E, *et al.* MicroRNA expression in zebrafish embryonic development. Science 2005; 309(5732): 310-1.

[180] Campagnolo L, Leahy A, Chitnis S, *et al.* EGFL7 is a chemoattractant for endothelial cells and is up-regulated in angiogenesis and arterial injury. Am J Pathol 2005; 167(1): 275-84.

[181] Soncin F, Mattot V, Lionneton F, *et al.* VE-statin, an endothelial repressor of smooth muscle cell migration. EMBO J 2003; 22(21): 5700-11.

[182] Wang Y, Zhou Y, He L, *et al.* Gene delivery of soluble vascular endothelial growth factor receptor-1 (sFlt-1) inhibits intra-plaque angiogenesis and suppresses development of atherosclerotic plaque. Clin Exp Med 2011; 11(2): 113-21.

[183] Psarros C, Lee R, Margaritis M, *et al.* Nanomedicine for the prevention, treatment and imaging of atherosclerosis. Nanomedicine 2012; 8 Suppl 1: S59-68.

[184] Winter PM, Neubauer AM, Caruthers SD, *et al.* Endothelial alpha(v)beta3 integrin-targeted fumagillin nanoparticles inhibit angiogenesis in atherosclerosis. Arterioscler Thromb Vasc Biol 2006; 26(9): 2103-9.

[185] Winter PM, Caruthers SD, Zhang H, *et al.* Antiangiogenic synergism of integrin-targeted fumagillin nanoparticles and atorvastatin in atherosclerosis. JACC Cardiovasc Imaging 2008; 1(5): 624-34.

[186] Almutairi A, Rossin R, Shokeen M, *et al.* Biodegradable dendritic positron-emitting nanoprobes for the noninvasive imaging of angiogenesis. Proc Natl Acad Sci U S A 2009; 106(3): 685-90.

[187] Lobatto ME, Fayad ZA, Silvera S, *et al.* Multimodal clinical imaging to longitudinally assess a nanomedical anti-inflammatory treatment in experimental atherosclerosis. Mol Pharm 2010; 7(6): 2020-9.

[188] Folkman J. Angiogenesis in cancer, vascular, rheumatoid and other disease. Nat Med 1995; 1(1): 27-31.

[189] McCarty MF, Liu W, Fan F, *et al.* Promises and pitfalls of anti-angiogenic therapy in clinical trials. Trends Mol Med 2003; 9(2): 53-8.

[190] An MM, Zou Z, Shen H, *et al.* Incidence and risk of significantly raised blood pressure in cancer patients treated with bevacizumab: an updated meta-analysis. Eur J Clin Pharmacol 2010; 66(8): 813-21.

[191] Chu TF, Rupnick MA, Kerkela R, *et al.* Cardiotoxicity associated with tyrosine kinase inhibitor sunitinib. Lancet 2007; 370(9604): 2011-9.

[192] Mourad JJ, des Guetz G, Debbabi H, *et al.* Blood pressure rise following angiogenesis inhibition by bevacizumab. A crucial role for microcirculation. Ann Oncol 2008; 19(5): 927-34.

[193] des Guetz G, Uzzan B, Chouahnia K, *et al.* Cardiovascular toxicity of anti-angiogenic drugs. Target Oncol 2011; 6(4): 197-202.

[194] Kurup A, Lin C, Murry DJ, *et al.* Recombinant human angiostatin (rhAngiostatin) in combination with paclitaxel and carboplatin in patients with advanced non-small-cell lung cancer: a phase II study from Indiana University. Ann Oncol 2006; 17(1): 97-103.

Subject Index

Atta-ur-Rahman and Muhammad Iqbal Choudhary (Eds)
Copyright © 2014 Bentham Science Publishers Ltd. Published by Elsevier Inc. All rights reserved.
10.1016/B978-0-12-803963-2.50011-9

D

E

V

W

X

Printed in the United States
By Bookmasters